The Principles of Green and Sustainability Science

Adenike A. Akinsemolu

The Principles of Green and Sustainability Science

Springer

Adenike A. Akinsemolu
The Green Institute
Ondo State, Nigeria

ISBN 978-981-15-2492-9 ISBN 978-981-15-2493-6 (eBook)
https://doi.org/10.1007/978-981-15-2493-6

This Springer imprint is published by the registered company Springer Nature Singapore Pte Ltd..
The registered company address is: 152 Beach Road, #21-01/04 Gateway East, Singapore 189721, Singapore

To Helen and Jerome.

Foreword

I think you should always bear in mind that entropy is not on your side.—Elon Musk

This quote is true for all of us. Every person, no matter how diligent or gifted, is doomed to experience a total failure of body processes. No system can be sustained forever. Systems fail, empires fall, and even the universe will succumb to total heat death. Sometimes, we expedite this process ourselves. Human intervention has rapidly deteriorated the sustainability of the environment.

However, there are some who resist.

Adenike Akinsemolu joins the ranks of men and women who fight for environmental sustainability, while others accept the diminishing state of the world. As a person active in promoting and investigating sustainability in my community and the world, who has spent decades focusing on sustainability in the field of energy and engineering, I recognize Akinsemolu's diligence and see its reflection in this work. She understands that the unprecedented and rampant growth in population during the past century presents daunting challenges—but also sees the opportunity in mass cooperation in efforts to move towards sustainability. The work is both informative in its content, a reminder of our relationship to the environment, and a call to action. However, there is still confusion about issues of sustainability and green living despite its importance. That is where this text comes in.

In the book, *The Principles of Green and Sustainability Science*, Akinsemolu provides a comprehensive and step-by-step guide to understanding sustainability. Beginning with the concept of Going Green, the text explores far-reaching implications in the fields of computer science, engineering, physics, and more. The book offers empirical evidence and examines best practices to create a holistic approach to more harmonious living. This understanding of sustainability in the totality of its reach can allow nations and individuals to set, and hopefully to achieve, more appropriate sustainability goals.

Today, we are far from meeting the 17 Sustainable Development Goals[1] (SDGs) set by the United Nations Development Programme (UNDP). Ironically, Akinsemolu first highlights the 17th goal: the need for partnership in turning these goals into our greatest accomplishments. Although some think of humanity as owning the planet, much of humankind has long viewed itself as stewards of the Earth. Yet, we have damaged the ozone layer, destroyed forests, littered the marine world, and threatened the animal kingdom. This text illuminates the path towards repairing this relationship with our environment by working together towards a more sustainable future. Akinsemolu takes an aspirational approach and reminds us that anyone can join the effort.

The book offers immediate actions that the individual can take. Akinsemolu adopts a practical approach which examines how everyday activities affect climate, ecosystems, and more. As such, the book presents both a top-down methodology and a grassroots vision. While covering broad topics, the text offers practices for readers to put into action in their everyday lives. It is an invitation to explore the mindset of sustainability and take action. A consistent source of inspiration and illustration throughout the text is the use of examples of sustainability contributions. Each chapter highlights a prime example of a single person's ability to bring about a greater balance with the environment.

This text is not a stand-alone effort. It is born within a culture of progress towards sustainability. The book enriches a global movement while highlighting efforts in Africa. Africa has seen examples of substantial initiatives towards sustainability and Green living. One such example is the Green Institute in Nigeria, which aims to "build the next generation of sustainability innovators through education, advocacy, and social entrepreneurship." This type of education will prove vital to global sustainability. Indeed, a 2016 Forbes report by Amy Jadesimi stated that Nigeria "lies on the cusp of greatness" as the country has an abundance of resources and an economy which has grown annually. One way this is becoming a reality is through education that challenges traditional forms of education, which at times have thwarted the progress of humanity. This text carries the same spirit of presenting an updated conversation on the topic of sustainability.

Akinsemolu views sustainability as progress. The benefits are manifold when social, economic, and environmental equilibrium is found. Applying sound principles of sustainability science is needed to avoid damaging Earth, our home, and the life and civilization on it. With this in mind, it is my hope that this text becomes a standard choice for inclusion in the education of readers globally on the challenges and threats to sustainability and the best practices to protect the environment and promote a balanced ecosystem.

The material and educational approach in this book are past due. At this stage, we must live in a more green and sustainable manner if we are to thrive. We must

[1] https://www.undp.org/content/dam/undp/library/corporate/brochure/SDGs_Booklet_Web_En.pdf.

change if we are to flourish. For Akinsemolu, sustainability is more than delaying the inevitable—it is the promise of a future.

Faculty of Engineering and Applied Science
University of Ontario Institute of Technology
Oshawa, ON, Canada
March 9, 2020

Professor Marc A. Rosen,
Ph.D., P.Eng., FEIC, FCSME,
FASME, FIEF, FCAE, FCSSE

Preface

"**What does it mean to go Green?**" I asked this question to my students in Nigeria. The response was silence and a sea of blank faces. I waited patiently, but not a single hand was raised. Not one student in the group of 250 bright minds was confident enough to talk about the most urgent issue facing our planet and species. I decided that the time for patience was over; I started working on this text which aims to inspire a green movement in the hearts and minds of readers of all ages.

The core purpose of this book is to use the breakthroughs in sustainability science to address the problems afflicting the environment and present measures for improving economies, societies, and human behavior by applying this emerging science through best practices. This book is both informative and prescriptive in its approach to sustainability, a combination missing in previous publications.

Sustainability in science leads to proactive measures geared towards the adoption of knowledge, skills, activities, and attributes necessary for using the existing resources in a way that does not compromise the ability of future generations to meet their own needs. There have been monumental sustainability texts produced prior to the publication of this book and I applaud their contributions. However, sustainability science is unique in its scope and accessibility. It acts as an introductory book for the average reader but can still be used as a condensed knowledge base for the more advanced. It is not just an academic text but a call to action for nations and individuals. It is an easy read filled with hard truths. It provides the basic know-how of sustainability science needed for the reader to start contributing today.

Everyday anthropogenic activities are responsible for the problems of our planet and there is a need to salvage the situation through creativity, innovation, and critical thinking. Now more than ever, there is great potential for sustainability science's incorporation into programs, policies, and cultural norms because of the common consensus on the need to develop proactive measures to protect and preserve the planet's finite resources. This level of common consensus is itself uncommon, and we must not squander this potential. To help accomplish these goals, this book is adapted to present the proposed ideas by using empirical evidence as the basis for achieving sustainability. This is a results-based approach which takes an honest look at unemployment, health, disease, unsustainable production, waste management, environmental ethics, and harmful anthropogenic activities. Whereas past literature has mainly examined sustainability as a purely environmental issue, the proposed book expands this conversation into other fields, including mathematics, biology, agriculture, computer science, engineering, and physics. By examining the application of sustainability in all areas of study, it illuminates the path to sustainability as defined by the World Commission on Environment Development in its Our Common Future report. The plethora of information this text offers combines intrigue with practicality and intellect with wisdom.

The text is also authoritative as it moves through each chapter, citing case studies of individuals with notable contributions to sustainability efforts. The achievements in sustainability from the likes of Ken Saro-Wiwa, Wangari Maathai, Jeffrey Sachs, and Elon Musk add an inspirational weight to the text as the reader learns the core concepts of sustainability. Finally, using examples from Africa and the Middle East adds a sense of grounding to the issue. These real-world examples of sustainability efforts show how the applications can lead to real progress and have dire consequences if ignored.

I invite you to view this book as a personal invitation to join the conversation on sustainability and actively engage in saving our planet for all its residents, present and future.

Ondo State, Nigeria
November 15, 2019

Adenike A. Akinsemolu

Acknowledgements

None of us, including me, ever do great things. But we can all do small things, with great love, and together we can do something wonderful—Mother Teresa

You would not be reading this book without the people in my life who supported me throughout its creation. My mentors, Prof. V.W. Ogundero, Prof. F.A. Akinyosoye, Prof. D.J. Arotupin, and Prof. D.S Olawuyi have inspired me and have been an endless resource for my constant questioning. Newton said that he saw further by walking on the shoulders of giants, and you have provided me the platform to see beyond the boundaries of my potential to reach new heights. I would also like to thank all the members of staff of the Green Institute, Ondo Kingdom. A special mention goes to Dr. Jason McSparren of the University of Massachusetts, Boston, for editing and reviewing this manuscript. You helped find the diamonds in the rough and are one yourself. The world is a better place with you in it.

Thanks to everyone on the Springer team who helped me and to Mei Han Lee, the ever-patient publishing manager, and John Martyn, a fantastic production editor who was always there when I needed him (which was often).

My most profound appreciation goes to my parents, Jerome and Helen Akinsemolu. Your support, encouragement, and constant love have sustained me throughout my life. Mom, you always make sure that I am well-fed whenever I am engrossed in writing. I know I am responsible for more than a few of your gray hairs. To my siblings, who have also given me their support throughout this period, I wholeheartedly appreciate you.

I want to also acknowledge the support of Olukoya Obafemi, Feyisayo Adeyemi, Kehinde Adenegan, Becky Olubunmi, Bankole Temitayo, Foluke Arijeniwa, and Olamide Mosuro for their love, guidance, and moral support during the period of writing this book. You reminded me of the light at the end of the tunnel, even when the mission was at its darkest. You taught me to never give up on my goals and I will carry this lesson with me for the remainder of my life and make sure it lives on through my students.

Special thanks to my amazing students whose class contributions and ideas served as the seeds that sprouted into this book. I appreciate your support, and I commit this book to the ambition I see in you.

Finally, I would like to thank the Supreme Being for life, strength, and provision to come this far.

The commitment to this book echoes Africa's commitment to sustainability; individual humans might be strong-willed but are ultimately insufficient in the face of goals of such enormity. We must turn to our fellow humans, and, together, a collection of individuals is capable of anything. We are human, and that is our legacy.

"*Dum inter homines sumus, colamus humanitatem.*"

Contents

Acronyms

3Rs	Reduce, reuse, recycle
AAR	African Air Rescue
AI	Artificial intelligence
AIDS	Acquired immune deficiency syndrome
ASAL	Arid and semi-arid land
ATP	Adenosine triphosphate
AU	African Union
BMW	Bayerische Motoren Werke
CEO	Chief executive officer
CGG	Commission on Global Governance
CH_4	Methane
CNN	Cable News Network
CO_2	Carbon dioxide
COD	Chemical oxygen demand
CSR	Corporate social responsibility
DAF	Dissolved air flotation
DC	Direct current
DCM	Direct current motor
DDT	Dichlorodiphenyltrichloroethane
DNA	Deoxyribonucleic acid
EIA	Environmental impact assessment
ELWA	Eternal Love Winning Africa
EMR	Electromagnetic radiation
EPA	Environmental Protection Agency
ERA	Environmental risk assessment
EU	European Union
FCEVs	Fuel cell electric vehicles
FCX	Fuel Cell eXperiment
FEPA	Federal Environment Protection Agency
FORNEWA	The Rwanda Green Fund
GBM	Green Belt Movement

GDP	Gross domestic product
GE	General Electric
GGE	Green growth and economy
GHSs	Greenhouse gases
GMOs	Genetically modified organisms
HEP	Hydroelectric power
HIV	Human immunodeficiency virus
IBM	International Business Machines
IEEE	Institute of Electrical and Electronics Engineers
IFPRI	International Food Policy Research Institute
IKEA	Ingvar Kamprad Elmtaryd Agunnaryd
Inc.	Incorporated
IPCC	Intergovernmental Panel on Climate Change
IR	Infrared ray
ISO	International Organization for Standardization
ITS	Intelligent transportation system
IUCN	International Union for Conservation of Nature
LDCs	Least developed countries
LED	Light-emitting diode
MCED	Ministerial Conference on Environment and Development
MDG	Millennium Development Goal
MEA	Millennium Ecosystem Assessment
MENA	Middle East and North Africa
MFC	Microbial fuel cell
MOSOP	Movement for the Survival of the Ogoni People
MSF	Medecins Sans Frontieres
NAS	National Academy of Sciences
NASA	National Aeronautics and Space Administration
NATO	North Atlantic Treaty Organization
NCDs	Non-communicable diseases
NESREA	National Environmental Standards and Regulations Enforcement Agency
NGO	Non-government organization
NIDA	National Institute of Drug Abuse
NIST	National Institute of Standards and Technology
NPP	Nobel Peace Prize
OSHA	Occupational Safety and Health Administration
OECD	Organisation for Economic Co-operation and Development
OPEC	Organization of the Petroleum Exporting Countries
PAHs	Polycyclic aromatic hydrocarbons
PDA	Personal digital assistant
PP	Precautionary principle
PPE	Personal protective equipment
PPP	Polluter Pays Principle
PV cells	Photovoltaic cells

RNA	Ribonucleic acid
SBSP	Space-based solar power
SDGs	Sustainable Development Goals
SMEs	Small and medium-sized enterprises
TED	Technology, Entertainment, Design
UK	United Kingdom
UPMC	University of Pittsburgh Medical Center
UN	United Nations
UNDP	United Nations Development Programme
UPP	User-pays principle
US	United States
USD	United States Dollars
US EPA	United States Environmental Protection Agency
UV	Ultraviolet radiation
WEMA	Water Efficient Maize for Africa
WHO	World Health Organization
WWF	World Wide Fund

Part I
Introductory Context and Principles of Green

Chapter 1
The Principles of Green

Abstract Going Green has increasingly become a global trend recently because of its potential benefits to the society, environment, economy, and future of life. Using case studies, theories, and various intertwined principles, Chap. 1 demonstrates the way Going Green entails embracing five fundamental principles which are important for making sure that we embrace sustainable lifestyles. Apart from the five principles, the chapter describes the various critical ways of going green at the individual, organizational as well as national level. The case studies of the Maasai in East Africa and Fulani in West Africa are used for bringing out the concept of the Tragedy of the Commons. In addition to the case studies, the first chapter also critically discusses the benefits related to going green, with the major pros identified being cost, environment, reputation, improved health, sustainable development, invention, and innovation. A notable contribution to the green initiative in this chapter is Wangari Maathai's Green Belt Movement. Also, future trends are identified to predict the future of going green.

Keywords Going green · Tragedy of commons · Termite architecture · Wangari Maathai · Conservation of resources

1.1 What Does It Mean to Go Green?

With the world facing a crisis due to scarce resources, there have been numerous attempts to ensure people use resources in a sustainable way. Conceptually, resources in this context refer to materials, energy, or other assets that can be used for benefiting society. This has led to the increased use of the term "Going Green," commonly defined as the adoption of environmentally friendly lifestyles through activities such as recycling, purchasing, reuse, reducing or minimizing, driving, and others. In project management, "Going Green" includes designing projects that meet environmental standards. The two definitions above suggest that Going Green involves engaging in activities that sustainably use resources. However, these definitions are determined by the specific activity. A more comprehensive definition of the term is "pursuing practices and knowledge that can result in ecologically

A. A. Akinsemolu, *The Principles of Green and Sustainability Science*,
https://doi.org/10.1007/978-981-15-2493-6_1

responsible and environmentally-friendly lifestyles and decisions, which can be beneficial for protecting the environment and sustaining natural resources for both the current as well as future generations."

Going Green emphasizes the need for countries and corporations to consider sustainable development in their activities. Sustainable development is defined as development that caters to the "needs of current generations without compromising the ability of future generations to meet their own needs" (Gardner and Abraham 2010). Through efficient production and deployment of Green technologies, organizations can reduce their carbon emissions and promote the protection of the ozone layer. They can also emit less hazardous waste through the 3Rs (reduce, reuse, and recycle), thus ensuring that the ecosystem continues providing its services efficiently. From an individual perspective, Going Green could be beneficial for encouraging people to develop positive ideas and use alternative resources (Cock 1991). This is crucial for avoiding the aspect of Tragedy of the Commons. Conceptually, this describes a situation where individual users of a resource act independently out of self-interest. This behavior contradicts that of common good users since it depletes and spoils that resource. Examples of Tragedy of the Commons are mining in South Africa and overgrazing among the Maasai of Kenya.

Going Green involves more than recycling alone and extends to living life in a way that is respectful and sustainable for the natural environment and the planet. It implies contributing positively towards the maintenance of the natural ecological balance as well as the preservation of the earth and its natural resources. It entails taking steps to reduce the harm ones does to the environment. This includes the carbon footprints that we leave behind because of living on this planet. Green also means (1) *Growth*, (2) *Renewable*, (3) *Efficient*, (4) *Economically Sustainable*, and (5) *Natural*. From these definitions and descriptions, Going Green involves the adoption of five fundamental principles in one's daily life to protect the environment and ensure sustainability: *Reduction of pollution, conservation of resources, conservation of energy, reduction of waste,* and *protection of the ecological balance of the Earth*. Consequently, it is crucial for us to adopt Green practices using these principles in order to make a difference. The five principles are further explained to enhance our understanding.

What steps have you taken in your life that can qualify as part of the "Going Green" initiative?

According to the first principle, Going Green means reducing pollution. Conceptually, pollution is the release of contaminants or toxic substances into the environment. Think about the cleaning detergent, soap, or shampoo that you like to use. Many of these products have chemicals that are washed down sinks and travel into rivers and the water table. Along these lines, think about the fast food lunch you consumed. During the production process of bread, burgers, salad, or pizza, there is the release of human-made fertilizers into the natural environment.

The wrappers that we use for packaging meals are disposed of in incinerators and landfills because they are difficult to be recycled. When the wrappers are burnt or buried, they release harmful gases and other wastes.

What are some of the wrappers that you use for packaging meals? List four effects of these wrappers on the ecosystem.

In addition, think about the car that you drive or the bus that you board while going to work. They emit greenhouse gases (GHG) that contribute to global warming and release toxic substances, including lead, into the atmosphere (Ashuri and Durmus-Pedini 2010). We have only scratched the surface with the forms of pollution produced by industries during the manufacturing of items we use daily.

In life today, it is difficult to leave no carbon footprint of toxic substances or have zero pollution. Consequently, we should understand that many of the things that we utilize every day create some form of a chemical by-product that could contribute to pollution. Going Green implies that we should do something about these activities. There are various things that people can do to minimize the effect of our activities on the planet. We should consider using more natural cleaners as well as personal products, including natural detergents and organic shampoos. These are made from natural substances, are biodegradable,, and impose minimal harm when produced or released into the environment (Eyraud et al. 2011). To support the Green initiatives, you could consider switching to organically grown food. This is the food that is produced without using synthetic fertilizers and pesticides that cause harm to the environment. Other suggestions that can reduce pollution include driving less, carpooling, using eco-friendly vehicles, and adopting Green driving practices.

Identify some organically grown food that you know.

The second principle suggests that Going Green entails conserving resources. While there are limited resources on the planet, our expanding human population is used to modern conveniences from technological advancements that increase the quantity of globally available goods. Therefore, many of us increase our consumption of goods that are eventually discarded as well as the natural resources used in manufacturing. This is an obvious problem in developing nations.

List some of the eco-friendly products that you use every week.

Deforestation, high energy use, over-mining, and intensive farming contribute to the depletion of our natural resources. Certain flora and fauna are becoming scarce resources and could become extinct unless we intervene and change our behavior. A World Wildlife Fund report (WWF 2016) highlights that people are using more natural resources than the planet can produce. Going Green can slow and eventually reverse this trend, but we have to use what we require rather than what we want. This includes learning to consume what is needed and to consider our consumption. In other words, we should remember that the Earth has to produce ecosystem services for all of us. These are defined as the numerous benefits that humans freely accrue from the natural environment and appropriately functioning ecosystems, some of which are aquatic, forest, and grassland ecosystems (Loreau et al. 2001). These benefits are collectively called ecosystem services and are crucial for providing clean drinking water, waste decomposition, and natural crop pollination. According to the 2005 Millennium Ecosystem Assessment (MEA) report, there are four main types of ecosystem services: supporting, provisioning, regulating, and cultural services.

Describe each type of these ecosystem services and provide at least three examples for each category.

One of the major proponents of the Green initiative was Kenya's Wangari Maatha, who founded the Green Belt Movement (GBM). As a renowned Kenyan, her environmental activism played a crucial role in the conservation of many trees in the Karura forest. Her contribution to sustainable development was recognized when she received the Nobel Peace Prize in 2004. Notably, she fought tirelessly, despite Kenya having a hostile regime, to ensure people conserved the environment. The Green Belt Movement was particularly instrumental in mobilizing Kenyans to plant numerous trees. It is estimated that at least 30 million trees were planted because of her efforts, thus serving as a model for other states including Malawi, Lesotho, and Tanzania. It is also worth noting that the movement's main objectives were resource and environmental conservation, tree planting, and women's rights. More information about Wangari Maathai's life can be found at the end of this chapter.

Do you believe that you are doing enough to conserve resources? Briefly explain. How can Africa emulate the achievements of Wangari Maathai? In what ways can we honor the legacy left by Wangari Maathai?

According to the third principle, Going Green means conserving energy. There are numerous ways to save energy that ensure that we go Green. By conserving energy, we contribute less to the creation of pollution.

The fourth principle argues that Going Green entails reducing waste and consumption. In this principle, the 3Rs cannot be overemphasized. The 3Rs stand for reduce, reuse, and recycle (Kranz and Picot 2011). The domino effect of reducing our consumption results in a lesser need for resources and energy, thus leading to less pollution from manufacturing industries. As we minimize our waste, we use fewer resources and energy for managing unwanted waste. Consequently, less pollution arises from incinerators and landfills. Reducing is essential for lowering the rate of raw material consumption and helps to minimize the waste we create. Recycling enables us to reuse materials that are found in unwanted materials for making new items. Finally, Going Green entails protecting the Earth's ecological balance . Ideally, the ecological balance is the equilibrium formed through the harmonious co-existence of organisms, including flora and fauna. Arguably, these five principles provide a platform for the best ways of Going Green and are further discussed in the next section.

How can the 3Rs (reduce, reuse, and recycle) be used in the home environment to conserve energy?

Case Study 1.1: The Fulani Herdsmen and Maasai overgrazing (tragedy of the commons)

Fig. 1.1 Maasai Cattle Overgrazing, *Source:* Irving (2016)

The Fulani pastoralists are nomadic herders who are mainly occupied in raising livestock. Pure Fulani pastoralists often engage in moving periodically with cattle based on the seasons and weather, whereas semi-nomadic pastoralists migrate back to their homes or camps. Historically, Fulani herdsmen are mainly found in the Sahel and semi-arid regions of West Africa, but because of climatic patterns, many of them have moved into the southern states, such as Senegal, Guinea, Mali, Niger, Burkina Faso, Cameroon, and Nigeria (Okeke 2014).

(continued)

Fulani pastoralists had long lived in relative harmony with sedentary agriculturalists before recurring droughts within arid and semi-arid land (ASAL) regions over the past five decades. The droughts have resulted in desertification, which minimizes the arable land for farmers and cattle to graze. As a result, the Fulani have migrated to the Guinea savanna and tropical areas leading to competition with other farmers (Okeke 2014). Overgrazing has been one of the contributing factors to the overuse of available resources, alongside farming and climate change.

Farmers accuse the herders of destroying their crops and contaminating watering-places. It is evident from this revelation that there could be a Tragedy of the Commons of not only land but also shared water resources. On the other hand, farmers are accused by the Fulani of denying them access to grazing lands when they cannot find other options (Okeke 2014). The conflict is mainly attributed to the increased competition for natural resources, which is a key characteristic of the Tragedy of the Commons concept.

Similarly, the Maasai of Kenya and Tanzania have been accused of engaging in overgrazing the land they occupy (see Fig. 1.1). The result is that they leave trails of destruction and are forced to move to other areas to find grazing lands. They consider cattle as a valuable aspect of their day-to-day lives. A few decades ago, there were few Maasai, probably due to the harsh environment, which deterred them from accessing clean water, shelter, and food. The Moran, a popular term in Maasai meaning warriors, often roamed the plains found in modern-day Ethiopia, Kenya, and Sudan searching for cattle to steal. They also waged war for water, grazing lands, and cattle. However, the life expectancy of the Moran was short because of low birth rates and high infant mortality. Today, because of significant changes, Maasai populations have substantially increased with improved lifespan and lower infant mortality. The factors that necessitated these changes are police control, laws and regulations, infrastructural development, and warring tribes. The Maasai are thriving but still hold onto their culture of cattle rearing. The increased Maasai population implies that there are a lot of cattle, which results in potential overgrazing and overpopulation, and subsequently, a decline in local wildlife. The Maasai used to live harmoniously with nature by cherishing Enkai, a higher spirit. Enkai would bring about life within the savannah and ensure that grass sprouted again for the livestock to feed after the dry seasons. Today, however, many of their stories and wisdom have disappeared and been overtaken by the desire to fit in and adapt to modern life. Overpopulation of the Maasai has made land sparse, while hunger has also become more frequent. The lack of knowledge over resource conservation has contributed to overgrazing that harms water resources and the land and has partly contributed to the destruction of the wilderness. This case study has provided insight into the importance of Mother Nature, and local communities play a crucial role in the conservation of such ecosystems.

1.2 Ways to Go Green

The first section examined the definitions of Going Green and the principles that encompass the concept of Going Green. The following ways of Going Green are among the most influential steps that individuals, organizations, and governments can take towards sustainable living and production. Many of these ideas can be implemented easily, are cost-effective, and can positively affect the environment as well as natural resources. Some of them are quite large, require time for implementation, and cannot be done overnight. To ensure that this section comprehensively covers all the ways of Going Green, it will examine strategies that can be deployed by individuals, communities, industries, and entrepreneurs or organizations.

1.2.1 Green Transportation

Green transportation is among the best ways of Going Green. The existing transportation system requires large quantities of energy to power vehicles. Many people acknowledge that automobiles result in excessive pollution, which affects the environment and contributes to health complications. Therefore, we must consider innovative solutions. However, before such innovations are developed, the world can use eco-friendly transportation modes that are already available. For example, the bicycle is the first mode of Green transportation available today. Commuting using a bicycle is an excellent way of reducing greenhouse gases (GHGs) found in the atmosphere. Despite walking also being an important mode of Green transportation, a bicycle is faster as well as affordable. Electric bikes are also great ways of Green transportation since they do not emit harmful substances into the atmosphere (Froehlich et al. 2009).

What do you think are the disadvantages of using a bicycle? Do these disadvantages affect the significance of bicycles in Green initiatives?

Green trains can also be an ideal way of Going Green. As many world governments embark on supporting Green transportation, trains have increasingly become Greener due to the deployment of environmentally friendly technologies and hybrid locomotives, which use similar technologies applied to hybrid cars (Froehlich et al. 2009). Today, modern electric trains use an electrified third rail that does not emit harmful gases.

Electric motorcycles also do not emit harmful emissions as they are battery-powered. Nevertheless, pollution can also occur because of the grid electricity generation used to charge electric bike batteries. The only disadvantage associated with electric motorcycles is the cost.

Hybrid cars depend on electricity and are designed such that they automatically recharge their batteries through energy conversion. Emissions in such cars are quite low compared to standard cars (Coad et al. 2009). Despite being important in reducing greenhouse gas emissions, they also have some weaknesses, particularly the batteries. In addition, new hybrid buses, including the Mercedes-Benz, Citaro G, BlueTec Hybrid Bus, and public transportation are essential Green transportation modes.

Carpooling can also be a great way of Going Green. Conceptually, carpooling entails multiple occupants sharing vehicles. Carpooling reduces the number of cars on the road, thus significantly minimizing pollution levels (Buliung et al. 2009). Carpooling is eco-friendly and a favorable Green transportation mode. For example, five friends can use one vehicle for commuting rather than each of them driving their car. This is not only economical but also ecologically sensible.

Can you identify ways that our community can improve carpooling? What measures can be put in place to ensure that carpooling becomes successful?

Finally, one cannot fail to talk about walking as a Green mode of transport. It involves zero-emission of GHG emissions, is free, and a great way of exercising. The other Green transportation modes that can be used today are service and freight vehicles. The use of electricity and biofuels instead of fossil fuels and by offering lots of travel alternatives will go a long way towards aligning the transportation sector to Green transportation.

1.2.2 Green Practices at Home

Everyone has a place that they call home, and as such, everyone can play a responsibility in Going Green by adopting Green practices at home. Instead of depending on artificial light, one can maximize the use of external lighting. Natural light offers a bright and warm feeling while also providing natural heat to homes. The best way of achieving this is by developing options that allow for great natural light solutions, including shutters and solar-powered roof windows made by the Velux Group. People with a home office should try to ensure that it is as Green as possible. For example, you can store digital file copies rather than hard and printed copies, unless it is mandatory that you have them in that form.

The other way of Going Green at home is reducing waste by embracing greener habits in the kitchen. A kitchen is among the major generators of waste. Consequently, we must educate family members on the benefits of minimizing our dependence on disposables.

What practices can we use in the kitchen to contribute towards a sustainable way of living?

The other ways that one can adapt to Green living at home are: installation of rain tanks to store water, switching off electricity when it is not needed, developing recycling initiatives, investing in a pressure cooker to conserve energy, minimizing laundry duties by washing clothes once in a week, and conserving water while bathing. Power is one of the best and easiest ways of Going Green at home. You can make use of environmentally friendly solar power at home to minimize power consumption from the grid and reduce reliance on fossil fuels. Also, consider environmentally friendly fuel sources such as gas cookers. Finally, you should eat local and organic foods as they do not have to be transported to distant markets, and so have less of a carbon footprint. Since food is one of the major aspects of our lives, it is among the most significant ways of Going Green.

Due to the nature of the business environment, it is quite difficult for organizations to go Green entirely. Notwithstanding, there are various ways that organizations can go Green. First, they can reduce their fuel consumption. With a barrel of oil being close to $100, this is a no-brainer. Regular engine services can be conducted to ensure that there is an efficient performance, maximization of gas mileage, and minimization of the excess exhaust.

What steps can be undertaken to reduce fuel consumption in engines? Briefly explain four steps that you deem important.

1.3 The Importance of Going Green

Going Green has wide-ranging benefits, including environmental, health, economic, and individual budgets. It is difficult for one to fail to notice the numerous advertising messages that emphasize the need for people to go Green. These messages underscore an aspect that humans should strive to embrace. Businesses have adopted this as a marketing strategy for convincing consumers that they are passionate about the environment. Notwithstanding, it is not just for organizational reputation, but the various benefits associated with Going Green.

1.3.1 *Cost Savings*

Green action is essential for saving costs. There are various ways which Going Green can enable us to save costs. For example, by going paperless, you can save money by paying bills online. It can also save organizations money associated with postage and paper costs. The use of compact fluorescent light bulbs can enable you to avoid expenses associated with energy costs. It is estimated that each fluorescent bulb substituted with an incandescent bulb will utilize 75% less energy (Eyraud et al. 2011). The United States Department of Energy estimates that replacing a 100-W bulb with a 25-W fluorescent bulb can save more than $100 in a year.

Admittedly, overhauling the traditional and unsustainable practices incurs extra costs upfront. However, these savings could be significant in the long-term. Using the energy bulb example, light-emitting diode (LED) bulbs are costlier than incandescent bulbs, but LED bulbs last longer and utilize 10% of the energy used by incandescent bulbs (Eyraud et al. 2011). Considering the energy bill savings, and need for fewer replacements, LED bulbs remain a cheaper alternative over an extended period. This also applies to renewable energy, while alternative sources of energy, including geothermal and wind, are costly in the short-term, they become cheaper with technological advancement. Furthermore, fossil fuel prices, mainly oil, are volatile and continually fluctuate. If you commit to renewable energy now, you could benefit with lower bills next month.

What types of bulbs or lighting system are used in your house? Estimate the amount of power that you consume utilizing these lighting systems. From this estimation, determine whether your actions can be said as "Green."

1.3.2 *Environmental Benefits*

Recycling and reusing products, instead of buying new ones, can bring in significant benefits. The United States Protection Environment Agency (US EPA, 2019) asserts that recycled products tend to use less energy for production than goods produced from raw materials due to the elimination of costs related to collecting raw materials. Consequently, recycled goods can contribute to reducing pollution and conserving natural resources.

In addition, using Greener energy can help maintain the available number of carbon sinks that are important for carbon sequestration, the process of removing carbon dioxide from the atmosphere. Carbon sinks are artificial or natural reservoirs that can store and accumulate materials with carbon compounds for an extended period. The main carbon sinks are soil, the ocean, and plants. Even the most basic acts can reduce the amount of trash one generates as well as the quantity contributed to landfills. Going Green is also a commitment to minimize your environmental impact. As such, you should calculate your carbon footprint to identify the aspects of your life that contribute most to environmental degradation. The carbon footprint calculator can be found at https://www.c2es.org/content/calculate-your-carbon-footprint/. A more detailed and personalized version (household unit) can be found at http://www.carbonindependent.org/.

1.3.3 Organizational Reputation

Large organizations emphasize the need to go Green because of the benefits they accrue. For example, restaurants like Starbucks and Chipotle are proud of using sustainable and organic ingredients. Likewise, Apple has just committed approximately $1 billion to solar power (Griskevicius et al. 2010). Such companies acknowledge that consumers value Green corporations. Consumers have gone beyond marketing strategies used by organizations that claim to practice sustainability and can now find endless information online and ascertain whether such claims about Going Green are valid. Companies that are genuinely Green have a better image and incur more trust.

Identify organizations in Africa that engage in Green practices. What is the general image/reputation of these organizations? Briefly explain some of their Green practices.

1.3.4 Innovation

Furthermore, Going Green can stimulate innovation. All organizations have to grapple with stagnation periodically by becoming stuck in one way of doing things for an extended period and never questioning whether the process is the most effective. An underappreciated benefit of Greenness is that it compels an organization to conduct an evaluation of all business aspects and devise better production means, including money-saving innovations (Griskevicius et al. 2010). Therefore, Going Green goes beyond the use of renewable energy. A complete audit of energy usage and emissions can unearth various opportunities for reducing costs as well as operating more efficiently.

1.3.5 Health

Sustainability is healthier for us. One of the principles of living sustainably is to eliminate processed foods. Consuming only whole and fresh nutrients enables us to reduce processed foods by minimizing the demand for meal components. It does not imply that you should become a full vegan but encourages us to look for our meat from the local butchers rather than factory options. Livestock that has been raised independently have fewer hormone and antibiotic injections. Healthier diets are tastier and healthier than mass-produced meats.

> What are some of the examples of healthier diets that can enhance our health while promoting Greenness?

Also, Greenness minimizes the probability of us dealing with health problems. This is because the fewer waste people produce, the less damage is done to the ozone layer. If we maintain our sustainability momentum, the ozone may be able to repair itself. The ozone layer is located in the stratosphere, which sits between 15–30 km above the planet. It plays a crucial role in absorbing the majority of the ultraviolet radiation of the sun, thus reducing the amount of radiation that reaches the Earth's surface. Since this radiation causes cataracts and skin cancer, the ozone layer plays a crucial role in protecting our health and preventing radiation from damaging animals, plants as well as materials. However, waste emissions lead to the depletion of this layer, thus hampering its efforts in insulating against solar radiation. The depletion of the ozone layer has been linked to the presence of chlorofluorocarbons as well as other halogen-source gases within the stratosphere. Many of the ozone-depleting substances are synthetic chemicals that are used in consumer applications, industrial applications, refrigeration, and fire extinguishers. The generation of waste from these chemicals harms the ozone layer.

1.3.6 Green Entrepreneurship

Furthermore, Going Green can promote Green entrepreneurship which is responsible for the creation of employment and the alleviation of poverty and hunger. For example, Wecyclers in Nigeria has been a source of employment for many young Green entrepreneurs. Started by Bilikiss Adebiyi-Abiola in 2012, the project sought to solve the problem of trash disposed off carelessly. The initiative has been essential for enabling poor communities to generate income from waste that occupies the streets. It is an emblem of community empowerment for wealth creation and a step towards sustainable livelihoods. Afrocentric Bamboo in Ghana is the other initiative founded by Winnifred Selby to create employment. It has become an effective way of alleviating poverty by employing the young in the community. Eco-Post Kenya has also been successful in creating employment for 300 Kenyans, especially marginalized women. An example of an initiative geared at achieving the sustainable development goals (SDGs) is the Green Campus Initiative in Nigeria that educates and empowers young people to be environmental and social change agents in their communities. Since inception by Adenike Akinsemolu, the initiative has trained thousands of youths in environmental sustainability. The initiative tackles climate change and environmental sustainability through the development of socio-environmental conscious leaders (Park 2016). Some of the strategies used for addressing some of the challenges are nurturing student leaders, promoting innovative academic research, and results-based eco-initiatives. Students are also encouraged to use bicycles on their campuses as a way of promoting Green growth and environmental stewardship.

1.4 Practical Application: Green Termite Architecture

The Eastgate Centre, located in Harare, Zimbabwe, is a perfect example of Green architecture (see Fig. 1.2). It is the nation's largest shopping and office complex that marvels many people because of the use of biomimicry principles. The building was designed by Mick Pearce and did not have any conventional air-conditioning. However, it remains regulated yearlong with significantly less energy use through innovative design methods. The inspiration for this building is the self-cooling mounds made by the African termites. African termites are great at building structures that are self-regulating and can maintain oxygen levels, humidity, and temperature (Nkandu and Alibaba 2018). It is this ingenuity that engineers and architects have adapted to the design of the Eastgate Building. The exterior of a termite structure appears like a pile of garbage. However, when sliced in half, you

Fig. 1.2 Eastgate Building, *Source:* Pearce (2016)

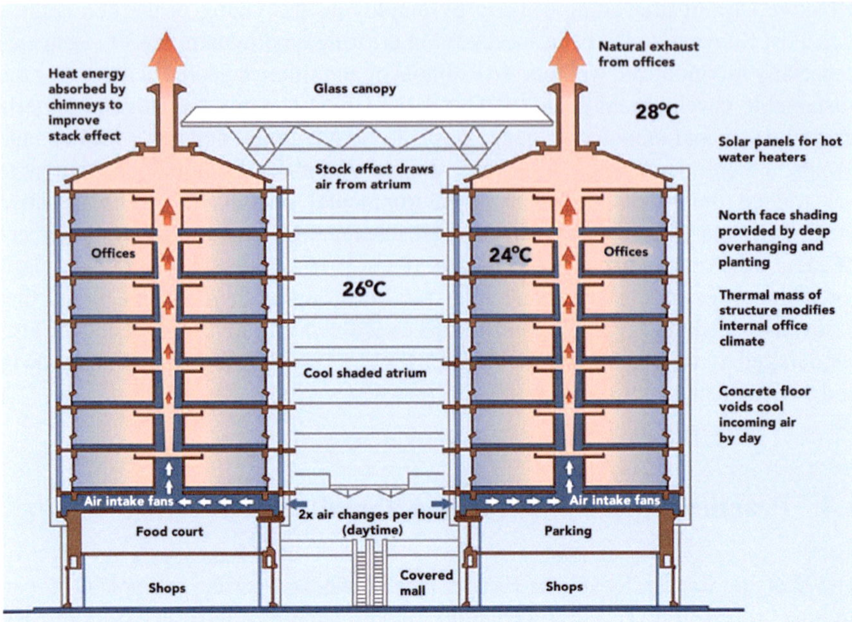

Fig. 1.3 Termitary Mechanism for Biomimicry Architecture, *Source:* Pearce (2016)

can notice that the above-ground mound has an outer wall with numerous holes that result in various tunnels with chimneys.

Eastgate Centre is mainly made of concrete and has a ventilation system that enables it to operate in a similar way to a termite mound (see Fig. 1.3). The air drawn in is cooled or warmed by the building depending on its state. The air is then ventilated into the floors and offices before exiting through the chimney at the top. The complex is also made of two buildings that are separated by an open space covered in grass. It is also open to local breezes, thus making the building conducive. Fans found on the first floor continuously draw air into the open spaces. The house design then enables it to move to the vertical supply sections found within the central spine of the two constructions.

Stale air is replaced by fresh air that rises and goes out through the exhaust ports found in each floor's ceilings. It ultimately gets into the exhaust section prior to being flushed out. Compared to a conventional building, the Eastgate Centre utilizes no more than 10% energy. The efficiency has been beneficial for the bottom-line. The Eastgate owners have saved about \$3.5 million from the low-cost air-conditioning system. Apart from being eco-friendly and eco-efficient, the savings are also beneficial to tenants who pay 20% less rent compared to occupants within the surrounding buildings (Pearce 2016). One would not have imagined that replicating designs created by termites would have provided a viable climate control solution. It is among the best applications of the Green initiative.

1.5 Individual Matter: Wangari Maathai

Maathai (see Fig. 1.4) was a Kenyan environmental activist and politician who served as the Assistant Minister in the Ministry of Environment, Natural Resources, and Wildlife. She was awarded a Ph.D. that made her the first woman in Central and East Africa to receive a doctorate (Napoli 2010). She won the Nobel Peace Prize because of her "holistic approach to sustainable development that embraces democracy, human rights, and women's rights in particular."

Maathai was born in 1940 in Nyeri, Kenya. Her father was a farmer who fully supported the family, while Kenya was a British colony. Her family surprisingly sent her to school, which was uncommon for Kenyan girls at the time. She excelled in primary school and furthered her schooling at Loreto Girls' High School. In 1960, she won a scholarship that enabled her to attend Mount St. Scholastica College in Kansas, United States (Napoli 2010). She earned an undergraduate degree in Biology before completing a Master's Degree (Biological Sciences) in 1966 at the University of Pittsburgh. She returned to Kenya and enrolled in veterinary anatomy. It was a historic moment when she became the first woman in East Africa to receive a doctorate. It propelled her to become university faculty and chair at university department in 1976; the first woman to achieve such a feat.

She sought to end the destruction of forests and lands in Kenya due to development. To achieve this, she founded the Green Belt Movement in 1977

Fig. 1.4 Wangari Maathai, *Source:* The Green Belt Movement (2019)

to spearhead the reforestation of the country, while empowering women. The movement successfully planted more than 30 million trees, inculcated new skills, and empowered about 30,000 women (Winter 2008). Maathai further challenged the government regarding its management of land and development plans. As an outright critic of the dictatorial President Daniel Arap Moi, she was beaten and arrested by the regime multiple times. A notable contribution to the environment was the famous actions she instigated in 1989. She organized a protest in Uhuru Park, Nairobi to deter the building of a skyscraper. The campaign elicited global attention, thus preventing the project from the beginning. The place of the demonstration was named Freedom Corner. The environmental movement also became political as she sought to deal with triggers of environmental degradation, particularly bad governance.

She remained vocal against the regime of President Moi until he lost the elections in 2002. In the same year, she was selected to the assistant ministry docket before receiving the Nobel Peace Prize (NPP) in 2004 for her contribution to peace, democracy, and sustainable development (Winter 2008). During her final years, she battled cancer and succumbed to it on September 25, 2011. Her contributions to the environment remain in the hearts and memories of not only Kenyans but also the entire world.

1.6 Future Trends

1.6.1 Africa

For Africa, Going Green will involve intensified organic farming, which entails the use of natural methods in animal rearing and crop cultivation. This will reduce the accumulation of waste and pollution, and can be ecologically sustainable since farmers avoid synthetic substances for maintaining soil fertility. There will also be a proliferation of eco-friendly home building techniques that can include things such as mud-bricks that can be easily recycled and created with mud. The continent could accrue benefits from electric cars, with the major benefit being that consumers will not depend on inconstant petrol prices because of the oil prices. Solar power could make it easier for the car batteries to be charged affordably and efficiently.

1.6.2 Asia

Also, going forward, green business trends will continue developing, with the foundation being green infrastructure, circular economy, B-2-B sharing economy, microgrids, and carbon recycling in Asia. These eco-friendly practices will become an opportunity for minimizing risks and raising revenue, thus opening the door for investors to invest in the sustainable growth of businesses.

1.6.3 Australia

In the future, Australia could go Green by producing 70% of its electricity needs from the tri-generation (low-carbon production method) (Cogentenergy 2019). One Central Park will be among the greenest as well as the most self-sufficient urban development. Being among the most urbanized countries globally, Australia will seek to have more zero waste cities and a circular economy.

1.6.4 Europe

In a matter of years, life cycle will become as crucial as lifestyle. To-date, recycling has been quite successful at eliminating large quantities of materials from the international waste stream. With the popularization of the cradle-to-cradle design concept (entails the creation of a zero-waste stream design system), product designers have increasingly begun considering not only the way to assemble the products but also reuse the materials. During the next two decades or so, recycling will be more sophisticated since retailers and manufacturers embed the concepts of return, recycle as well as reuse into the production of products and sales. One of the emerging trends is the electronic recycling adopted by Best Buy and compostable chip bag introduced by SunChips. In the future, biomimicry will also be a common aspect along with permaculture (designing human habitats that are respectful and healthy to the environment as they mimic relationships found within nature). Bioclimatic architecture will also become a common aspect (comprises of the design of buildings that account for climatic conditions and tapping on the available natural resources). Bioclimatic houses could significantly minimize the use of energy and even eliminate the need for it.

1.6.5 North America

Regarding green transportation, there is a likelihood of the adoption of lighter vehicles in the future. It is predicted that reducing the weight of a car by close to 220 pounds could raise its energy efficiency by more than 3%, thus lowering greenhouse gas emissions (Serrenho et al. 2017). It is the reason magnesium is being considered, as it is the lightest metal on the planet. The Kinetic Energy Recovery System technology could ultimately be utilized within passenger cars. The system is used for transferring braking energy into electrical energy that could be stored for later use.

1.6.6 South America

There will be a surge in driverless vehicles that are fuel-efficient. Gridlocked traffic tends to increase the number of carbon emissions, and improving traffic speeds can minimize greenhouse gases. South American citizens will continue becoming enlightened on environmental issues. With such awareness, it is anticipated that the public will oppose more projects that cause pollution. Public outcry will result in more private sector and public sector collaboration. More importantly, green businesses and entrepreneurship ventures will become prevalent.

1.7 Summary

- Going Green involves adopting five fundamental principles in life: resource conservation, pollution reduction, energy conservation, waste reduction, and protecting the planet's ecological balance.
- These five principles are essential for ensuring that we adopt sustainable lifestyles.
- There are various ways of Going Green at the individual, organizational, and national level.
- Green transportation and practices at home have been identified as major ways that can be used for promoting Greenness.
- Organizations are better placed to influence the Green movement because of the contribution they have on society and the quantity of non-Green materials they use or produce.
- This chapter also explored the Maasai in East Africa and Fulani in West Africa as case studies for the Tragedy of the Commons.
- This chapter has discussed the benefits associated with Going Green.
- The main benefits highlighted are cost savings, environmental benefits, organizational reputation, innovation, better health, and sustainable development.
- Wangari Maathai is among the significant contributors to the Green initiative. The Green Belt Movement she founded was beneficial to the conservation of the environment, the empowerment of women, and sustainability.
- Africa will be characterized by intensified organic farming in the future, while biomimicry and sophisticated recycling will be predominant in Europe.
- North America and South America will adopt lighter vehicles, while Asia will increasingly focus on a circular economy and B2B sharing. Trigeneration will be used as the most preferred way of producing electricity in Australia.

1.8 Discussion Questions

1. What benefits have you reaped from incorporating elements of "Green"?
2. Which of the five discussed principles of Going Green do you think is the most important?
3. List three words that are important in the definition of "Green."
4. What are alternative "Green" technologies that you would propose for the transport sector?
5. How is going "Green" beneficial to learners?
6. Africa has been cited as the next frontier of the Green movement. What are some of the Green projects that you have seen or heard of in Africa? Which project was exciting to you? Why? Briefly explain.
7. Explain the meaning of the term "carbon footprint." Why is it an important element in Going "Green"?
8. People, the community, organizations, and the state are important contributors to the "Green" movement. Which is the most important among the identified stakeholders? Identify and briefly discuss three reasons for your answer.

References

Ashuri, B., & Durmus-Pedini, A. (2010). An overview of the benefits and risk factors of going green in existing buildings. *International Journal of Facility Management, 1*(1), 1–15.

Buliung, R., Soltys, K., Habel, C., & Lanyon, R. (2009). Driving factors behind successful carpool formation and use. *Transportation Research Record: Journal of the Transportation Research Board, 2118*(1), 31–38.

Coad, A., De Haan, P., & Woersdorfer, J. S. (2009). Consumer support for environmental policies: An application to purchases of green cars. *Ecological Economics, 68*(7), 2078–2086.

Cock, J. (1991). *Going green: People, politics and the environment in South Africa*. Oxford: Oxford University Press.

Cogentenergy (2019). *Delivering a low carbon energy network for city of Sydney*. Retrieved October 10, 2019: https://cogentenergy.com.au/delivering-a-low-carbon-energy-network-for-city-of-sydney/.

Eyraud, L., Wane, A. A., Zhang, C., & Clements, B. (2011). *Who's going green and why? Trends and determinants of green investment*. Washington, D.C: International Monetary Fund.

Froehlich, J., Dillahunt, T., Klasnja, P., Mankoff, J., Consolvo, S., Harrison, B., et al. (2009). UbiGreen: Investigating a mobile tool for tracking and supporting green transportation habits. In *Proceedings of the SIGCHI conference on human factors in computing systems* (pp. 1043–1052). New York: ACM.

Gardner, B., & Abraham, C. (2010). Going green? Modeling the impact of environmental concerns and perceptions of transportation alternatives on decisions to drive. *Journal of Applied Social Psychology, 40*(4), 831–849.

Griskevicius, V., Tybur, J. M., & Van den Bergh, B. (2010). Going green to be seen: status, reputation, and conspicuous conservation. *Journal of Personality and Social Psychology, 98*(3), 392.

Irving, A. (2016). *Meeting the Maasai cattle*. Retrieved November 11, 2019: https://earthstar.blog/2016/08/24/meeting-the-maasai-cattle/.

Kranz, J., & Picot, A. (2011). Why are consumers going green? the role of environmental concerns in private green-IS adoption. In *ECIS* (p. 104).
Loreau, M., Naeem, S., Inchausti, P., Bengtsson, J., Grime, J. P., Hector, A., et al. (2001). Biodiversity and ecosystem functioning: current knowledge and future challenges. *Science, 294*(5543), 804–808.
Millennium Ecosystem Assessment (MA). (2005). *Ecosystems and human well-being: Synthesis* (155pp). Washington: Island Press.
Napoli, D. J. (2010). *Mama Miti: Wangari Maathai and the Trees of Kenya*. New York: Simon and Schuster.
Nkandu, M. I., & Alibaba, H. Z. (2018). Biomimicry as an alternative approach to sustainability. *Architecture Research, 8*(1), 1–11.
Okeke, O. E. (2014). Conflicts between Fulani herders and farmers in central and southern Nigeria: Discourse on proposed establishment of grazing routes and reserves. *AFRREV IJAH: An International Journal of Arts and Humanities, 3*(1), 66–84.
Park, J. (2016). Clean Energy Entrepreneurship in Sub-Saharan Africa. In *Global Entrepreneurship: Past, present & future* (pp. 257–277). Bingley: Emerald Group Publishing Limited.
Pearce, M. (2016). *Of termites and architecture*. Retrieved November 08, 2019: http://www.mickpearce.com/biomimicry.html/.
Serrenho, A. C., Norman, J. B., & Allwood, J. M. (2017). The impact of reducing car weight on global emissions: The future fleet in Great Britain. *Philosophical Transactions of the Royal Society A: Mathematical, Physical and Engineering Sciences, 375*(2095), 20160364.
The Green Belt Movement. (n.d.). *Wangari Mathaai*. Retrieved November 05, 2019: http://www.greenbeltmovement.org/news-and-events/media-resources/.
United States Environmental Protection Agency (2019). *Reducing and reusing basics*. Retrieved October 24, 2019: https://www.epa.gov/recycle/reducing-and-reusing-basics/.
Winter, J. (2008). *Wangari's trees of peace: A true story from Africa*. Boston: Houghton Mifflin Harcourt.
World Wildlife Fund, WWF (2016). *Living planet report 2016: Risk and resilience in a new era*. WWF International: Gland, Switzerland.

Chapter 2
Basic Ecology

Abstract The processes that happen at the ecosystem, population, and community levels are important for understanding ecology, weather and climate, and biogeochemical cycles of the Green World. With vagaries that are adversely impacting the way people live, weather patterns, food production, spread of diseases, inception of diseases and other phenomena, it is pertinent to understand basic terms including ecology, organisms, biosphere, and community. The chapter has three major sections, with the first part exploring concepts of ecology (organism, population, community, biosphere, ecosystem) and carrying capacity. The second part explores the weather and climate, while the third section explores the various biogeochemical cycles (carbon, nitrogen, sulfur, and phosphorous cycles) of the Green World. Using practical applications (particularly experiments) and individual contribution of James Hansen, this chapter is a key role in acting as the foundation for understanding the various processes in our ecology and planet, and their contributions to the Green World.

Keywords Biogeochemical cycles · Ecology · Mountain Everest · James Hansen · Climate

2.1 Ecology

Ecology is defined as the study of the relationship between organisms and the environment. It can also be defined as the study of the abundance and distribution of living things, and the way they are influenced by their interactions with their environment. From a broader angle, ecology considers the processes which happen in the ecosystem, population as well as community levels. The primary focus of ecology is on biodiversity. Myers et al. (2000) define biodiversity as "the variability among living organisms from all sources." Ecology is studied at various levels of organization: organism, population, community, ecosystem, and biosphere.

A. A. Akinsemolu, *The Principles of Green and Sustainability Science*,
https://doi.org/10.1007/978-981-15-2493-6_2

2.1.1 Organisms

Plants, animals, and microbes are organisms. The common definition in literature is that an organism is an individual life-form that can grow, reproduce, and is made up of one or more cells. Furthermore, an organism can respond to stimuli and maintain homeostasis. In ecology, the study of organisms explores their physiology, evolution, and behavior in relation to environmental factors.

What is your understanding of the term organism?

2.1.2 Community

This is the interactions between living organisms that occupy or live in the same area. In ecology, a community is perceived as a collection of populations of the various species. Community ecology explores the relationships among species. It factors the way such interactions influence community structure as well as species richness and abundance patterns. In a community, species interact in three major ways: mutualism, predation as well as competition. Typically, competition leads to a double negative, whereby both species do not benefit during the interaction. Predation is a win or lose situation with one of the species winning. Contrarily, mutualism entails both species engaging in some way but with both species winning. It also defines a group of more than two populations of organisms from various species that inhabit the same location during the same period. Communities tend to consist only of biotic factors, whereas abiotic factors only constitute the ecosystem.

Differentiate between biotic and abiotic factors. In reference to Charles Darwin of Natural Selection, is competition healthy or unhealthy in an ecosystem? Identify mechanisms used by organisms to adapt to their communities and environments?

2.1.3 Population

In ecology, population is defined as the characteristics of a group of organisms. A population is also perceived as a group of organisms or individuals of the same species living in the same area and may be defined at several spatial scales. The

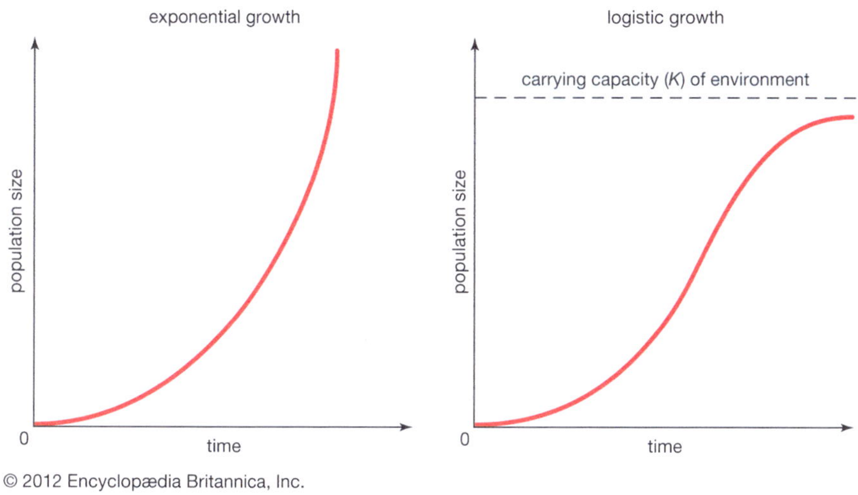

Fig. 2.1 Exponential versus logistic population growth, *Source:* Encyclopædia Britannica (2012a)

main challenge experienced in population ecology is the derivation of population characteristics from individual characteristics (Fig. 2.1).

What is the meaning of derivation in the above sentence? Identify the major characteristics of the term population.

2.1.3.1 Carrying Capacity

This refers to the maximum number of individual organisms that a specific environment can sustain during a particular period and is defined in terms of population, species, time, and space. A population's carrying capacity can change during resource fluctuations, such as changes in water levels, nutrients, biotic interactions, and habitat structure (Daily and Ehrlich 1992). From these elements, it is defined as the features of a single population, single species at a single place, and during a single period. Due to the range of factors and potential fluctuations an environment may experience, it is difficult to measure a true factor for carrying capacity in an environment.

What are your ideas about the ecological carrying capacity of the human environment? How can the carrying capacity of the human population be done? Identify three terms that must be present in any definition of carrying capacity for it to have a complete meaning. Differentiate between biological and cultural carrying capacity. Why do people ignore issues of carrying capacity? What are some of the activities that we can engage in to ensure that we do not strain the carrying capacity?

2.1.4 Ecosystem

An ecosystem is a community of living organisms (animals, plants, and microorganisms) that interact with each other in a specific area and the non-living environment. The components of the living environment include the climate, soil, and weather. For example, when foxes eat more rabbits, the plants increasingly begin thriving due to the fewer number of rabbits eating them. At the basic functional level, the ecosystem has varieties of primary producers and consumers. The producers harvest from the sun through photosynthesis. The energy flow is then facilitated by the food chain, where producers attract consumers and consumers seek out producers. Consumers can be primary or secondary consumers and feed on captured energy. Whatever remains is further consumed by decomposers (Brooks et al. 2006). Decomposers are considered the bottom of the food chain. There are three types of ecosystems: freshwater, terrestrial, and ocean. Freshwater ecosystems are categorized into pond ecosystems, river ecosystems, terrestrial ecosystems, forests, and grasslands.

Identify some of the goods and services provided by the ecosystem? Which of these goods is most important for life sustenance?

2.1.5 Biosphere

In ecology, biosphere refers to parts of the Earth that are occupied by life. Bio is a Greek word meaning "living," while sphere refers to "globe" which encompasses the varieties and differences seen in the environment. Some authors suggest that the biosphere consists of the lithosphere, hydrosphere (water), and atmosphere (air). These spheres are essential for the sustenance of life. The lithosphere consists of a crust and supports the biosphere. Conventional currents in the mantle lead to a shift of tectonic plates, leading to the formation of mountains, earthquakes, and volcanic eruptions (Runcorn 1962). These can be disastrous in the short-term, but long-term benefits include the development of new plant life, new habitats, and better adaptation. The lithosphere also acts as the source of all resources and has numerous elements, including calcium, aluminum, and copper, that has been used by humans since the Ancient Egypt period. When the lithosphere and biosphere interact, organic compounds embedded in the crust can be excavated as natural gas or oil and used as sources of fuel. The lithosphere, hydrosphere, and atmosphere combine to provide a crucial source of nutrients that support botanical life, which produces glucose used by higher organisms for sustenance. Similarly, the biosphere is an integral part of the Green World since it plays a crucial role in providing water and air to the planet.

What is the major role of the biosphere? Differentiate between the lithosphere, atmosphere, hydrosphere, and biosphere. Which of these spheres is the most important?

2.2 Weather and Climate

Weather is defined as the state of the atmosphere during any specific period; this includes precipitation levels (rain, snow, and sleet), temperature (warm at 75°F and freezing at 0°F), along with other factors, such as wind and humidity. Weather is mainly caused by air pressure, moisture, as well as temperature differences in various locations (Lydolph 1985). On the other hand, climate is defined as the average weather conditions over a prolonged period. Whereas the weather may change in only a few hours, climate may take numerous years. Some of the factors that affect climate are terrain, altitude, latitude, and proximity to water bodies.

How do terrain, altitude, latitude, and proximity to water bodies affect climate?

James Hansen is among the major contributors to climate and climate change topics. He directs the Program on Climate Science, Awareness and Solutions, and is famous for his research in climatology (Hansen 2013). His work helped raise awareness about climate change encouraging advocacy in avoiding dangerous climate change. Despite not directly contributing to weather and climate, his work is influential for highlighting our understanding of climate and climate change trends. Topography influences snowfall and rain. Mountains particularly play a crucial role in precipitation patterns. For instance, topographic barriers, like hills and mountains, compel prevailing winds over their slopes (Lydolph 1985). Cooler air can hold less water vapor compared to warmer air. As the air rises, it cools and condenses, leading to the deposition of rain or snow. As it is commonly understood, the higher the altitude, the colder it becomes. Mountain ranges usually act as barriers to airflow across the Earth's surface. When warm air travels from a low-lying area to the mountain's foot range, it rises along the mountain's slope. As it rises, it cools, leading to the formation of clouds and precipitation on the windward slope as well as at the summit. In meteorology, windward is a technical name for the sides of a mountain. The windward side is the one facing the prevailing wind or upwind, while the leeward side is the side that the mountain's elevation shields. The windward side is responsible for the enhancement of precipitation in the surrounding of mountain ranges, while the leeward side is responsible for withholding it. For example, these precipitation patterns can be seen in the Sierra Nevada mountain range, which usually traps moisture as it travels off the Pacific Ocean. Topography is also instrumental for the creation of regional winds.

What is the difference between altitude, topography, and slope?

Experiment 2.1: The influence of topography on weather and climate

Objectives

- Create a "rainstorm."
- Explore the way topography affects weather and climate.

Required Materials

Small branches, metal bowl or pan water, clay, and sponge

Background of the Experiment

An area's topography (or relief) can affect weather and climate. An area close to a water body tends to have milder climates. Mountainous regions have more extreme weather events due to the barrier the mountains pose on-air movements. The leeward or dry side of the mountain can make the weather of the region to be characterized by prolonged periods of drought. On the other hand, the windward side could result in frequent rainfalls. Mountains may also create a physical barrier to clouds that form rain. In this experiment, students will create rainclouds and a mountain to illustrate how one side of the mountain has less vegetation than the other.

(continued)

Procedure

1. Give a group a metal tray or pan and clay.
2. Give each of the group instructions to come up with a mountain scene. Small branches can be used to represent trees, while aluminum foil can be used to create a lake's appearance.
3. Give every group a sponge and ensure that the sponge is wet. Note: Students should assume that it is a cloud filled with water and ready to change to rain or snow.
4. When a cloud is wet, it is often heavy, and as the winds force it up, it becomes cool. It does not have the ability to become lighter and rise over the mountain. As this happens, there is a little amount of water left on the other side. In case the mountain's altitude is high enough, a desert will be created on the other side.
5. Students should experiment with the mountain and sponge.

Conclusion

Students should be able to observe the effect of topography on climate and weather.

Topography is closely related to the concept of altitude. With an increase in altitude, climatic conditions change and become colder. Places found in high latitudes or far away from the equator tend to experience significantly less sunlight compared to those at low latitudes. Latitude provides the location of a place south or north of the equator and is usually expressed through angular measurements that range from 0° at the equator to 90° at the poles. Different latitudes receive a varying quantity of sunlight and play a crucial role in determining the climate of a region. For instance, the high latitude of a specific place or the farther it is from the equator, the sharper the angle of the rays of the sun that reach it. This means that the sun's rays spread across a wider area. In this regard, higher latitudes usually receive less heat compared to lower latitudes found closer to the equator. The quantity of sunlight and precipitation influence the kinds of living organisms that may be found in a place. There are various ecosystems found in climates experienced in varying latitudes. The highest quantity of sunlight is often experienced at the equator, thus making the area quite warm. Some of the ecosystems that highlight the influence of latitude on climate are rainforests, deserts, savannah, chaparral, taiga, tundra, and grasslands. Temperatures tend to drop when an area is farther away from the equator because of the Earth's curvature. In areas that are close to the poles, temperatures are usually cooler because of the high amount of energy loss. Distance from the sea is another crucial element that affects climate. Oceans tend to heat up and cool down at a slower rate compared to land. This means that coastal locations tend towards being cooler during the summer and warmer during the winter. For instance, Glasgow, Scotland occupies latitude similar to Moscow, Russia, but tends to be hotter during

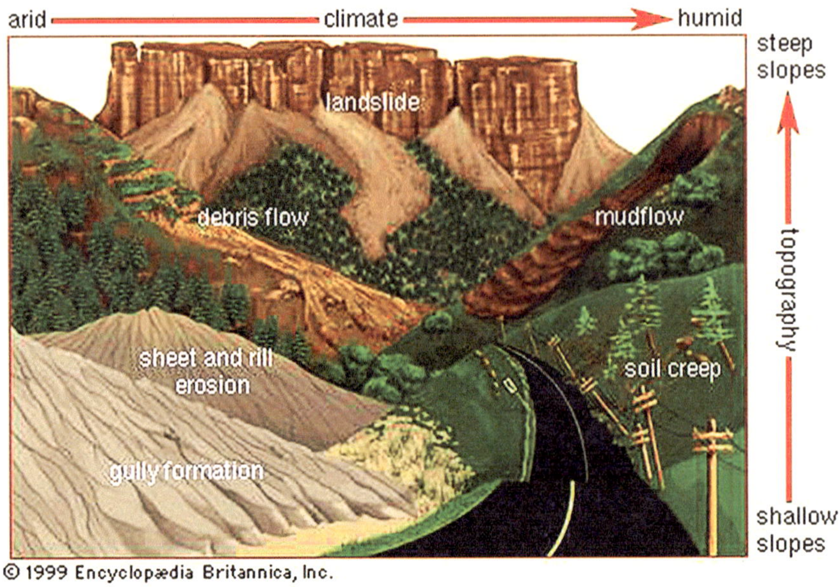

Fig. 2.2 Effect of topography and climate on water-induced soil erosion, *Source:* Encyclopædia Britannica (1999)

the winter due to its proximity to the coast. Figure 2.2 shows the effect of topography and climate on water-induced soil erosion.

Weather and climate are major factors that affect the distribution of species. Weather falls under the abiotic factors that influence biological processes, including germination. Some of the weather components that influence biological processes are water, sunlight, wind, soil, and rocks.

2.2.1 Temperature

Temperature is the degree of hotness or coldness of an object. It is among the critical and changeable ecological factors and can vary in climatic regions as well as temporally. Temperature infiltrates all areas of the biosphere and significantly affects all life forms as it can lower or increase important biological activities, including embryonic development and metabolism.

Identify the units for measuring temperature.

2.2.2 *Humidity*

Humidity is the quantity of water vapor in the atmosphere and is among the gaseous states of water. It depicts the probability of fog, dew, and precipitation. There are three major humidity measurements: specific, relative, relative and absolute (Lydolph 1985). Absolute humidity refers to the content of water in the air that is expressed in grams per kg/m^3. Relative humidity, on the other hand, is a percentage that measures the existing absolute humidity compared to the maximum for that temperature, while specific humidity refers to the weight of water vapor in a unit weight of air expressed as grams of water vapor per kilogram of air.

Which is the commonly used measurement of humidity?

2.2.3 *Wind*

This refers to the flow of gases, and on the Earth's surface, it comprises the movement of air. Simply put, wind is the movement of air molecules. It is attributed to the uneven heating of the surface of the Earth. Outer space also experiences wind; for example, the solar wind is depicted as the movement of gases from the Sun (Lydolph 1985). Typically, winds are categorized according to their types, spatial scale, the force that creates them as well as the regions where they occur. They are characterized by velocity, the density of the gas involved, wind energy, or content. In ecology, wind significantly influences the various stages of plant life, including growth, death, distribution, reproductionas well as evolution. Wind has further numerous impacts, including the destruction of forests, crops, and erosion of the soil. While plants are still seedlings, slight breezes enable them to grow sturdier. However, winds at high speed can damage and blow down the strongest trees. Depending on the place you live, this storm damage can occur at any time. Winter winds are particularly damaging since plants cannot replace the water they lose.

What triggers the wind to blow? Define a dust devil. What is the significance of the wind to the planet?

2.2.4 Precipitation

Precipitation is defined as any form of water that forms in the atmosphere and drops onto the Earth's surface. Precipitation is an important aspect of the water cycle. It falls to the surface of the Earth as rain and snow and eventually evaporates into the atmosphere in a gaseous state. People depend on precipitation for getting freshwater used to wash and drink and to carry out irrigation for crops. The most common forms of precipitation are rain, hail , hail and snow.

Rain is defined as the precipitation that gets to the ground in the form of water droplets. Ice pellets are a form of rain that falls from the clouds, but freezes before reaching the ground (Lydolph 1985). Hail is formed within intense storm clouds when cold water droplets freeze and turn solid. On the other hand, snow is precipitation which falls to the ground in the form of ice crystals. Ideally, all these forms of precipitation happen when some part of the atmosphere is saturated with water vapor. Finally, precipitation plays a crucial role in ensuring that freshwater is deposited on the Earth's surface.

In which months does your local area experience heavy amounts of precipitation? Identify the seasons if possible.

2.2.5 Sun's Radiation

The Sun is the major heat source that reaches the Earth. Conceptually, the Sun's radiation is a combination of electromagnetic waves (EV) that range from ultraviolet rays (UV) to infrared rays (IR) and influences weather systems as well as photosynthesis in plants. Electromagnetic radiation (EMR) is deposited mainly in infrared and visible wavelengths. As the Earth is heated by EM radiation, the Earth also radiates heat into space. The energy balance of the Earth and the climate relies on a balance between the EM energy that gets into the Earth and the energy emitted from the Earth. When the total quantity of energy leaving the planet is equivalent to that getting in, the global climate may remain unaffected. The sun's angle causes seasonal weather changes.

Experiment 2.2: Self-experiment for a Tornado

Objectives To create a tornado at home.

Required Materials
Tall drinking glass, table salt, long stirrer, and a bottle of carbonated water.

Procedure

1. Pour the water or soda into a glass until 2/3 full.
2. Use stirrer to stirring the liquid as fast as you can without it spilling over, and add some salt.
3. Watch the results as carbon dioxide bubbles develop into a tornado.

Conclusions
Tornadoes develop when the moist and warm air combines with the cool and dry air. They are linked to the storm clouds that have vortexes. As the surrounding air is sucked in by the vortex, more of the air moves in at high velocity to fill the void.

Experiment 2.3: Measuring wind speed

Required Materials
Stopwatch, paper plate, colored tape, a pencil with eraser, colored tape, four paper cups, double-sided tape, and thumbtack that has a plastic end.

Procedure

1. Take one of the paper cups and wrap the colored tape on the exterior.
2. Using a pencil, draw a "y" on the plate. Ensure that the center of "y" is close to the center of the plate.
3. Carefully attach the tape to the side of all the cups. Attach one of the cups to the plate's edge. Ensure that the opening faces left.
4. Next, rotate the plate and similarly attach all the remaining cups.
5. Pin the thumbtack in the center of the plate, where "y" is. Pin the eraser and the plate together then hold the cups and plate to enable them rotate in the breeze.
6. Locate a website that provides you with the speed of wind in your locality. Count the number of times that the marked cup rotates in 30 s and jot it down.

(continued)

7. Go through the process at different periods for various days. You should be able to convert the numbers into wind speeds. After a short while, you will not have to check the speed of wind of the area using the Internet.

Conclusion

An anemometer measures wind speed. It does so by spinning discs that are connected to an electronic device. The device counts the rotations and converts them into speeds.

2.3 Biogeochemical Cycles of the Green World (Planet Earth)

2.3.1 Water Cycle

As shown in Fig. 2.3 below, water cycle describes the way water evaporates from the Earth's surface, gets into the atmosphere, and cools into rain before falling to the Earth's surface as precipitation. Water cycling is an integral part of weather

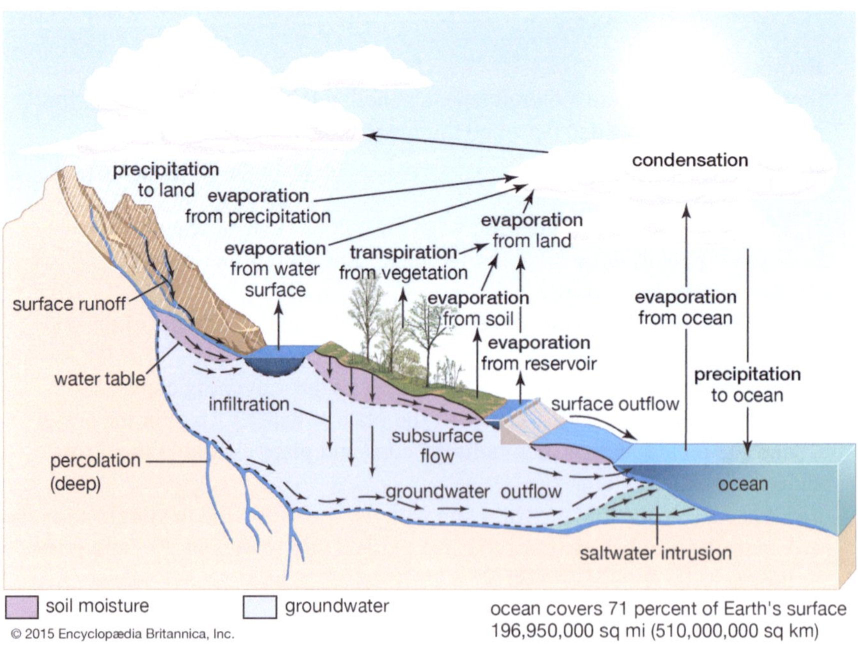

Fig. 2.3 Water cycle, *Source:* Encyclopædia Britannica (2015)

patterns. The water cycle plays a crucial role in the sustenance of life on Earth and has six major steps:

- Step 1: Evaporation
- Step 2: Condensation
- Step 3: Sublimation
- Step 4: Precipitation
- Step 5: Transpiration
- Step 6: Runoff
- Step 7: Infiltration

2.3.1.1 Step 1: Evaporation

Water can be found in plants, lakes, oceans, soil, and swamps. Heat from the sun excites the water molecules causing them to spread out. Evaporation is mainly defined as the loss of density and causes water to form water vapor clouds in the air (Barry and Chorley 2009). Water evaporation usually happens when water reaches its boiling point, which is about 100 °C. Contrarily, in places where humidity and air pressure are lower, less heat energy is required for evaporating water due to the lower pressure holding the water molecules. Water evaporating from oceans is non-salty as the salt is too heavy for it to rise along with the vapor. This is the reason water found in lakes and rivers is not salty. Ice and snow can become water vapor without necessarily turning into water. The process is referred to as "sublimation" and is attributed to dry winds and low humidity. Sublimation happens typically at mountain peaks where there is low air pressure, thus implying that less energy is required for sublimating the ice into vapor. Mountain Everest is among the highest peaks on the planet with the necessary components for sublimation.

What is the role of evaporation in availing water in water bodies? Is it a positive or negative thing?

2.3.1.2 Step 2: Condensation

Water vapor that rises into the atmosphere cools when it meets the cooler air found higher in the sky. Consequently, the vapor becomes a cloud that winds and moving air currents push around the world. The water vapor starts condensing on the surface of tiny particles that have escaped during the evaporation process. The tiny droplets begin falling into each other and merge to produce a larger droplet. Gravity pulls down a droplet that is large enough at a rate that is more than the force in the cloud, resulting in rainfall or precipitation (Barry and Chorley 2009). In case precipitation happens within conditions that are specifically cold or with low air pressure, such water droplets can usually crystallize and freeze, making a waterfall in the form of solid ice or snow.

2.3.1.3 Step 3: Sublimation

Sublimation is defined as the conversion from a solid to a gaseous state of matter without necessarily going through the liquid stage (see Fig. 2.4). In the water cycle, sublimation is usually utilized for describing the process of ice and snow changing into water vapor without going through the liquid state. It is quite challenging to see sublimation happening. However, one of the ways that you can use for seeing the sublimation results is hanging a wet shirt outside on a freezing day. The ice found in the shirt typically disappears. Furthermore, the best way of seeing sublimation is to ensure that you do not use any water, but carbon dioxide. An example is the loss of snow cover from Mt. Everest because of Chinook winds that trigger the process of sublimation. Mount Everest is the highest mountain in the world, located in the Himalayas (Barry and Chorley 2009). Chinook winds have been named after the Chinook Native Americans who lived within coastal areas of Oregon and Washington close to the Columbia River. Chinook winds usually develop when the warm and moist air blows from the Pacific Ocean towards the Rocky Mountain range, and can be around ten degrees Celsius (10 °C). Sublimation happens when some weather conditions are present, including dry winds and low relative humidity. It also often occurs at higher altitudes where there is lower pressure than that found at lower altitudes.

Mention two mountains that experience the sublimation process.

Case study 2.1: Sublimation on Mountain Everest

Fig. 2.4 Sublimation on Mountain Everest

(continued)

Mountain Everest is the highest mountain on the planet, reaching more than 8840 m in height. However, one wonders what processes of the water cycle are at work at a place where there is no liquid water. The mountain is cold most of the time. At more than 29,000 feet, temperatures at the mountain's peak are quite low, hence making it difficult to rain (Yang et al. 2011). However, it does get some snow. Snow melts but rarely goes above freezing. With this in mind, how do you think that snow accumulates forever on the mountain? As the figure at the end of the case study highlights, winds blow large amounts of snow from the peak. The sublimation process is functioning. In the same way, evaporation converts liquid water into vapor, sublimation converts frozen water into vapor gas, thus bypassing the melting phase. The heat of the sun also causes sublimation, and on top of the mountain on any clear day, the sun is quite intense and offers energy for sublimation despite being below freezing. However, sublimation does not happen at the peak of the mountains. In case snow is present and ice accumulates in your neighborhood, sublimation is also happening on a dry and sunny day. Mountain Everest loses snow because of sublimation.

2.3.1.4 Step 4: Precipitation

Precipitation refers to the several forms of water present within the atmosphere, typically falling in the form of sleet, hail, and rain. After condensation, the size of water droplets becomes larger as more water particles meet each other. The force of gravity is higher than the wind holding it, causing the water to fall in the form of rain, sleet, hail, or snow (Barry and Chorley 2009). Particulate matter plays a fundamental role in the development of water droplets. They act as a nucleus where the water will settle. It is the reason rainwater is not often pure since it can contain several minerals. The water quantity falling from the sky varies significantly. While some regions always have rain, others have very little.

2.3.1.5 Step 5: Transpiration

Soil absorbs some of the water from precipitation leading to the transpiration process. It is similar to evaporation, but the main difference is that water is released from plant leaves. Plant roots first absorb water and push it towards the leaves. The water then moves through the plant by capillary action. This water goes through the transpiration process, which entails its evaporation from flower petals and leaves (Barry and Chorley 2009). It should be noted that only the extra water that has not been used for photosynthesis moves out through the stomata. Water then gets into the biosphere as a gaseous state.

How does relative humidity influence the rate of transpiration? Identify and explain four major factors that influence the rate of transpiration.

2.3.1.6 Step 6: Runoff

After the waterfalls and the soil saturates or the snow melts, the water follows gravity or other inclines in order to join or form rivers through a process called runoff. This is the process through which water reaches the lakes and the ocean. The direction the water flows is called stream flow.

How does surface runoff impact the quality of water? What factors influence the prevalence of runoff?

2.3.1.7 Step 7: Infiltration

Conceptually, infiltration is the process through which precipitation soaks into the soil and flows into rocks through pore spaces and cracks. The larger quantity of rainwater ends up infiltrated. The soil may absorb the water and can remain in it for an extended period after evaporation (Huntington 2006). In case of substantial vegetative cover, infiltrated water can be absorbed through the roots, and later transpire. It should be noted that infiltration happens within the upper ground layers. However, it can also go on further downwards into the water table. Figure 2.5 shows phase change of matter.

How long does it take for rainwater to infiltrate into the ground?

The level of ground saturation determines whether water may proceed downwards for replenishing aquifers and water tables, a process called percolation. In the case of close water bodies, infiltrated water may end up within water bodies. The infiltration rate relies on factors like the precipitation amount, soil types, quantity of vegetative cover, pre-saturation levels, land topography, and evapotranspiration levels within the region.

Activity Take a walk along a riverbank to find evidence of surface water. As you walk, stop at several places where surface water gets into the river. Identify the likely sources of the water. Write down your observations.

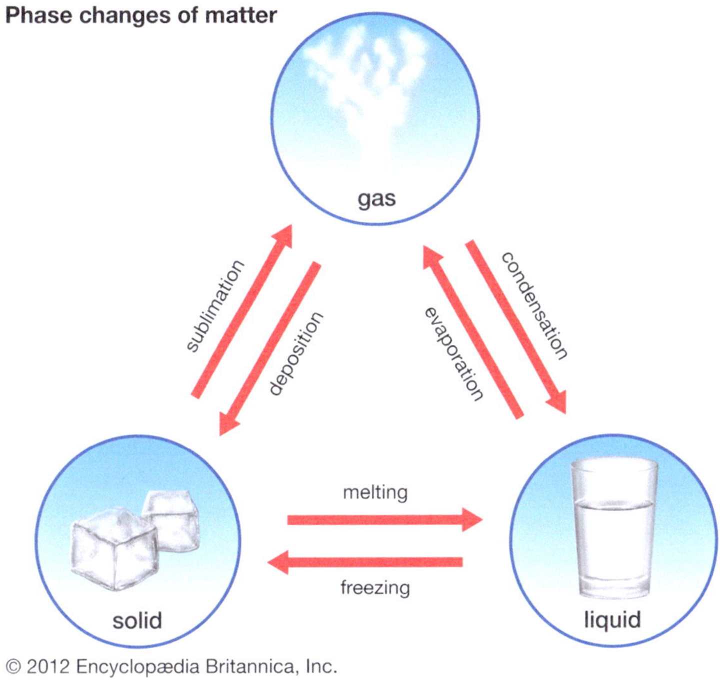

Fig. 2.5 Phase change diagram, *Source:* Encyclopædia Britannica (2011a)

2.3.2 *Carbon Cycle*

The carbon cycle is a type of the biogeochemical cycle and is a term used for describing the ways carbon moves between the atmosphere, biosphere, hydrosphere, and lithosphere (see Fig. 2.6). Carbon is critical for the sustenance of various functions of the planet. It is a significant component of various minerals and living beings. These spheres act as carbon sinks or reservoirs in the short-term or long-term. As the Earth is a dynamic environment, processes like respiration, decomposition, and erosion continually shift carbon between reservoirs (Prentice et al. 2001). Carbon enters is stored and leaves the various Earth spheres using various methods and quantities.

In the atmosphere, carbon is stored in the form of carbon dioxide (CO_2) and methane (CH_4); greenhouse gases that absorb and retain heat. Living organisms release CO_2 into the atmosphere through respiration. Photosynthesis and dissolution are essential processes for removing this gas from the atmosphere. Methane is released into the atmosphere through processes such as decomposition, combustion of fossil fuels, and animal emissions.

In the biosphere, all organisms have organic carbon by removing it from atmospheric CO_2 through photosynthesis or the consumption of other organisms (Prentice et al. 2001). Often, carbon is retained until an organism decomposes to

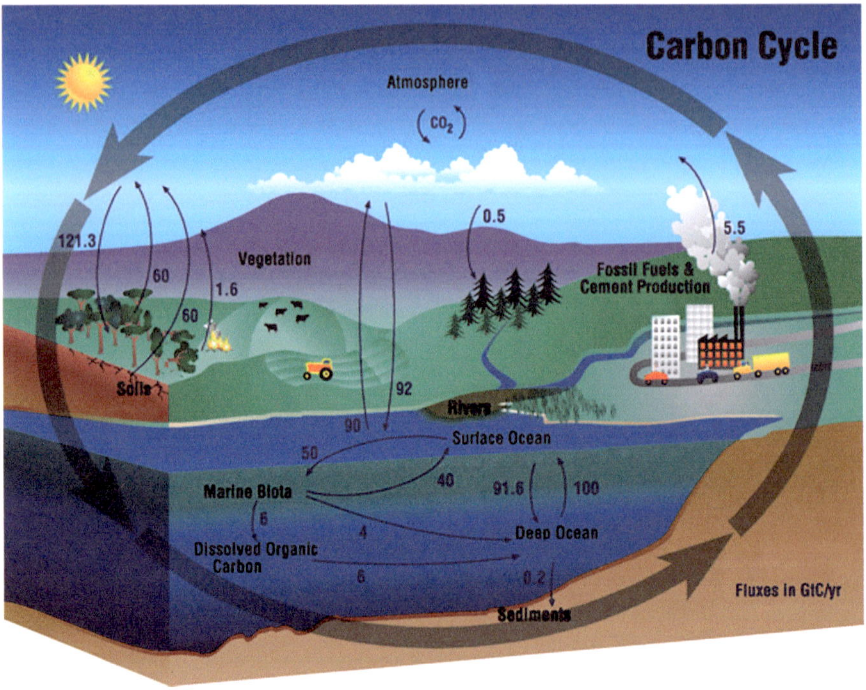

Fig. 2.6 The carbon cycle, *Source:* Science Kids (n.d.)

release carbon into the lithosphere or atmosphere. Regarding the hydrosphere, the upper ocean layers hold a large quantity of dissolved organic carbon, whereas the lower ocean waters have vast amounts of dissolved inorganic carbon.

Furthermore, the soil holds carbon within the lithosphere as both organic and inorganic carbon. Soil respiration can make the carbon leave and carry it into oceans or rivers where it gets into the hydrosphere. In the Earth's crust, a large quantity of carbon is stored within kerogens and limestone (Prentice et al. 2001). Organics consist of compressed living matter. Such forms of carbon are quite stable and could remain within the lithosphere for numerous years. Furthermore, reservoirs can be categorized as carbon sinks or sources.

Today, humans have a significant effect on the carbon cycle. Human activities have led to the elimination of carbon by directly adding it into the atmosphere. The major human impacts on the carbon cycle are the combustion of fossil fuels, land-cover change, and land use (Prentice et al. 2001). Ideally, land use changes, particularly deforestation affects the carbon cycle in two significant ways: elimination of plants and replacement of the built environment, thus leading to a reduction in the carbon store. Built environments are aspects of the surroundings that humans build. It includes the buildings and urban spaces, such as infrastructure and parks, that support human activity. In addition, the built environment incorporates all building forms as well as civil engineering infrastructure.

What is the role of the primary producer in a carbon cycle? Does the carbon cycle begin with photosynthesis in plants? Briefly explain your answer. Are plants the primary consumers of carbon dioxide? Explain. Does carbon dioxide dissolve readily in water? Is all the carbon available always recycles within the carbon cycle? In what way does carbon get into the soil?

2.3.2.1 Concepts of the Carbon Cycle

1. Carbon presence ascertains whether something is organic or inorganic.
2. Carbon moves through the ecosystem in various ways, including respiration, photosynthesis, weathering, and geological processes.
3. Numerous factors, including anthropogenic activities and seasons, affect carbon concentration in the atmosphere.

2.3.3 Nitrogen Cycle

The nitrogen cycle has three major steps: nitrogen fixation, nitrification, and denitrification. It is defined as a biogeochemical cycle through which nitrogen is converted into multiple chemical forms as it circulates via terrestrial, marine, and atmospheric ecosystems (see Fig. 2.7). Nitrogen conversion can take place through physical and biological processes. Some of the significant nitrogen cycle processes are fixation, nitrification, denitrification, and ammonification (Gruber and Galloway 2008). A large proportion (78%) of the Earth's atmosphere is made up of nitrogen. Despite this, there is little atmospheric nitrogen available for biological use, thus resulting in a scarcity of usable nitrogen. The nitrogen cycle is crucial to ecologists since nitrogen availability can influence ecosystem processes such as decomposition and primary production. Nitrogen exists in the environment in various chemical forms, including ammonium nitrate, organic nitrogen, nitrous oxide, and nitric oxide (Gruber and Galloway 2008). Many of the processes within the nitrogen cycle are carried out by microbes.

Conceptually, nitrogen fixation refers to the conversion of nitrogen gas into nitrites and nitrates through biological, industrial, or atmospheric processes. Nitrogen assimilation refers to the absorption of ammonium or nitrate from the soil through plant roots. On the other hand, ammonification is beneficial for providing living organisms with nitrogen (Gruber and Galloway 2008). Ammonification occurs because of decomposers that convert plant and animal cells into simpler components, hence availing nutrients within the ecosystem. Nitrification is defined as the conversion of ammonium nitrate by nitrifying bacteria. Furthermore, denitrification refers to the conversion of nitrates into nitrogen gas. The process is facilitated by

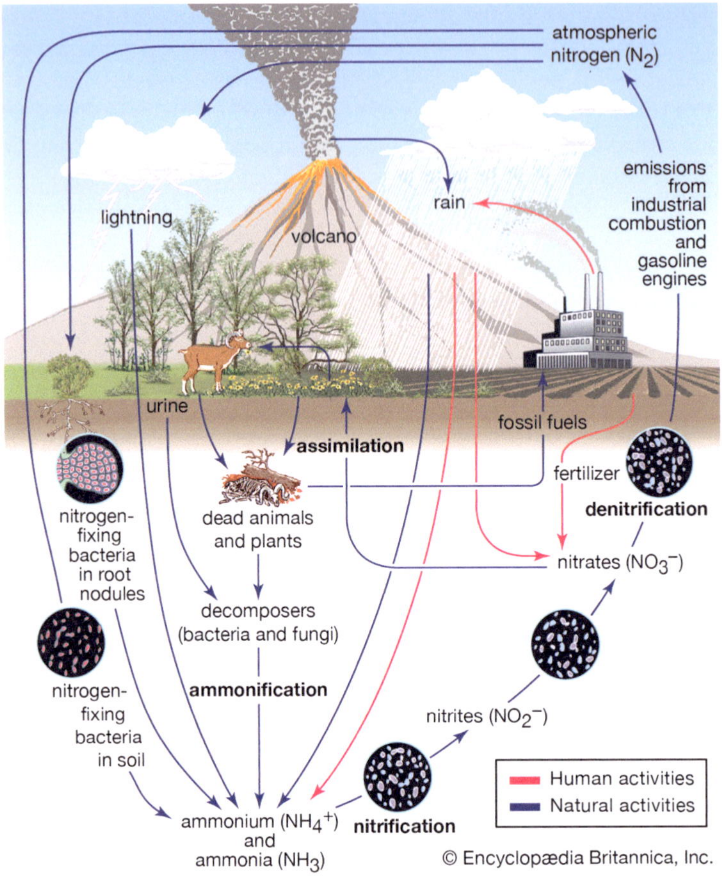

Fig. 2.7 The nitrogen cycle, *Source:* Encyclopædia Britannica (2011a)

bacterial species, including Clostridium and Pseudomonas. The process takes place in anaerobic conditions, including waterlogged soils.

Many human activities significantly influence the nitrogen cycle. The combustion of fossil fuels, nitrogen-based fertilizer applications, and other activities can drastically increase the quantity of available nitrogen. Since nitrogen availability impedes the primary productivity of numerous ecosystems, substantial changes in nitrogen availability can result in severe changes in the nitrogen cycle within terrestrial and aquatic ecosystems (Gruber and Galloway 2008). Anthropogenic activities, including the combustion of fossil fuels and the deployment of artificial nitrogen fertilizers, have significantly influenced the global nitrogen cycle. Adding nitrogen to the terrestrial ecosystems could lead to a nutrient imbalance in trees and the decline of biodiversity.

Increased nitrogen availability often leads to a change in carbon storage, thus influencing processes. Within agricultural systems, fertilizers are utilized extensively for increasing plant production while available nitrogen may undergo leaching and get into rivers and streams (Gruber and Galloway 2008). High nitrogen concentration in water streams can result in anoxia, altered biodiversity, and degradation of habitats. A typical result of increased nitrogen is the proliferation of harmful algal blooms that contributes to a high mortality rate in fish. Some of the hazardous impacts of nitrogen on aquatic ecosystems are water acidification, eutrophication, and toxicity of the water. Eutrophication results typically in low levels of dissolved oxygen within the column of water, which can lead to the demise of aquatic fauna. An example of the effects of eutrophication is the oceanic dead zones found close to the mouth of Mississippi within the Mexican Gulf (McIsaac et al. 2001). Other visible effects that result from the deposition of nitric acid rain are Hudson Highlands, Catskills, and the New York Adirondack Lakes. They result in the death of fish and numerous aquatic species.

Even in the absence of such economically catastrophic impacts, the addition of nitrogen may result in biodiversity changes and composition of species, thus negatively affecting the overall functioning of the ecosystem. Aerosols and nitrogen gases can be directly poisonous to some plant species, and negatively affect the growth of plants close to large sources of nitrogen pollution. There could also be alterations in plant species interactions as the accumulation of nitrogen compounds heightens its availability within a particular ecosystem (Canfield et al. 2010). Ammonia can also result in higher toxicity towards sensitive plant species, especially ones that use nitrate as the nitrogen source, leading to poor shoot and root development. Higher nitrogen deposition also results in soil acidification that increases base leaching within the soil. Changes that result from high nitrogen deposition can result in environmental vulnerabilities to ecological disturbance and stress. This makes the environment less resilient to processes or situations that would have had little effect on its prolonged vitality. Some suggest that changes to the nitrogen cycle can result in a heightened susceptibility to infectious and parasitic diseases among wildlife and humans.

What are the negative impacts of the nitrogen cycle? Explain two major roles of the nitrogen cycle in life sustenance.

2.3.4 Sulfur Cycle

Sulfur is among the components that constitute vitamins and proteins. Proteins are made up of amino acids, of which some are made up of sulfur atoms. It is beneficial for both plants and animals and is a crucial element for living things'

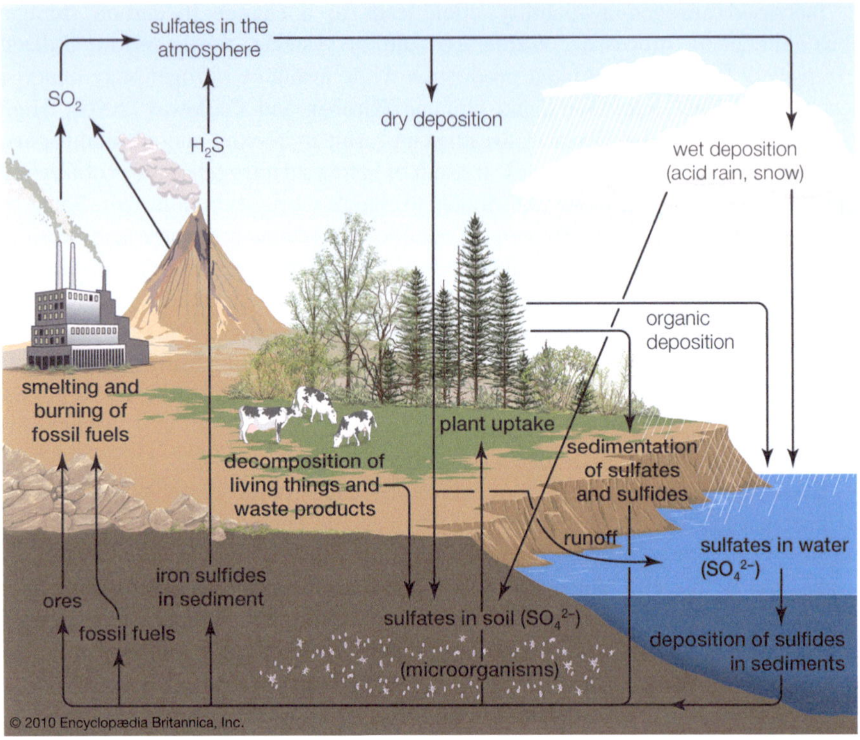

Fig. 2.8 The sulfur cycle, *Source:* Encyclopædia Britannica (2010)

macromolecules. Its release into the atmosphere is a result of fossil fuel combustion, such as coal. Sulfur plays a pivotal role in the development of disulfide bonds in proteins which assist in ascertaining their three-dimensional folding patterns. As Fig. 2.8 shows, sulfur moves between the atmosphere, land, and oceans. In the atmosphere, sulfur is in the form of sulfur dioxide and gets into the atmosphere in three ways: decomposition, volcanic activity, and combustion of fossil fuels.

Atmospheric sulfur dioxide becomes available to marine and terrestrial ecosystems when it dissolves in the form of weak sulfuric acid. Rock weathering also avails sulfates to terrestrial ecosystems. The decomposition of living organisms returns sulfates into the atmosphere, soil, and ocean. On the land surface, sulfur can be deposited through four main ways: rock weathering, precipitation, geothermal events, and directly from the atmosphere. As rain falls, sulfur dissolves into weak sulfuric acid (Haneklaus et al. 2003). Sulfur can also fall from the atmosphere directly as fallout. The weathering of rocks that contain sulfur can also contribute to its release. Terrestrial ecosystems can then use the soil sulfates and after the decomposition of organisms, release it back into the atmosphere in the form of hydrogen sulfide gas.

Human activities have been instrumental in alteration of the balance of the sulfur cycle. The combustion of vast amounts of fossil fuels from coal leads to the release of high quantities of hydrogen sulfide gas into the air (Haneklaus et al. 2003). When rain combines with this gas, it creates acid rain, which can lead to environmental deterioration by lowering the pH of lakes. It also leads to chemical degradation of the built environment, especially buildings, by changing the appearance of buildings. For instance, many of the marble monuments, including Washington's Lincoln Memorial, have suffered the disastrous effects of acid rain. These examples highlight the various impacts of human activities on the environment.

Identify and explain the significance of the sulfur cycle. How is sulfur dangerous to humans?

2.3.5 *Phosphorus Cycle*

The phosphorus cycle refers to the circulation of phosphorus in several forms (see Fig. 2.9). Phosphorus is one of the scarcest of all elements recycled within the biosphere. It is crucial for life since it enhances the transfer of energy and acts to

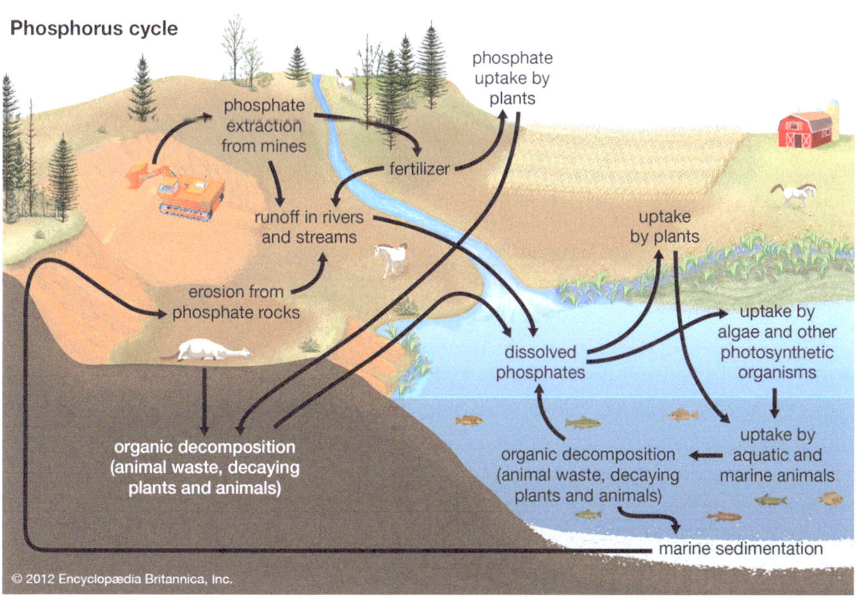

Fig. 2.9 The phosphorus cycle, *Source:* Encyclopædia Britannica (2012b)

hold genetic information in the form of ribonucleic acid (RNA) and deoxyribonucleic acid (DNA) (Kümmerer 2004). Also, phosphates are a pertinent component of adenosine triphosphate (ATP), which organisms use as energy. Similar to calcium, phosphorus is beneficial for vertebrates, and 80% of it is found within the bones and teeth.

The major difference between the phosphorus cycle and other cycles is that it lacks a gas phase. However, it is worth noting that small phosphoric acid quantities can enter the atmosphere to produce acid rain. Sedimentary rock is the major reservoir of phosphorus, where the cycle starts. During precipitation, the process of weathering removes phosphates from rocks and distributes them throughout the water and soil. Plants absorb phosphate ions from the soil, and they move to animals when herbivores consume plants or when carnivores eat herbivores or plants. Phosphates absorbed via consumption ultimately come back to the soil through feces and urine excretion. The above process also happens in the aquatic ecosystem (Kümmerer 2004). Since phosphorus is less soluble, it mainly reaches waters through runoff. Phosphates also get into the water streams through fertilizer runoff, natural deposition of minerals, and other industrial process wastes. Phosphates have a tendency of settling on the bottom of lakes and ocean floors. Phosphorus benefits many biological processes, but a high concentration of it is perceived as a pollutant. Phosphates often promote the growth of plants and excessive growth consume large quantities of dissolved oxygen, thus leading to suffocation of marine animals and fish. It also blocks the available sunlight from reaching bottom-dwelling species; an aspect called eutrophication. Human beings may also affect the phosphorus cycle in numerous ways, including deforestation and the application of agricultural fertilizers. The recycling of nutrients supports rainforest ecosystems, while agricultural runoff increases the proportion of phosphates in waterways.

2.4 Individual Matter: James Hansen

James Hansen is a respected American professor, who currently lectures at Columbia University in the Earth and Environmental Sciences' department (see Fig. 2.10). His fame is attributed to him being called the "father of global warming" due to his early warnings about global warming.

He was born in 1941 in Iowa. He drew his inspiration from Red Giants and White Dwarfs, authored by Jastrow (1990), during his high school years, which inspired him to venture into a scientific career. Hansen was trained in physics and astronomy from the University of Iowa, where he graduated with a bachelor's degree in Mathematics and Physics with distinction in 1963. Hansen further earned his master's degree in 1965 and a Ph.D. degree in Physics from the University of Iowa. He participated in the National Aeronautic and Space Administration (NASA) graduate traineeship program and attended Tokyo University (Llavador et al. 2015). His first job was on radiative transfer models to get acquainted with the Venusian atmosphere. He later used the knowledge obtained while in school to understand

Fig. 2.10 James Hansen, *Source:* Columbia University (2018)

the Earth's atmosphere, particularly the impacts of aerosols and trace gases on climate. His deployment and development of various earth climate models have been crucial for enhancing the understanding of the planet's climate. He was the leader of NASA Goddard Institute for Space Studies in between 1981 and 2013 (Niemeier 2017). He has been the director of Columbia University Earth Institute's Program on Climate Science since 2014. The program is dedicated to the advancement of basic climate science to the propagation of public awareness regarding policy actions for the prevention of global warming. One of his career highlights is his research in climatology and testimony about climate change in 1988. His major achievements are being elected to the National Academy of Sciences (NAS), the Dan David Prize in 2007, and the authorship of numerous research publications and books (Llavador et al. 2015). As a pioneer of sustainable future, his voice remains pivotal for the combat of harmful impacts of climate change today.

2.5 Future Trends

2.5.1 Africa

Higher temperatures and increased variability in rainfall related to climate change will undermine not only the livelihoods but also food security of Africa's farmers

(Girvetz et al. 2019). Predictions of climate change will become better with the availability of adequate data, unlike now (Kotir 2011). Africa will remain among the continents that will be most vulnerable to the negative impacts of climate change due to the anticipated change and the perceived lack of ability of Africans and their governments to withstand the changes.

2.5.2 *Antarctica*

With the current climate change trends, there is a likelihood of Antarctica's melting ice changing the weather around the world. Melting ice will result in more trapped heat within some regions as well as colder temperatures within others.

2.5.3 *Asia*

Climate change will lead to higher temperatures, intense storms, lower crop yields, the collapse of coral reefs, and erratic rainfall patterns. Many countries in the continent are the most vulnerable of plunging deeper into disaster and poverty in case climate change mitigation and adaptation interventions are not robustly and promptly rolled out.

2.5.4 *Australia*

Projections from climate change in Australia department indicate in the coming years, Australia will experience: more increase in temperatures, with many extremely hot days, increased acidification of oceans and warming, longer-lasting marine waves, reduced cool-season rainfall, more high fire weather danger days, and fewer tropical cyclones.

2.5.5 *Europe*

Considering the current trend, temperature will rise from 3 to 5-°C by the end of the twenty-first century (European Environment Agency 2019). With increased exploitation of fossil fuel resources, the rise in temperature will be significantly higher. The average temperature will rise by about 0.2 °C every 10 years for the next 20 years.

2.5.6 *North America*

In a single generation, the climate in numerous North American cities will change to that of locations found miles away or to another climate, unlike the one experienced in this continent today. By the 2080s, even with limits being imposed on emissions, North American urban areas will have completely different climates (University of Maryland Center for Environmental Science 2019). In case conditions continue uncontrolled in the entire century, the urban areas of North America will have the contemporary climate like that of the South. For example, within the Eastern part of the US, almost all the urban areas, including Philadelphia, New York, and Boston will resonate with the contemporary climates of the southwest and south.

2.5.7 *South America*

South America will experience reduced rainfall in the south and east of the Amazon region as well as the coast of South America. There is also a likelihood of droughts intensifying in some areas and seasons because of lower precipitation and higher levels of evapotranspiration. Heatwaves will increase in frequency, intensity, and length.

2.6 Summary

- This chapter explored the definitions of major Basic Ecology terms, such as ecology, biosphere, ecosystem, community, population, organisms, and carrying capacity.
- The second part of the chapter discussed weather and climate, with a focus on temperature, humidity, wind, precipitation/rainfall, and the sun's radiation.
- The third part of this chapter examined the biogeochemical cycles of Green World. The primary processes of the water cycle were explored, including evaporation, condensation, sublimation, precipitation, transpiration, runoff, and infiltration. The main processes of the carbon cycle explored were dissolution and photosynthesis. One of the findings is that the presence of carbon determines whether a substance is organic or inorganic. Various factors, especially human activities and seasons, influence the presence of carbon in the atmosphere.
- This chapter has also explored three steps of the nitrogen cycle: nitrogen fixation, denitrification, and nitrification. In addition, the sulfur and phosphorus cycles have been explored comprehensively.
- From this chapter, a major point to note is that concepts in Basic Ecology play a crucial role in Greenness.

- James Hansen's research, publications, and voice remain crucial for understanding the effects of climate change today.
- The future trends in the various continents mainly show that there could be adverse impacts associated with more climate change vagaries.
- In the future, Africa will be characterized by food insecurity because of the high temperatures, while Europe and North America will experience extreme temperatures
- South America and Asia will have reduced rainfall, heatwaves, lower crop yields, and more droughts
- Australia's future trends could be increased acidification, fewer tropical cyclones, and unpredictable weather patterns

2.7 Discussion Questions

1. Differentiate between organism, population, community, biosphere, and ecosystem.
2. Describe community ecology in terms of mutualism, predation, and competition.
3. Which is the most effective way of protecting biodiversity?
4. Describe ecosystem services and provide examples.
5. What is the importance of wetlands in maintaining biodiversity?
6. Discuss the components of an ecosystem.
7. Differentiate between ecosystem diversity, genetic diversity, biodiversity, and invasive species.
8. Which is the major threat to biodiversity today?
9. Outline four benefits of biodiversity.
10. What could change a certain habitat's carrying capacity?
11. Describe factors that determine the carrying capacity of (a) a country and (b) a population.
12. What is the difference between population and community ecology?
13. Briefly describe the water cycle.
14. What factors could lead to the disruption of the water cycle?
15. How does energy limit the number of steps involved in a food chain?
16. In what way is the sun the engine of the motorcycle?
17. Describe the carbon cycle.
18. Describe the way photosynthesis acts as the major consumer of carbon dioxide.
19. What are fossil fuels, and what is their effect on the environment?
20. Describe the four spheres of the Earth and their importance.

References

Barry, R. G., & Chorley, R. J. (2009). *Atmosphere, weather and climate*. Abingdon: Routledge.

Brooks, T. M., Mittermeier, R. A., da Fonseca, G. A., Gerlach, J., Hoffmann, M., Lamoreux, J. F., et al. (2006). Global biodiversity conservation priorities. *Science, 313*(5783), 58–61.

Canfield, D. E., Glazer, A. N., & Falkowski, P. G. (2010). The evolution and future of Earth's nitrogen cycle. *Science, 330*(6001), 192–196.

Columbia University. (2018). *James Hansen*. Retrieved December 2, 2018, from http://www.mediafire.com/file/cdcybew6bbi7f6q/Hansen_boulder_photo.jpg/file/.

Daily, G. C., & Ehrlich, P. R. (1992). Population, sustainability, and Earth's carrying capacity. *BioScience, 42*(10), 761–771.

Encyclopædia Britannica. (1999). *Effect of topography and climate on water-induced soil erosion*. Retrieved November 04, 2019, from https://www.britannica.com/science/topography-geology#/media/1/599655/19386/.

Encyclopædia Britannica. (2010). *Sulfur cycle*. Retrieved November 02, 2019, from http://www.britannica.com/science/sulfur-cycle#/media/1/572740/111671/.

Encyclopædia Britannica. (2011a). *Nitrogen cycle*. Retrieved October 01, 2018, from https://www.britannica.com/science/nitrogen-cycle#/media/1/416271/118424/.

Encyclopædia Britannica. (2012a). *Carrying capacity; exponential versus logistic population growth*. Retrieved November 10, 2019, from https://www.britannica.com/science/carrying-capacity/media/97118/158213/.

Encyclopædia Britannica. (2012b). *Phosphorus cycle*. Retrieved November 02, 2019, from https://www.britannica.com/science/phosphoruscycle#/media/1/457621/139360/.

Encyclopædia Britannica. (2015). *Water cycle*. Retrieved October 01, 2018, from https://www.britannica.com/science/water-cycle/media/278858/112177/.

European Environment Agency (2019). *How is the climate going to change in the future?* Retrieved October 24, 2019, from https://www.eea.europa.eu/themes/climate/faq/how-is-the-climate-going-to-change-in-the-future/.

Girvetz, E., Ramirez-Villegas, J., Claessens, L., Lamanna, C., Navarro-Racines, C., Nowak, A., et al. (2019). Future climate projections in Africa: where are we headed? In *The climate-smart agriculture papers* (pp. 15–27). Cham: Springer.

Gruber, N., & Galloway, J. N. (2008). An Earth-system perspective of the global nitrogen cycle. *Nature, 451*(7176), 293.

Haneklaus, S., Bloem, E., & Schnug, E. (2003). The global sulphur cycle and its links to plant environment. In *Sulphur in plants* (pp. 1–28). Berlin: Springer, Dordrecht.

Hansen, J. E. (2013). Climate urgency. *Simulation & Gaming, 44*(2–3), 232–243.

Huntington, T. G. (2006). Evidence for intensification of the global water cycle: review and synthesis. *Journal of Hydrology, 319*(1), 83–95.

Jastrow, R. (1990). *Red giants and white dwarfs*. New York: WW Norton & Company.

Kotir, J. H. (2011). Climate change and variability in Sub-Saharan Africa: a review of current and future trends and impacts on agriculture and food security. *Environment, Development, and Sustainability, 13*(3), 587–605.

Kümmerer, K. (2004). Resistance in the environment. *Journal of Antimicrobial Chemotherapy, 54*(2), 311–320.

Llavador, H., Roemer, J. E., & Silvestre, J. (2015). *Sustainability for a warming planet*. Boston: Harvard University Press.

Lydolph, P. E. (1985). *Weather and climate*. Lanham: Government Institutes.

McIsaac, G. F., David, M. B., Gertner, G. Z., & Goolsby, D. A. (2001). Eutrophication: nitrate 1flux in the Mississippi River. *Nature, 414*(6860), 166.

Myers, N., Mittermeier, R. A., Mittermeier, C. G., Da Fonseca, G. A., & Kent, J. (2000). Biodiversity hotspots for conservation priorities. *Nature, 403*(6772), 853.

Niemeier, D. (2017). Building a wall on the imaginary line. *Environmental Science & Technology, 51*(3), 1053–1053.

Prentice, I. C., Farquhar, G. D., Fasham, M. J. R., Goulden, M. L., Heimann, M., Jaramillo, V. J., et al. (2001). *The carbon cycle and atmospheric carbon dioxide*. Cambridge: Cambridge University Press.

Runcorn, S. K. (1962). Convection currents in the Earth's mantle. *Nature*, *195*(4848), 1248.

Science Kids. (n.d.). *Carbon cycle*. Retrieved November 11, 2019, from http://www.sciencekids.co.nz/pictures/-earth/carboncyclediagram.html/.

University of Maryland Center for Environmental Science (2019). Climate of North American cities will shift hundreds of miles in one generation: New web application helps visualize climate changes in 540 North American cities. *ScienceDaily*. Retrieved February12, 2019, from www.sciencedaily.com/releases/2019/02/190212120044.htm/.

Yang, X., Zhang, T., Qin, D., Kang, S., & Qin, X. (2011). Characteristics and changes in air temperature and glacier's response on the north slope of Mt. Qomolangma (Mt. Everest). *Arctic, Antarctic, and Alpine Research*, *43*(1), 147–160.

Chapter 3
Everyday Activities in the Green World

Abstract Anthropogenic activities play a crucial role in determining the state of the Green World. To this extent, understanding the role played by everyday activities is crucial for ascertaining their likely impacts on the environment and the ecosystem. Each activity plays a crucial role in sustainability and improvement of the Green World, but this chapter will mainly focus on organic and sustainable agriculture, energy, science and technology, socio-political systems and rural-urban migration, house-to-house activities, and transportation. Emphasis is placed on the five principles identified in the first chapter, which were resource conservation, pollution reduction, energy conservation, waste reduction, and protecting the earth's ecological balance. Sky Greens Singapore is the case study embedded in the chapter to provide insight on everyday activities. Using an experiment on energy and the individual contributions of Vandana Shiva, this chapter lays the foundation for understanding the contribution of our activities to the environment and society. This chapter lays the foundation for understanding the Going Green movement or subject, particularly the contribution of our activities to the environment and society.

Keywords Ecosystems · Vertical farming · Vandana Shiva · Sustainable agriculture · Industries

3.1 Introduction

Various activities play detrimental roles in the Green World. In this chapter, the "Green World" will refer to activities that promote the conservation, protection, and sustainability of the environment. They emphasize on the five aspects identified in Chap. 1 about Going Green: **G**rowth, **R**enewable, **E**fficient, **E**conomically Sustainable, and **N**atural. These activities also involve the adoption of the five major principles identified in Chap. 1: Conservation of resources, reduction of pollution, conservation of energy, reduction of waste, and protection of the ecological balance of the earth. In discussing the various sub-chapters, the focus will be on the contribution of each towards a greener world. The negative impacts of these sectors will be explored in Chap. 4.

A. A. Akinsemolu, *The Principles of Green and Sustainability Science*,
https://doi.org/10.1007/978-981-15-2493-6_3

3.2 Sustainable Agriculture

Sustainable agriculture is the major form of agriculture that plays a crucial role in the Green World and is defined as the production of animal or plant products, including fiber and food, using farm methods that conserve and protect the environment, human communities as well as the animal welfare (Altieri 2018). The main objective of sustainable agriculture is to meet the food and textile needs of the present society without necessarily compromising the capacity of future generations (Definition adapted from the Brundtland Commission 1987). Sustainable agriculture practitioners strive to incorporate three primary objectives in their practices: economic profitability, healthy environment, and socioeconomic equity. All stakeholders within the food system, including food processors, growers, distributors, consumers, retailers, and waste managers, play a crucial role in making sure that there is a sustainable agricultural system.

There are many Green practices utilized by people who work in sustainable agriculture. Growers can utilize methods for promoting soil health, minimizing water usage, and lowering the levels of pollution on the farm. Similarly, retailers and consumers who value sustainability can use values-based foods that are typically grown through methods that promote the well-being of farm-workers, protect the environment, or enhance the local economy (Altieri 2018). Today, the sustainable agriculture movement has increasingly gained support in mainstream agriculture, which has been welcomed by the Green movement that has advocated for sustainable practices. This is because sustainable or Green agriculture tackles numerous social and environmental issues, and provides innovative as well as economically viable opportunities for the consumers, growers, and laborers.

As the Fig. 3.1 above shows, sustainable agriculture embeds three fundamental objectives which can be viewed as more significant roles in enhancing Greenness: environmental health, economic profitability, and social equity. An example of such sustainable agriculture is vertical farming, which involves the production of medicinal plants and food in stacked layers as well as inclined surfaces. Many modern concepts of vertical farming tend to utilize indoor farming methods and controlled environment agriculture. The main benefits of vertical farming are better human health and urban growth. The other benefits are weather disruption, energy sustainability, mass extinction, conservation, and reduction of poverty (Benke and Tomkins 2017). Some of the renowned case studies of vertical farming are Food Field (Detroit, Michigan), The Distributed Urban Farming Initiative (Bryan, Texas), and Sky Greens (Singapore). A case study of Sky Greens is presented towards the end of the chapter to offer more insight on vertical farming. In Africa, Kenya and Uganda are some of the states that practice vertical farming. In Uganda, the lack of adequate financial resources for building a modern vertical farm and the limited access to water and land have made farmers venture into vertically stacked wooden crate units. These units are made of a vermicomposting chamber and use water bottles to irrigate crops continuously. The advantage of these stacks is that they consume less water and enable urban farmers to plant crops such as kale. Kampala

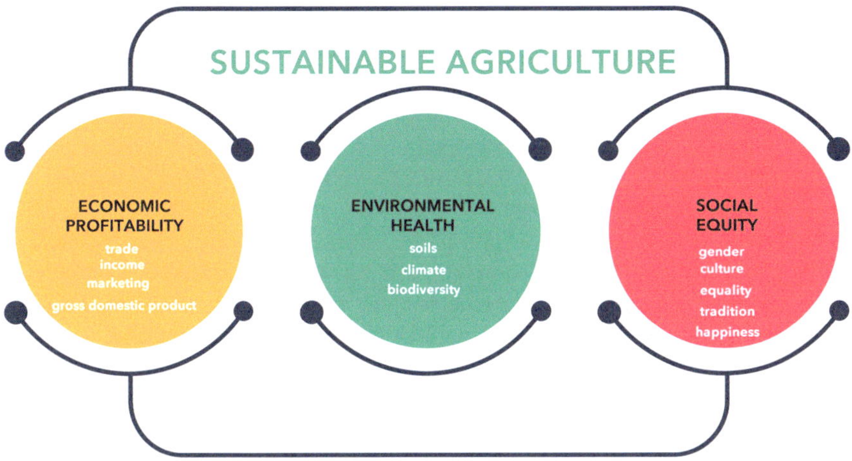

Fig. 3.1 Sustainable agriculture

has 15 such farms with the hope more will be created. In Kenya, sack gardens are effective forms of vertical farms. They are made from sisal fibers at a low cost with each sack estimated to cost $0.12. More importantly, sack gardens use fewer resources and local materials but provide substantial yields similar to those achieved by other farmers globally. For example, in Kibera, Kenya, many people have turned towards sack gardening with more than 22,000 households farm on sacks. Despite this, the continent still has numerous unique opportunities for vertical farms. It is time for African innovators and entrepreneurs to invest in building and designing vertical farms.

Furthermore, sustainable agriculture plays a crucial role in promoting the stewardship of human and natural resources. The human resource stewardship entails considering social responsibilities, including the laborers' living and working conditions, rural community needs, and consumer safety, while land and natural resource sustainability involve the maintenance or enhancement of this vital resource base (Altieri 2018). Every stakeholder in the sustainable agriculture sector plays a crucial role in the development of a sustainable food system. This is an element of stewardship that seeks to ensure that every participant contributes and is responsible for their actions, which makes the sustainable agriculture community stronger.

Moreover, sustainable agriculture ensures that the production of fiber and food is done sustainably. For instance, it acknowledges that water has to be used appropriately as it is the primary resource that enables the society and agriculture to prosper. Sustainable production practices involve various approaches that consider the climate, available local resources, soil characteristics, and topography (Glick 2018). Furthermore, sustainable agriculture enhances the conservation of resources by deploying proactive and preventive mechanisms. The early adoption of preventive strategies can minimize inputs and develop a sustainable production system.

Sustainable resource use is one of the indispensable roles played by sustainable agriculture in the Green movement. For instance, sustainable agriculture emphasizes diversity as an economically viable option (Glick 2018). Diversified farms are often ecologically and economically resilient. While monoculture farming has benefits regarding efficiency and management, crop loss in a single year can disrupt the stability of the community. By growing various groups, farmers spread their economic risk and are less vulnerable to significant price fluctuations that are related to supply and demand changes. Appropriately managed diversity is also crucial for buffering a farm. For instance, crop rotation can also be utilized for suppressing pathogens and weeds. Furthermore, maximum diversity can be obtained through the integration of livestock and crops within the same farming operation.

Sustainable agriculture promotes the efficient management of resources. One of the prevailing philosophies of sustainable agriculture is that a healthy environment and soil are crucial ingredients of sustainability. In other words, healthy soil can lead to the production of healthy crop plants that have maximum vigor and are more resistant to pests. Appropriate water, nutrient, and soil management can be crucial for preventing specific pest issues that are associated with crop stress and nutrient imbalance (Altieri 2018). Sustainable systems consider the soil as a fragile living medium that should be nurtured and protected to ensure that it remains stable and productive for an extended period. Some of the methods used by sustainable agriculture for protecting and enhancing soil productivity include the reduction of tillage, the use of cover crops, avoidance of traffic on wet soils, and the maintenance of soil cover. By advocating for efficient utilization of inputs, sustainable agriculture plays a crucial role in enhancing Greenness. Many of the practices and inputs that conventional farmers use are also deployed with the only difference being that sustainable farmers maximize their dependence on renewable, natural, and less intensive energy inputs.

Finally, consumers play a crucial role in the development of a sustainable food system. The purchases they make send strong messages to the retailers and producers about the crucial aspects. Consumers often consider the nutritional quality and cost of food while making their decisions. By purchasing foods produced using environmentally friendly choices, they play a crucial role in ensuring that producers sell food that protects the quality of the environment (Gold 2016). This creates a foundation for ensuring that the agricultural chain is sustainable and responsive to the needs of both the present and future generations. Overall, sustainable agriculture plays numerous roles in ensuring that the production, distribution, purchase, and intake of food promote a greener world. It offers healthy food for consumers while simultaneously protecting human health and the environment.

Consumers determine the availability of products on the market through supply and demand. How can they use this power to ensure that producers bring sustainable products to the market?

3.2.1 *Organic Farming*

Organic farming emerged during World War II due to technological advances that led to input-intensive agriculture as well as industrialization, which resulted in the availability of many fertilizers. However, farmers globally believed that input-intensive agriculture would result in land degradation and unsustainable production. Consequently, they rejected the idea and commenced studying and developing methods that increased sustainable production of their farms (Meng et al. 2017). This became referred to as organic farming as farmers became farm stewards. Organic farming had the same principles as sustainable agriculture, including natural plant fertilization, ecological balance, natural pest management, biodiversity, sustainability, and soil integrity. However, organic products can be produced unsustainably on farms that have not been certified organic. For instance, some organic dairy farms raise cows within large facilities but can meet the requirements for certification. Despite organic farming not being overly sustainable, many of its practices can be perceived as sustainable. When shopping, you should consider the organic label as a valuable tool since it enables you to get food that is free from pesticides and artificial hormones (Meng et al. 2017). The easiest way of avoiding the confusion is to buy whole fruits, vegetables, and dairy products from an organic producer. In case you seek to purchase processed commodities, get them from local and independent sources as they have a less likelihood of having additives and preservatives.

Discuss two other specific roles of organic farming.

3.2.2 *Specific Roles of Sustainable Agriculture*

3.2.2.1 Environmental Preservation

Sustainable farms often raise animals and produce crops without depending on synthetic fertilizers, toxic chemical inputs, genetically modified seeds, or any other practices that pollute natural resources. Through the growth of various plants, the use of techniques such as conservation tillage, and livestock farming based on pasture and crop rotation, sustainable farms play a crucial role in protecting biodiversity and fostering the development of healthy ecosystems (Gold 2016).

3.2.2.2 Health Protection

Sustainable agriculture acknowledges that food production should never be done without considering its impact on human health. Sustainable crop farms tend to avoid the use of hazardous pesticides, and can therefore grow vegetables and fruits that are safe for workers, the consumers, and surrounding communities. Similarly, sustainable livestock farmers raise their animals without adopting dangerous practices, including the use of non-therapeutic antibiotics. Responsible livestock management ensures that humans are protected from toxins and pathogens. Genetically modified organisms (GMOs) are one of the models for sustainable agriculture through the farming practices used in their production (Gold 2016). These methods help in reducing the environmental footprint and preserving natural resources. Today, crop biotechnology enables GMOs to reduce the losses related to extreme weather events. Biotechnology is already providing solutions to significant challenges experienced globally.

3.2.2.3 Sustaining Vibrant Communities

A crucial aspect of sustainable agriculture is its capacity to be economically profitable by offering farmers, food processors, and other individuals employed within the food system fair working conditions. Sustainable farms also enhance economies, create jobs, and build healthy communities.

3.2.2.4 Upholding Animal Welfare

Sustainable ranchers and farmers treat animals appropriately and implement appropriate livestock husbandry practices that protect the health and well-being of animals. By feeding animals pasture, sustainable farmers allow animals to move freely, engage in instinctive behaviors, eat a natural diet, and avoid the problems associated with confinement, including the spread of diseases.

3.2.2.5 Sustainable Food Future

The transition to sustainable food will take some time, but its importance for a Green World cannot be disregarded. The critical aspect of stakeholders in sustainable agriculture is to provide resources and tools that can enable consumers to make informed food choices (Kalantari et al. 2018). One can also contribute to sustainable agriculture by learning to shop, cooking sustainably, and eating the appropriate foods.

3.2.3 Demerits of Sustainable Agriculture

Whereas sustainable agriculture has numerous benefits, it is crucial to understand some of its disadvantages or challenges that would have to be overcome. First, sustainable agriculture makes use of natural fertilizers that take time to break down. Organic materials usually break down at varying rates, thus making the content of the fertilizer to be inconsistent. Also, many of them are too smelly to use on indoor plants. Thirdly, collecting natural materials, including leaves and seaweed, to add to the compost pile is time-consuming and labor-intensive. Finally, it is also a misconception for one to believe that since fertilizer is organic, it should be deemed as automatically safer. In case organic fertilizer is not applied correctly, it could result in groundwater and surface contamination, thus leading to a nutritional imbalance within the soil and causing salt burns.

On the other hand, chemical fertilizers are manufactured from synthetic ingredients that are designed for stimulating the growth of plants. The advantage of chemical fertilizers is that they are predictable and reliable because the blending of formulations is done with accuracy. They also have a balanced distribution of nitrogen, potassium, and phosphorus—the essential nutrients required for optimal growth of plants. The formulations allow a farmer to know the exact nutrients that they are giving their plants instead of guessing as maybe the case with organic formula composition.

Case Study 3.1: Sky Greens, Singapore

Fig. 3.2 Vertical farming. *Source:* Sky Greens (2014)

Due to Singapore's small size and lack of resources, it imports a large proportion of its food from the USA, Europe, China, and Indonesia. The Sky Greens project as shown in Fig. 3.2 above emerged as a technique for

(continued)

feeding the expanding population. Locally grown vegetables contributed to approximately 7% of consumption and showed how vertical farming used in Sky Greens was the first water-powered low-carbon hydraulic urban vertical farm (Nichols 2014). It reduced the size of land and the quantity of energy needed for traditional farming methods.

Sky Greens was designed by engineer Jack Ng and runs on a Sky Urban Vertical Farming System. It is a modern innovative idea that makes use of gravity and rainwater. A water-pulley system is used in the farm with 38 growing troughs revolving around an aluminum tower of about nine meters tall. The role of the rotating troughs is to ensure that there is an even distribution of sunlight for all the plants. The water used for turning the troughs is also essential for plant nourishment (Despommier 2013). The flooding method is used for irrigating and fertilizing the plants, thus eliminating the need for sprinklers. This is crucial for eliminating water and electricity wastage because of runoffs. It is estimated that about only half a liter of water is needed for rotating the entire vertical structure. This water is in an enclosed reservoir system and is often reused. Also, only 40-W electricity is required for powering one tower. From this case study, it is apparent that vertical farming offers an environmentally sound solution that can be replicated globally (see Fig. 3.3). With the mission of providing enhanced agricultural solutions with less impact on energy, water, and land resources, this can enable cities that experience food scarcity to achieve food security and enhance a low-carbon footprint in agriculture.

Fig. 3.3 Sky Greens facility. *Source:* Sky Greens (2014)

3.3 Industries

Environmental and development experts believe that Green extends beyond color and also describes activities that are beneficial to the environment. Such activities emphasize the careful utilization of the planet's finite resources. Despite 20 years elapsing after the Rio Summit's first session, the planet continues facing a twin challenge: the expansion of economic opportunities for all in a growing population and tackling environmental pressures that can undermine the capacity to seize opportunities. Environmental pressures and unprecedented population growth may be perceived as problems, but can provide the impetus for a greener world. Green industrialization is about fostering the growth of industries while making sure that natural assets continually offer environmental services and resources that act as the foundation for our well-being. Green industries are particularly important for promoting the Going Green initiative, including wind and solar power plants.

The Green industry acknowledges that in a world where resource scarcity is becoming common, economic growth should depend on efficient and effective production processes. The Green industry has various attributes that distinguish it from other industries: sustainable production, low utilization of virgin materials, recycling of the solid waste streams, toxin-free production processes, methods that reduce materials and water usage, longevity and durable products, and lower emissions. Green industrialization does not substitute sustainable development but offers a flexible and practical approach for attaining measurable progress across the environmental as well as economic pillars while considering the social repercussions of greening (Gold 2016). Furthermore, they play a pivotal role in ensuring that the available natural resources and assets can sustainably deliver their economic potential. However, some are critical of how they can lessen the environmental carbon footprint. For instance, the hybrid trucks used by Walmart play a crucial role in ensuring lower atmospheric emissions.

Identify and explain three practical industries that use fewer resources and reduce the emission of hazardous gases into the atmosphere.

The importance of Green industrialization has been heralded because it is the Holy Grail of socioeconomic transformation. The infusion of Green initiatives into value-chain activities can cure economic stagnation, but also promote the sustainable utilization of resources (Gold 2016). It is because of this benefit that recent economic forums in the African continent, particularly in Abuja, Nigeria, and Addis Ababa, Ethiopia, as well as Rabat, Morocco, have primarily supported Green industrialization. Some have termed it as the only way that Africa can attain inclusive and sustainable growth. Green industrial initiatives will play a crucial role in moving Africa from the periphery to the center of the global economy, and these efforts have mainly borrowed from the 2015 Paris Climate Change Agreement and the Sustainable Development Goals. In addition, African nations that are rolling out Green initiatives will not have to undergo all the intermediary stages of technology but can access the latest ones available on the market. Numerous African nations are planning or already rolling out Green projects, including Solomon Corporation, Green Institute Nigeria, Eco-Post Kenya, and WeCyclers. At the same time, they are encouraging their industries to engage in Green initiatives. The central role of Green industries, perhaps, is that they play a crucial role in reducing the emission of gases and other pollutants that could harm the ozone layer and contribute to further global warming. With a planet that is staring at environmental crises, this is necessary for ensuring that finite resources are conserved and used appropriately. It is also beneficial for ensuring industrial activities are carried out with due consideration for the planet's carrying capacity. The general belief among many development experts in Africa is that Going Green through industrialization is a socioeconomic imperative.

Identify three Green buildings in Africa. How do they ensure that their building standards/practices are sustainable?

3.4 Energy

Green energy has been heralded as a vital promoter of Green growth as it considers the five principles of "Going Green" highlighted in Chap. 1. Whereas non-renewable energy is a source of environmental issues, renewable energy has increasingly emerged as a platform for Green growth and sustainability. The major roles of renewable energy in Green growth are stable energy prices, decreased global warming, better public health, inexhaustible energy, resilience and reliability, jobs as well as other economic roles (Kuzemko et al. 2016). Regarding global warming, it is worth noting that human activity is pressuring our atmosphere with emissions and gases that trap heat. This results in substantial and harmful effects, including drought, rising sea levels, and extinction. Unlike non-renewable energy sources, renewable energy produces minimal global warming emissions. Even when one includes the life cycle emissions of clean energy, the emissions related to renewable energy are quite low. Life cycle emissions refer to the emissions that emanate from the respective stages of the life of a technology, including manufacturing, installation, and decommissioning (Kuzemko et al. 2016). This comparison is evident when you examine the statistics and numbers behind renewable and non-renewable energy. For example, the burning of natural gas to obtain electricity leads to the release of between 0.6 and two pounds of CO_2 equivalent per kWh, and coal leads to the emission of between 1.4 and 3.6 pounds. In contrast, wind energy leads to 0.02–0.04 pounds on a life cycle basis. Solar, geothermal, and hydroelectric also contributes to less than 0.5 pounds of emissions. Figure 3.4 shows that renewable

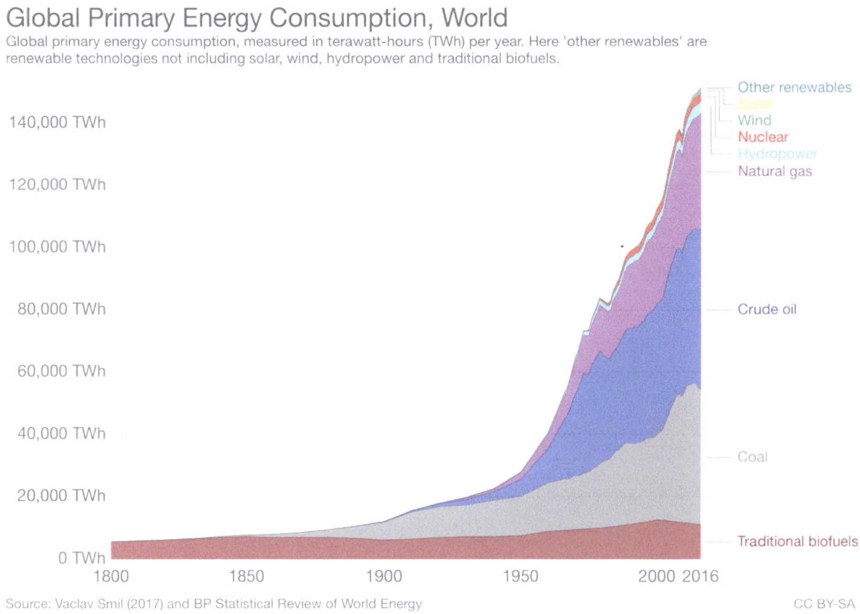

Fig. 3.4 Global primary energy. *Source:* Smil (2017); British Petroleum Company (2017); Ritchie and Roser (2019)

energy accounts for less than 5% of global energy consumption, thus the journey to transition from non-renewable sources of energy to a low-carbon one is progressive.

How can Africa spearhead the transition to low-carbon industrial growth?

Green energy reduces the number of emissions released into the atmosphere and water bodies, thus improving public health. Furthermore, they do not produce any adverse health effects. Solar and wind energy sources, for example, generate electricity and do not have any resultant air pollution emissions. Despite biomass and geothermal systems emitting some air pollutants, the cumulative air emissions are lower than those emitted by natural gas-operated industries. Thirdly, Green energy enhances the sustainable utilization of resources (Kuzemko et al. 2016). For instance, solar and wind energy do not need any water for operation, and therefore do not result in polluting water resources and disrupting water ecosystems as well as other ecosystems. This ensures that these ecosystems continue providing goods and services. This is in contrast to fossil fuels that can have a substantial effect on water resources. Furthermore, Green energy plays a crucial role in providing employment and other economic benefits. They are a source of Green jobs since they mainly require human labor, unlike non-renewable energy sources that require machines. The resilience and reliability of renewable energy sources also make them lucrative

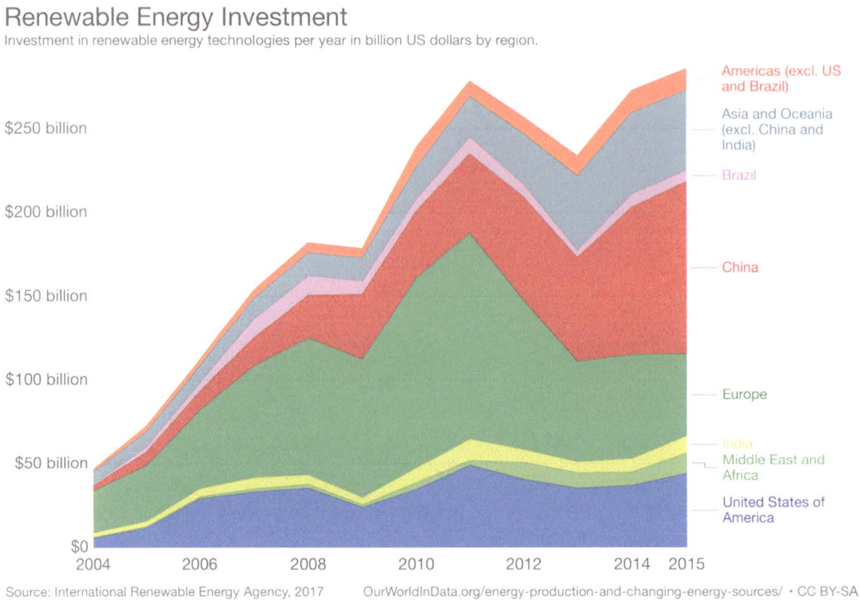

Fig. 3.5 Renewable energy investment. *Source:* International Renewable Energy Agency (2017); Ritchie and Roser (2019)

and attractive. As such, Green energy plays numerous crucial roles in Green growth across the world (See Fig. 3.5)

Some analysts tout the Lamu Coal Project in Kenya as a dangerous and disastrous project. Evaluate the trueness of this statement in light of sustainability. Are there any other projects that degrade the environment in Africa? Briefly discuss.

Experiment 3.1: Testing solar/photovoltaic (PV) cells

Required Materials

- Small Photovoltaic (PV) cells
- Several sheets of transparency film (multicolored).
- Aluminum foil
- Protractor
- Direct circuit (DC) ammeter

(continued)

- Two electrical leads having alligator clips
- Direct circuit (DC) voltmeter
- Goggles
- Direct sunlight/desk lamp/flashlight
- Heat gun/hairdryer

Experiment Setup

Fig. 3.6 Equipment setup. *Source:* Tyukhov et al. (2018)

Procedure

A: Construction of the PV Energy System

1. Using goggles, attach 15 cm of wire to the nodes of the photovoltaic cell. The cell needs to have hooks or clips where the wire can manually be twisted.
2. Follow the safety instructions provided by your teacher. Attach the red wire to the red lead on the ammeter.
3. In the same way, attach black wires from the photovoltaic cell to the ammeter's black lead.
4. Use the light source to see where you are receiving a current reading. In case the ammeter shows no current, re-check the wire connections. See Fig. 3.6 above for the equipment setup.

(continued)

B: Carrying Out the Light Activity

1. Maintaining constant sunlight, cover the photovoltaic cells with a colored transparency film piece. Redo with the other transparency film colors and use only direct sunlight. Make a recording for the current generated for all the colors that have been tested and the direct light in Table 3.1.
2. With only one photovoltaic within the circuit, shade 25% of the photovoltaic cell using a piece of paper and take a reading. Next, shade $\frac{1}{4}$, $\frac{1}{2}$, and then the whole photovoltaic cell. Jot down the results in Table 3.2.

Table 3.1 Impact of wavelength color on cell

Filter colour	Current
No filter	

Table 3.2 Impact of shading on the current of the cell

Amount of shade	Current
No shade	
$^1/_4$ covered	
$^1/_2$ covered	
$^3/_3$ covered	
All covered	

C: Connecting and Powering the PV

1. Connect the photovoltaic cells in series. Take a reading of what you observe. Next, completely shade one cell and take a reading. Then cover all the photovoltaic cells before taking another reading.
2. Make a parallel connection between the photovoltaic cells and observe the reading. Completely shade one cell and record the reading. Cover all the photovoltaic cells and record the reading.
3. Place the photovoltaic cells directly pointing at the sun. Determine the angle using the protractor and slant the photovoltaic cells at 15° intervals. Make a recording of the generated amps at each degree change.

(continued)

4. Using aluminum foil, design a light reflector for the photovoltaic cell to focus the light shining on it. Use the attached reflector to measure the new current and make a recording.

Discussion Questions from the Experiment

1. Which colors led to the generation of the highest quantity of electricity?
2. Explain what is observed when you shade the cell.
3. What happens to the current when: a. Photovoltaic cells are shaded and connected in series; b. Photovoltaic cells are connected and shaded in parallel?
4. What takes place when you heat the photovoltaic cell?
5. In what way does temperature influence a photovoltaic cell's temperature?
6. More light leads to the production of more current. Evaluate the trueness of this statement.

3.5 Science and Technology

Currently, there has been an increased dependence on technology to solve environmental issues globally. This has been brought about by the realization that many governments, especially in Africa, are unwilling to make necessary socio-political changes that minimize consumption and growth in production. Many renewable energy sources are currently being exploited because of technological advances that are cognizant of environmental changes. Technology relies on society to shape Green growth initiatives. Science and technology have mainly been crucial in ensuring that many development forms are sustainable (Kates 2010). As such, many programs incorporate sustainable development policies that seek to alter the economic growth nature instead of limiting it. These are based on the perception that continued growth in a world with finite resources can be spearheaded by technological powers, which can enable humans to find better sources or offer alternatives in case-specific resources begin to dwindle. In other words, technology plays a vital role in helping us use and reuse what is left efficiently. Sustainable development tools and technologies aim to bring about changes, including the minimization of wastes, material substitution, recycling, pollution control, efficient resource use, and changed processes in production.

How does biotechnology enhance sustainability? Discuss three African projects that make use of the concept of biotechnology.

Furthermore, science and technology play a crucial role in promoting the 3Rs: reduce, reuse, and recycle, which are essential components for a sustainable planet. Resource use can be facilitated through the reduction of wastage and recycling. Recycling, through the redesign of products, low-waste technology,, and conservation, can interrupt the movement of waste, thus making it a key feature of sustainable development (Kates 2010). Therefore, science and technology play crucial roles in ensuring that finite resources are used efficiently and that the generated waste can be reused, reduced, and recycled. Since waste is among the major impediments towards Green growth, science and technology are crucial in tackling this recurrent problem.

3.6 Socio-Political Systems

A socio-political system is one that consists of political and social elements and the link between them. Political systems rely on the existing social system, and the interaction between the two systems is crucial for influencing Green growth and development. The design of policies, implementation, evaluation, and adoption are critical components in determining the sustainability of development. Today, development is among the major sources of pollution and degradation of the planet. As such, policies that aim to ameliorate such impacts play a crucial role in promoting Green growth and the "Going Green" initiative. For example, governments that aim to conserve and protect available resources play a crucial role in promoting sustainable growth (Kuzemko et al. 2016). The creation of awareness by civil societies on environmental conservation may also lead to more informed choices that can be important in ensuring that nations, states, and their citizens develop steps necessary for bringing about environmental stewardship.

Social systems, especially cultural, play a vital role in determining whether a particular society conserves its resources. Some cultures advocate for environmental protection. Indigenous conservation strategies have been heralded as the most effective as people take responsibility for ensuring that the environment stays clean and that they use resources in a sustainable manner (Kuzemko et al. 2016). An example is the Aboriginals or indigenous Australians who have consistently been at the forefront of environmental conservation. Traditional knowledge founded by observing nature, including trees and birds, enables communities, especially nomad communities, to determine seasonal migration. Preservation of the ecosystem's balance is often the indigenous way of life. Even within some tropical African forests, where some of the communities utilize wood products for building settlements, they engage in these practices without causing harm to the environment.

Discuss three African cultures that are detrimental to the conservation of finite resources? From a general perspective, has the African culture been beneficial for conserving resources?

Social systems act as the first opponents of deforestation. For example, in Chad, communities' seasonal movements are crucial for fertilizing the soil and the creation of a natural buffer against deforestation. This is an act of nomadic pastoralism whereby the communities migrate when there is depletion of pasture for their animals. The movement is determined by seasons, thus allowing the land to regain its capacity to produce more pasture before the communities can come back. Similarly, pastoralism has played an important role in the success of the Sahel Green Great Wall Initiative, a flagship initiative by Africa to tackle the impacts of climate change and desertification. Spearheaded by the African Union (AU), its major aim is to change the lives of many people through the creation of productive landscapes across the Sahel and North Africa. Pastoralism ensures that available resources are used responsibly and pastoralist communities are at the forefront of protecting resources as their livelihoods depend on them (Olson 2017). They view resources as their heritage that is crucial for sustenance. For a long period, indigenous communities have been at the forefront of combating desertification and regenerating vegetation. Overall, the socio-political system plays an indispensable role in environmental conservation and promoting the Green initiative.

3.7 Gender

Gender equality and empowerment is heralded as being among the major determinants of sustainability. Gender equality remains an important aspect of the world today. Research suggests that societies that value gender equality tend to have slightly better progress compared to those that are still characterized by gender inequality (Shiva and Mies 2014). Reduction of maternal mortality and attainment of universal access to reproductive health has been identified as crucial for sustainability. According to Starbird et al. (2016), meeting the needs of women by planning their pregnancies would result in lower population growth. Consequently, there would be a reduction of global carbon emissions by about 8–15%, which is equal to the stoppage of all the current deforestation (Starbird et al. 2016).

Women play a pivotal role in resource conservation, protection, and stewardship because of their unique relationship. For instance, women are the collectors of resources in some parts of Africa, including firewood for fuel and water for domestic purposes. As they are responsible for the management of these resources, they play a crucial role in ensuring their sustainability (Kassinis et al. 2016). In Africa, there are numerous examples of women who play crucial roles in protecting natural resources. An example is Ilaje women who are best-suited to protect their water bodies as they require them in the future. The Ilaje women occupy Ondo State in Nigeria, which has been adversely affected by resource degradation (see Fig. 3.7). Since they use these resources for obtaining water and food, they are best placed to ensure that people do not misuse them as it would directly negatively impact their lives.

Fig. 3.7 Ilaje women

Women are also great stewards of not only resources but also the environment. The best illustration of environmental conservation was the activities spearheaded by Wangari Maathai of Kenya. Her activities, under the Green Belt Movement, were critical in ensuring that women were at the forefront of promoting environmental conservation. Maathai's actions resonate with the concept of ecofeminism that advocates for a sustainable environment. Ecofeminism contends that capitalistic societies have resulted in a disastrous impact on the environment and the people. Some of the ecofeminism personalities and movements that have played a crucial role in promoting sustainability are Karen Warren, Maria Mies, and Vandana Shiva (Shiva and Mies 2014). Vandana Shiva is an environmentalist who has heavily criticized biotechnology and conventional agriculture. Notably, her work has been summarized as a "one-woman movement for peace, sustainability, and social justice," and is further described in detail in the Individual Matter Section at the end of this chapter.

Economic empowerment of women also has the potential of reducing poverty, which is one of the sources of pollution in water streams and other natural resources (Shiva and Mies 2014). The implication of providing women with equal chances, voices, and choices is crucial for their lives and the future of our planet. If we are to achieve sustainable development and Greenness, women must remain part of the system and decision-making.

Discuss three women groups that have been important for environmental conservation in Africa. From these discussions, what are the roles of ecofeminism in environmental sustainability?

3.8 Rural-Urban Migrations

Migration is an important factor for economic growth; however, it is also a major source of pollution when it occurs in large numbers. Notwithstanding, rural-urban migration can offer opportunities for nature conservation, especially in regions where deforestation is brought about by subsistence agriculture. The rural population is usually a key driver of deforestation. Rural-urban migration and planning concerning land use could favor nature conservation with little effect on the urban areas. Despite the existence of clear benefits attributed to this form of migration on the natural ecosystems, migrants may increase socio-environmental issues within urban areas. It is challenging to make sure that the benefits accrued in the countryside are not washed away by socio-environmental costs within urban environments. In other words, rural emigration can contribute substantially to native forests' conservation in the absence of any specific policy. Recent land-planning legislation may also facilitate the protection of forest resources by constraining future deforestation. The bottom line is that rural-urban migration ensures that the finite resources left at the local level are used in small quantities. It eliminates the possibility of the Tragedy of the Commons phenomenon (see Chap. 1) as many people migrate into urban centers to find suitable and alternative livelihoods. With a considerable number of people left in the countryside, there could be a possibility for more responsive and sustainable means of using available resources. However, quantifying this remains difficult, thus calling for the need for further studies. In addition, rural-urban migration can strain resources in urban areas via over-reliance and could stretch them beyond their carrying capacity. This is disastrous to the fragile ecosystem. Inadequate planning for such a large population could also lead to adverse health, environmental, and human impacts.

3.9 Transportation

Contemporary modes of transportation have been at the forefront of promoting environmental pollution and contamination. However, Green transportation has increasingly been adopted as a way of reversing this trend of pollution. Many people recognize that existing automobiles lead to excess pollution, which negatively impacts the environment. As such, Green transportation has emerged as a solution mainly because it promotes the use of minimal energy (Santos and Ribeiro 2013). The second role is that they reduce the emission of gases into the atmosphere.

The use of vehicles that run on alternative fuels and advanced technologies leads to less environmental harm when compared to the contemporary internal combustion engines. Green transportation also plays a crucial role in encouraging behaviors that support environmental protection. For example, carpooling is one of the best ways of Going Green that people can rely on. It reduces costs and minimizes pollution by reducing the number of vehicles on the road. Finally, the reliance on Green forms of transportation, such as hybrid cars, ensures that they convert

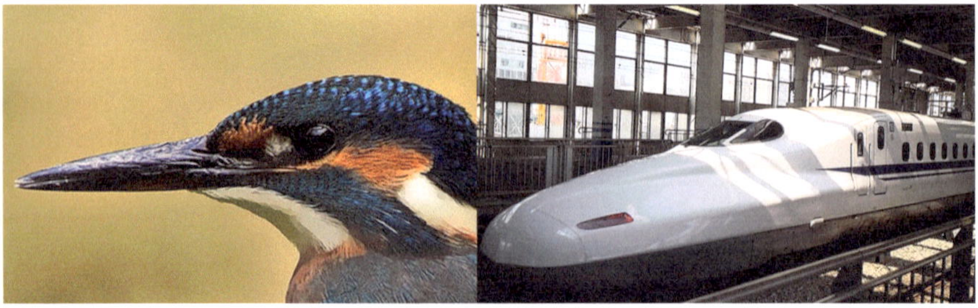

Fig. 3.8 Kingfisher-inspired bullet train

energy without releasing harmful substances. An example of Green innovation is the kingfisher-inspired bullet train in Japan called the Shinkansen bullet train (see Fig. 3.8). The modification was as a result of sound pollution that previous trains brought about. The engineer who designed the train was a birdwatcher and used the body of the kingfisher, which creates little splashes while diving in water, as inspiration.

Explain the positive effects of the Shinkansen bullet trains in Japan.

3.10 House-to-House Activities

Domestic life plays a crucial role in determining the future of the planet. House activities determine the level of pollution. Therefore, the use of biodegradable house commodities helps to conserve the environment, whereas the use of non-biodegradable commodities leads to the degradation of the environment. For instance, the deployment of more natural cleaners and personal products can lead to minimal environmental harm. Responsible house activities also play a role in conserving resources. Thirdly, house activities play a crucial role in energy conservation, which also contributes to less pollution. The fourth role is that house activities that reduce waste and consumption promote Greenness. The resultant effect of minimizing our consumption, especially goods, is that we use fewer resources at home, thus lessening the generation of waste, the usage of energy, and the environmental impact of consumption. When less waste is generated, it means that only a small quantity of waste is taken into landfills. Consequently, house activities geared toward reducing the quantity of waste plays a crucial role in sustainability.

3.11 Individual Matter: Vandana Shiva

Vandana Shiva was born on November 5, 1952, and is an Indian environmental activist as well as a renowned author (see Fig. 3.9). She has supported many traditional practices and is a member of many global solidarity movements. Her criticism of genetically modified organisms (GMOs) has made her famous in the environmental realm and enabled her to appear in many rallies that oppose GMO. She is famous for arguing that the introduction of genetically modified seeds within India has resulted in the deaths of numerous poor farmers who were enticed by false GMO promises. For instance, it is estimated that 270,000 farmers in the Indian seed market committed suicide.

After completion of her studies, Shiva founded the Navdanya organization. The website of the organization states that its mandate is to provide protection to the integrity and diversity of native seeds and living resources as well as enhance organic farming (Navdanya n.d). Under her leadership, the organization has become a national movement (Shiva and Mies 2014). Furthermore, Shiva is an opponent of globalization and agricultural genetic engineering. In this regard, she has authored more than 20 books on these topics. Perhaps, it could be stated that many of her efforts have mainly been premised on ecofeminism.

She has played an important role in the international ecofeminist movement. In her 2004 article titled "Empowering Women," she argued that sustainable agricultural production could only be attained through the reinstatement of a

Fig. 3.9 Vandana Shiva. *Source:* Frank Schwichtenberg via Encyclopædia Britannica (n.d.)

farming system that revolves around women (Shiva and Mies 2014). According to her, such a system would lead to positive change. In addition, industrial catastrophes and destruction of the environment threaten the day-to-day life, and that it is the responsibility of women to maintain such issues. Arguably, many of her works have been spurred change and raised awareness about environmental issues. Whereas some are controversial, there is a need for such a debate to ensure that people become custodians of the environment through their activities.

3.12 Future Trends

3.12.1 Africa

Demonstration farms for teaching sustainable and organic agricultural technologies and techniques will become more common and help in accelerating game-changing innovations. Through these farms, farmers will be able to learn new ways of doing things without necessarily having to do it on their lands/farms. The green energy sector within sub-Saharan Africa could become 100% renewable by 2050, resulting in lower costs of energy, environment, and better health as well as increasing the number of green jobs from about 1.2 million to more than five million (Osmundsen 2019). Increased scaling of wind, decentralized solar PV energy, electrification of heating, transport, and cooking, and use of batteries along other storage technologies will phase out fossil fuels by 2050 (Osmundsen 2019). In North Africa, almost all countries have sufficient sunlight for supporting an industry that can thrive. The region could tap on the high-quality winds found in the North African coastal areas to construct wind farms. As a result, an increasing number of countries across Middle East and North Africa (MENA) are setting ambitious official targets for renewable power, while creating laws to cultivate the industry. Morocco will become the leading producer of solar power in the region with the capacity to power at least one million homes upon completion (Stephen 2019).

3.12.2 Antarctica

Antarctica in 2070 will be characterized by greenhouse gas emissions remaining unchecked; the climate will continue warming (Rintoul and Chown 2018). Under such a scenario, Antarctica and the Southern Ocean will be characterized by rapid and widespread change, with international repercussions. Ocean warming and global warming will lead to significant loss of significant ice shelves. It will lead to more ice loss from the Antarctic ice sheet as well as the acceleration of sea-level rise to rates that have not been, since the demise of the last glacial period. Ocean acidification will also substantially change the marine ecosystems and proliferation of invasive species.

3.12.3 Asia

India and China will play a crucial role in transitioning towards a circular economy for creating socioeconomic and environmental benefits. The renewable energy cost will continue dropping, thus making them a competitive and viable alternative to fossil fuels. With falling prices and increasing investments in the Middle East, the renewable energy sector will experience more expansion in the coming decade. The likely future trends will include the deployment of hybrids, mini-grids, building integration, more use of artificial intelligence, and utility-scale battery storage. The Middle East will particularly witness more investment in solar and wind power.

3.12.4 Australia

Clean energy will provide 35% of the total electricity needs, thus underlining the pace at which solar power will transform the energy market (Ison 2019). There will be a rise in precincts of the renewable industry whereby the heat-intensive industries could be able to access renewable heat from renewable energy production, bioenergy, and solar thermal production. They will also be the major locations for renewable export industries of Australia. The continent's renewable exports will continue supporting countries like Indonesia, South Korea, and Japan to shift towards 100% renewable energy (Ison 2019).

3.12.5 Europe

Europe will experience an increase in the number of electric vehicles and car-sharing schemes because they do not need refueling. The transition to electromobility will begin within cities. They could provide parking spaces for visitors, enabling people to drive in with combustion vehicles, then switch to electrified forms of public or personal transportation in the towns.

3.12.6 North America

Herbs, vegetables, and fruits will occupy a substantial share of the vertical farming market. Tilapia will dominate the vertical farming market since it is currently the fastest-growing species. North America will continue dominating the international vertical farming market, and the industry is posed to continue growing because of increased demand for food. Also, more buildings will be converted into vertical farms.

3.12.7 South America

South America will continue having a dynamic renewable energy market, and scaling up its contribution to the future energy mix will remain a major priority. Wind and solar power generation have great potential throughout the continent. Public transportation will continue being widely used, leading to the creation of opportunities for electrifying larger vehicles that can run the whole day. National and city-wide efforts for improving air quality and tackling climate change will remain the main driving forces for the increasing interest in electric mobility. In Bogota, Colombia financial incentives and more electric vehicle models will spur electronic vehicle sales. Within Santiago, Chile strong electric mobility and climate change goals will provide long-term direction for the markets whereas green vehicle taxes could be instrumental in leveling the playing field for the Electric Vehicles (EVs). Generally, programs introducing electric taxi and government fleets will lead to more EVs on the road.

3.13 Summary

- This chapter seeks to explore the role of various activities in the Green World.
- Sustainable agriculture is crucial for environmental preservation, public health protection, sustenance of vibrant communities, upholding animal welfare, conservation of finite resources, and enhancing a sustainable food future.
- Organic agriculture also plays the same roles in promoting Green growth.
- Industries promote the appropriate use of finite resources, sustainable production, recycling solid waste streams, low emissions, and the use of fewer virgin materials. Perhaps, the significant role may be the triple-bottom phenomenon.
- Green energy incorporates the five principles of Going Green. It contributes to less global warming, the creation of employment opportunities, and improved public health. Also, it minimizes the number of atmospheric emissions.
- The success of Vandana Shiva and Wangari Maathai highlights that there is a need for more women to raise voices against environmental vices. The chapter further explains how women can be empowered to become active participants if not leaders in environmental stewardship debate and change.
- Science and technology have been beneficial for ensuring that development efforts are sustainable. Technological changes lead to a reduction of waste, recycling, control of pollution, and efficient use of resources.
- Socio-political systems are instrumental in the development of proactive and responsive policies that emphasize the need for environmental conservation and protection. These systems also establish mechanisms for ensuring less harm to the natural environment.
- Rural-urban migration may lead to environmental degradation or conservation, depending on their nature and as such, ascertaining their exact roles is quite tricky.

- Green transportation reduces the quantity of fuel used, gas emissions, and supports the adoption of activities that cause less environmental harm.
- House-to-house activities also determine the level of pollution and resource use depending on the commodities used and the quantity of waste released.
- Africa will experience an increase in the number of demonstration farms while Europe will have more electric vehicles and car-sharing schemes, particularly in the urban centers.
- It is also projected that North America will continue having a dynamic vertical farming market whereas South America will increasingly have a dynamic renewable energy market.
- Australia's clean energy shift will continue being among the best while Antarctica will report substantial loss of ice shelves.

3.14 Discussion Questions

1. Explain the differences between organic, inorganic, and sustainable agriculture. From these differences, which is the best form of agriculture in ensuring sustainability?
2. What is the major challenge that impedes the shift to Green energy?
3. What steps can be undertaken to ensure that agricultural activities embrace Greenness?
4. Explain why gender is an important factor in ensuring that the shift towards Greenness remains on course.
5. Science and technology have been heralded as a prerequisite for less resource use. Discuss.
6. The Chipko movement was crucial for environmental conservation. Discuss.
7. Elon Musk is one of the contributors to Greenness, primarily through transportation. Explain how his discovery and vision have been useful in driving the world towards Green transportation.
8. How can industries ensure that their activities support the Going Green project?
9. Identify three examples of projects in Africa that highlight the significance of Green industries in Green growth.
10. Explain the challenges that prevent developing states, especially industries, from achieving sustainable growth.
11. Does Africa have any examples of vertical farming? Identify and discuss any two critical examples.
12. With an increasing urban population and scarcity of food, explain how vertical farming could help in sustainably promoting food security.
13. Ecofeminism has been heralded as counterproductive. What reasons explain this position?
14. What future trends do you anticipate in terms of everyday activities globally?

References

Altieri, M. A. (2018). *Agroecology: The science of sustainable agriculture*. Boca Raton: CRC Press.

Benke, K., & Tomkins, B. (2017). Future food-production systems: Vertical farming and controlled-environment agriculture. *Sustainability: Science, Practice and Policy, 13*(1), 13–26.

British Petroleum Company, BP. (2017). *Statistical review of world energy*. Retrieved November 10, 2019, from https://www.bp.com/en/global/corporate/energy-economics/statistical-review-of-world-energy.html/.

Brundtland, G. (1987). Report of the world commission on environment and development: Our common future. United Nations General Assembly document A/42/427.

Despommier, D. (2013). Farming up the city: The rise of urban vertical farms. *Trends in Biotechnology, 31*(7), 388–389.

Encyclopædia Britannica. (n.d.). *Vandana Shiva*. Retrieved February 09, 2019, from https://www.britannica.-com/biography/Vandana-Shiva/images-videos#/media/1/541204/125949/.

Glick, B. R. (2018). Soil microbes and sustainable agriculture. *Pedosphere, 28*(2), 167–169.

Gold, M. V. (2016). *Sustainable agriculture: The basics* (p. 1). Boca Raton: CRC Press.

International Renewable Energy Agency. (2017). *Investment in renewable energy technologies*. Retrieved November 05, 2019, from http://resourceirena.irena.-org/gateway/dashboard/.

Ison, N. (2019). Australia could hit 100% renewables sooner than most people think. *The Guardian*. Retrieved August 8, 2019, from https://www.-theguardian.com/commentisfree/2019/jan/14/australia-could-hit-100-renewables-much-sooner-than-most-people-think/.

Kalantari, F., Tahir, O. M., Joni, R. A., & Fatemi, E. (2018). Opportunities and challenges in sustainability of vertical farming: A review. *Journal of Landscape Ecology, 11*(1), 35–60.

Kassinis, G., Panayiotou, A., Dimou, A., & Katsifaraki, G. (2016). Gender and environmental sustainability: A longitudinal analysis. *Corporate Social Responsibility and Environmental Management, 23*(6), 399–412.

Kates, R. W. (2010). *Readings in sustainability science and technology* (No. 213). Boston: Center for International Development at Harvard University.

Kuzemko, C., Lockwood, M., Mitchell, C., & Hoggett, R. (2016). Governing for sustainable energy system change: Politics, contexts and contingency. *Energy Research & Social Science, 12*, 96–105.

Meng, F., Qiao, Y., Wu, W., Smith, P., & Scott, S. (2017). Environmental impacts and production performances of organic agriculture in China: A monetary valuation. *Journal of Environmental Management, 188*, 49–57.

Navdanya (n.d). *Navdanya*. Retrieved April, 17, 2018 from http://www.navdanya.org/site/.

Nichols, M. (2014). Vertical farming and urban agriculture. *Practical Hydroponics and Greenhouses, 149*, 14.

Olson, R. (2017). Building sustainable agriculture: A new application of farming systems research and extension. In *Integrating sustainable agriculture, ecology, and environmental policy* (pp. 53–66). Abingdon: Routledge.

Osmundsen, T. (2019). *Africa's energy future: 100 percent renewable is feasible*. Retrieved November 5, 2019, from https://www.esi-africa.com/industry-sectors/future-energy/%EF%BB%BFafricas-energy-future-100-renewable-is-feasible/.

Rintoul, S., & Chown, S. (2018). *What will Antarctica look like in 2070?* World Economic Forum. https://www.weforum.org/agenda/2018/06/antarctica-has-lost-3-trillion-tonnes-of-ice-in-25-years-time-is-running-out-for-the-frozen-continen/.

Ritchie, H., & Roser, M. (2019). *Energy production & changing energy sources*. Retrieved November 06, 2019, from https://ourworldindata.org/energy-production-and-changing-energy-sources/.

Santos, A. S., & Ribeiro, S. K. (2013). The use of sustainability indicators in urban passenger transport during the decision-making process: The case of Rio de Janeiro, Brazil. *Current Opinion in Environmental Sustainability, 5*(2), 251–260.

Shiva, V., & Mies, M. (2014). *Ecofeminism*. London: Zed Books.
Sky Greens. (2014). *Sky greens*. Retrieved November 10, 2019, from http://www.sky-greens.com/.
Smil, V. (2017). *Energy transitions: Global and national perspectives*. Retrieved November 10, 2019, from http://vaclavsmil.com/2016/12/14/energy-transitions-global-and-national-perspectives-second-expanded-and-updated-edition/.
Starbird, E., Norton, M., & Marcus, R. (2016). Investing in family planning: Key to achieving the sustainable development goals. *Global Health: Science and Practice, 4*(2), 191–210.
Stephen, C. (2019). *Morocco prioritizes world's largest solar plant*. Retrieved November 5, 2019, from https://www.petroleum-economist.com/articles/low-carbon-energy/renewables/2019/morocco-prioritises-worlds-largest-solar-plant/.
Tyukhov, I., Schakhramanyan, M., Strebkov, D., Mazanov, S., & Vignola, F. (2018). *Combined Solar PV and Earth Space Monitoring Technology for Educational and Research Purposes Proceedings Solar 2008*, American Solar Energy Society Conference, San Diego (2008). Retrieved November 08, 2019, from https://www.researchgate.net/publication/258831586/.

Chapter 4
Effects of Everyday Activities on the Ecosystem

Abstract Everyday activities significantly influence the weather, climate, ecosystem, and cycles. Such activities result in climate change, global warming, ocean acidification, pollution, seasonal changes, high rates of disasters, and extinction. To this end, this chapter seeks to explore the impacts of these activities on the ecosystem. To achieve these objectives, experiments, examples, case studies, and individual matter have been embedded into this chapter to particularly understand the impact of climate change, weather and climate, global warming, ocean acidification, deforestation, acid rain, pollution, high disaster rates, extinction, and seasonal changes on the ecosystem. The Individual matter section elaborates on Paul Berg's contribution on the subject and provides insight on how other personalities can contribute to maintenance and protection of the ecosystem. Future trends in various continents are then provided to offer a glimpse of the impacts that everyday activities may have on the ecosystem.

Keywords Pollution · Global warming · Weather · Paul Berg · Extinction · Climate change

4.1 Climate Change

4.1.1 Ecosystems

Climate change refers to prolonged weather events that bring about changes in climatic conditions and weather. Climate change has adverse impacts, particularly on various ecosystems. Unchecked global warming can negatively affect the terrestrial eco-regions; for example, an increase in global temperatures can lead to significant changes within the ecosystem (Bonan 2015). Today, some species, including Cetti's warbler that has been moving to the north of the United Kingdom, little egret (that has colonized Britain), and comma butterfly that is moving to Edinburgh, are being forced to migrate from their habitats due to climate change; others are facing possible extinction due to these changing conditions and include the koalas, orange-spotted filefish, quiver tree in Germany, and the polar bear. At the same time, some

A. A. Akinsemolu, *The Principles of Green and Sustainability Science*,
https://doi.org/10.1007/978-981-15-2493-6_4

are flourishing due to these adverse weather conditions. The flourishing organisms are the mainly adaptable ones, including many pests and weeds. Cold-sensitive species, including the Burmese python in Florida, are also benefitting from climate change. The secondary impacts of global warming, like snow cover, rise in sea levels, and changes in weather, could influence human activities and the ecosystem. Studies have established that during the last three decades, global warming as a result of human activities has altered numerous biological and physical ecosystems.

What do you think could happen to our ecosystems by 2050 at the current climate change levels?

Various studies suggest that climate change will lead to an extinction of numerous species and a minimization in ecosystem diversity (Cahill et al. 2012; Archer and Predick 2008). A rise in global temperature could also result in the disappearance of many coral reefs in the world (Archer and Predick 2008). Cahill et al. (2012) argue that there could be higher rates of extinction of vulnerable species. Many of the vulnerable species are Antarctica and Arctic animals, including emperor penguins and polar bears. Climate change also interacts with other human activities, called stressors, to affect ecosystems. Stressors are defined as biological or chemical agents, or simply environmental conditions, or external stimulus that cause stress to organisms (Hoegh-Guldberg and Bruno 2010). Events that could trigger environmental stress responses include incredibly high or low temperatures, high sound levels, overcrowding, or over-illumination. While some of the stressors could lead to only minor effects when acting independently, their combined effect could result in dramatic ecological changes. For example, climate change could aggravate the stress placed by land development on fragile coastal ecosystems. In addition, forest lands that have been recently logged could be vulnerable to erosion if climate change leads to heavy rainstorms.

Identify some of the vulnerable species in Africa.

Experiment 4.1: Global warming as evidence for climate change

Objectives To demonstrate global warming phenomenon.

Required Materials 2 thermometers, 6 cups of water, 2 glass containers (each with a lid), 1 plastic bag, and 12 cubes of ice.

(continued)

Background of the Experiment This experiment seeks to demonstrate the process of global warming and the impact it has on the Earth's temperature. The results will help in developing better reasoning capacity, and critical thinking skills will be developed. Fundamental scientific knowledge will also be gained by comprehending the global warming process, thus easily connecting the concept of climate change. Greenhouse gases, such as carbon dioxide, are found within our atmosphere and are referred to as greenhouse gases since they make the atmosphere act like greenhouse glass. In other words, the rays of the Sun pass through the atmosphere, heat the earth, and greenhouse gases ensure that the heat does not leave. Human activities, such as the use of fossil fuels, increase the number of greenhouse gases found in the atmosphere. Excess greenhouse gases add another layer that traps even more heat, thus warming the planet.

Procedure

1. Take a jar and place 3 cups of water and 6 ice cubes. Repeat the procedure for the other jar and cover the lids.
2. On one of the jars, completely cover it with a plastic bag. (greenhouse jar)
3. Strategically position both jars within bright sunlight outdoors for 60 min. Make sure that they are in a secure and safe environment.
4. After the stipulated period, use a thermometer to measure the temperature of the jars.
5. Record the temperature of each jar and use it for answering the Discussion Questions

Discussion Questions

- Which of the thermometers recorded higher temperature?
- Why do you think this thermometer recorded a higher temperature?
- Explain how the plastic bag acted in a similar way to greenhouse gases.

4.1.2 Biogeochemical Cycles

Climate change impacts the precipitation and evaporation stages of the water cycle. It can speed up some stages of the water cycle as the rise in global temperatures also raises the evaporation rate. On average, more evaporation leads to higher precipitation. Today, people are already witnessing the effects of precipitation rates and evaporation, and these impacts are anticipated to rise over this century. Higher rates of precipitation and evaporation are not uniformly distributed globally, and some areas could experience higher precipitation. Droughts could make some areas to become vulnerable because of the shifting of the rain belt and desert locations in response to climate change. Some models predict that there will be higher rates of rainfall within coastal areas, whereas central parts of continents will experience prolonged dry spells (Joos et al. 2001).

Similarly, Joos et al. (2001) predict that there could be more rainfall and evaporation over the ocean bodies rather than the lands. As some of the regions increasingly become susceptible to droughts, they are vulnerable to wildfires and failure of crops to nourish within dry conditions. In other areas where the rainfall is becoming increasingly common, there is vulnerability that they may experience more damaging floods.

Do you think that the water cycle is the most affected by climate change? Provide reasons for your answer.

Regarding the carbon cycle, the combustion of fossil fuels and alteration of landscapes, including industrial agriculture and deforestation, lead to the disruption of the carbon cycle, thus contributing to climate change. Today, anthropogenic activities affect the broader aspect of the carbon movement. After being released into the atmosphere, greenhouse gases (GHGs), including methane and carbon dioxide, can alter the atmosphere's thermal behavior, the chemistry of seawater, along with the movement of oceans. It is a significant role reversal within the way the systems of the planet naturally operate (Pataki et al. 2011). Consequently, modern climate change is an event that significantly influences the contents of the carbon cycle. The overall image is that the carbon content of the atmosphere is rising steadily while the global oceans are slowly absorbing this carbon.

How does methane affect global climate change?

Altogether, global land absorbs and stores a substantial quantity of carbon produced by human activities. Forests and other related wood products account for a larger number of these carbon sinks.

Identify the other carbon sinks that you know.

When biogeochemical cycles are altered by climate change, there is higher susceptibility of disruptions to food security, biodiversity, the health of humans, and quality of water. Notwithstanding, managed shifts of these biogeochemical cycles could be instrumental in limiting climate change rates.

4.1.3 Weather and Climate

The rise in temperatures has resulted in numerous changes to the Earth's climate and weather. Droughts and heatwaves are some of the most common evidence of climate change that humans are experiencing today. Anthropogenic climate change has already led to a rise in the strength of some of these events. For instance, during the last five decades, the United States (US) has witnessed prolonged periods of high temperatures, severe droughts, floods, and torrential downpours. Heatwaves are defined as periods of abnormal levels of hot weather that could last from days to weeks. Recently, there has been an increase in the number of heatwaves experienced globally that analysts say is due to human-induced climate change (Mori et al. 2014). For instance, heatwaves were reported in Texas in 2011 and the Midwest in 2012. Higher temperatures result in higher evaporation rates. Even within the zones that do not show a decrease in precipitation, reports indicate that the rise in surface evaporation and evapotranspiration results in higher drying of the soil. With this, a larger percentage of incoming heat leads to heating of the soil, resulting in drier summers. For example, the 2011 drought in the United States made many of the areas in Texas experience temperatures higher than 100 °F for more than 100 days.

Can you identify two such examples in Africa?

Flooding is the other effect of climate change on weather and climate. A flood is a high flow or overflow by water that may result in or threaten damage. Floods result from human and weather factors. Some weather factors are prolonged precipitation, hurricanes, hurricane storm surges, and debris jams. Some human factors are altered drainages, alterations of the land cover, and structural failure of levees or in some cases, dams (Schmittner et al. 2008). The 2017 Solai Dam Tragedy in Nakuru County, Kenya, is evidence of flooding as a result of structural failures of dams. The tragedy led to the loss of approximately 44 lives and the displacement of more than 100 families. Finally, the rise in hurricanes is the other impact on weather and climate. For instance, the North Atlantic hurricanes have risen in duration, frequency as well as intensity, and is mainly attributed to an increase in surface temperatures of the sea within regions that the ocean flows through.

4.2 Global Warming

4.2.1 Ecosystems

Many of the effects associated with global warming are related to climate change. As such, this section will examine the impacts that have not been previously discussed.

Regarding ecosystems, forest destruction as a result of fires and severe winters has been commonplace. For example, in Columbia, South America, pine forests have been destroyed by pine beetle infestations that have continually expanded, partly because of harsh winters. With global warming increasing, and habitats as well as species reducing, the probabilities of ecosystems adapting naturally are dwindling (Bonan 2015). Global warming can also have numerous effects on fish as well as other aquatic species as some water bodies become too hot for these species to survive.

Identify some of the water bodies in Africa where this effect is being felt (at least three).

Global warming could also alter the water's chemical composition, thus leading to a decline in oxygen quantity, and subsequently an increase in salinity levels as well as pollution. Today, it is estimated that at least 20% of the freshwater fish in the world are becoming extinct, threatened, or endangered (Bonan 2015).

Briefly identify and explain how some fish species are being threatened by global warming.

Global warming has contributed to the rise in sea level from 3.0 ± 0.4 millimeters annually during the 1993–2017 period (Nerem et al. 2018). This rise in sea level has contributed to the loss of mangroves and wet prairies. Increased salinity is attributed to rising salt water, and it can damage freshwater ecosystems that have slough and saw-grass. Plants will have little capacity to adapt to global warming. Furthermore, many animal and plant species are migrating northward or towards higher altitudes because of the warming temperatures. They follow comfortable temperatures and move away from the rising global temperature. Because of this, many animals may not be able to compete in the new climate regime and may go extinct.

Also, migratory birds and insects currently arrive in their nesting and feeding grounds days earlier than they did during the twentieth century (Cahill et al. 2012). Warmer temperatures are further anticipated to increase the disease-causing pathogens that were restricted to tropical areas, leading to the death of animal and plant species that were insulated from these diseases. If left unchecked, such effects of global warming are poised to contribute to the disappearance of 50% of the plants and 30% of animals by 2080.

4.2.2 Biogeochemical Cycles

The water cycle will also be affected by the change of precipitation along with other climate conditions. In case global warming prolongs, patterns of rainfall, flood cycles, storms, and drought will happen, thus causing rapid changes in plant distribution globally. Evaporation leads to the loss of freshwater from oceans; however, it will precipitate over the land and get back into the sea. It is expected that global warming will intensify this cycle, thus resulting in a rise in river runoff, called river discharge (Pataki et al. 2011). River runoff has been increasing by an estimated 1.5% annually, and the global water cycle is intensifying as a result of global warming (Pataki et al. 2011).

Similarly, global warming will also increase the carbon cycle content because of the large quantities of carbon in the atmosphere. On the sulfur cycle, there have been no documented effects of global warming on this cycle (Refer to Chap. 2: Sulfur Cycle section). However, recent evidence suggests that the sulfur cycle could play a role in reducing the effects of global warming. Regarding the nitrogen cycle, climate change may make it more difficult controlling it. Changing global weather patterns could aggravate the creation of ozone. Extreme weather events have a likelihood of driving up the use of fertilizers by making it harder for farmers to raise crops.

Explain three more effects under this section.

4.2.3 Weather and Climate

An apparent impact of global warming is the rise in global temperatures. Higher global temperatures contribute to extreme weather events. Africa is the most susceptible of all continents due to poverty and lack of preparedness. Some parts of the African continent cannot withstand the hazardous weather that results in severe damage and loss. For instance, East Africa has had numerous droughts. Ethiopia is a case example where a prolonged drought in 2016 compelled the government to mobilize resources to save lives. In 2017, Liberia and Sierra Leone were affected by flash floods as well as mudslides, where hundreds of individuals lost their lives.

Global warming is further expected to contribute to the formation and intensity of hurricanes. Whereas some experts predict that the frequency of hurricanes will be reduced, while some places will experience frequent and intense hurricanes (Mori et al. 2014). It is because hurricanes are facilitated by the temperature difference between the cold upper atmosphere and the warm tropical ocean. Intense hurricanes, including the 2013 Philippines' Typhoon Haiyan could become more destructive.

Discuss the effects of the Philippines' Typhoon Haiyan.

Global warming also leads to the rise of flash floods and lightning. Forecasts of climate models also predict that global warming will lead to significant changes in global climate patterns. Also, due to the likelihood of high levels of GHGs remaining in the atmosphere for numerous years, these changes will last for various decades.

Case Study 4.1: Africa and climate change

Currently, climate change effects are being felt across Africa. Evidence suggests that temperature changes affect food productivity, livelihoods, water availability, and general security of the African people. The 2017 Climate Change Vulnerability Index suggests that seven of the ten states that are most vulnerable towards climate change are found in Africa (Verisk 2016). Africa has witnessed a decrease in the quantity of rainfall over large parts of Sub-Saharan Africa, while the central parts have had an increase in the amount of rainfall. During the last three decades, weather-related disasters, including droughts and floods, have significantly increased in Africa. This has led to an increased spread of infectious diseases, thus increasing the number of deaths. On weather patterns, climate change has led to an increase in flooding, thus making it become the most frequent disaster in Central, East, and South Africa. In the northern part of the continent, the 2001 disaster within northern Algeria led to approximately 800 deaths as well as an economic loss of approximately $400 million (Bonan 2015). The other effect of climate change has been the prolonged drought reported between July 2011 and mid-2012 when a severe drought swept Kenya, Tanzania, Rwanda, and Burundi (East Africa nations) and was reported as the worst drought experienced in 60 years. Some of the apparent impacts of climate change on the water ecosystems in Africa are drought, flooding, rainfall distribution changes, receding of water bodies, melting of glaciers, and drying up of both permanent and seasonal rivers. For example, some parts of West Africa have also suffered because of the decrease in the volume of water in some rivers. Rivers act as sources of water, and some countries use it for the generation of hydroelectric power. The reduction of the volume of water has affected some places. For instance, Ghana has become entirely dependent on hydroelectric power generated by the Akosombo dam found on River Volta, while Mali relies on River Niger to get food and water. Climate change has also contributed to the gradual disappearance of glaciers found

(continued)

on Mountain Kilimanjaro. Glaciers often act as a water tower, and various rivers are currently drying up. By 2001, Intergovernmental Panel in Climate Change (IPCC) reported that about 82% of the ice that initially capped Mt. Kilimanjaro had disappeared.

4.3 Ocean Acidification

This refers to the ongoing reduction in pH of the Earth's oceans, attributed to the uptake of carbon dioxide (CO_2) absorbed from the atmosphere. Ocean acidification takes place when oceans absorb carbon dioxide to produce acid. The acid is a result of the chemical reactions that result in more concentration of hydrogen ions. The increase leads to more acidity of seawater and makes carbonate ions to become abundant. It is estimated that about 30–40% of the CO_2 from human activity that enters the atmosphere gets into water bodies (Andrews and Schlesinger 2001). For chemical equilibrium, some of this gas reacts with water and forms carbonic acid. The dissociation of carbonic acid molecules into hydrogen ions and bicarbonate ions thus increases ocean acidity.

What world oceans have been affected by ocean acidification?

4.3.1 Ecosystems

Higher acidity potentially harms marine organisms, including the immune responses of some organisms and causes coral bleaching. The net reduction in carbonate ions quantity could make it difficult for marine calcifying organisms, including planktons and coral, to form calcium carbonate. Such structures are susceptible to dissolution (Andrews and Schlesinger 2001). In addition, a high drop of seawater pH can make it difficult for organisms to build their skeletons or shells. Lower pH of seawater can weaken the metabolic processes of various organisms, including reproduction and respiration. Whereas it is difficult to predict the changes in ocean ecosystems due to higher acidity, these ecosystems become less productive and resilient.

Identify some of the ecosystems that have become less productive because of ocean acidification.

4.3.2 *Biogeochemical Cycles*

Ocean acidification accelerates the carbon cycle. When the ocean absorbs atmospheric CO_2, the reaction leads to the formation of carbonic acid and subsequently acidifies the water. Since the Industrial Revolution, scientists report that oceans have gained 30% more acidity. Unless the emission of carbon is reduced in the future, scientists estimate that oceans may become 170% more acidic at the turn of the next century. Studies, where CO_2 has been added to marine habitats, suggest that there is a reduction in dimethyl sulfide that phytoplankton produce (Doney et al. 2007; 2009). It is quite interesting since it is the primary source of biological sulfur that gets into the atmosphere where sulfur compounds are instrumental for cloud formation. Sulfur facilitates the creation of cloud condensation nuclei. Because cloud cover influences the number of sun rays reflected into the atmosphere, this could potentially affect climate.

How does cloud cover affect climate?

4.3.3 *Weather and Climate*

Ocean acidification leads to a rise in water temperature and sea-level rise. Ocean acidification will also contribute to higher atmospheric warming as GHG concentrations in the atmosphere remain high. The reaction of oceans to higher CO_2 levels will have a significant impact on the way global warming affects other people. Oceans absorb a large quantity of gas released into the atmosphere (Cao et al. 2007). High levels of sulfur in the atmosphere are poised to minimize the quantity of solar energy that reaches the surface of the Earth, thus resulting in a cooling effect. In this case, acidification reduces the emissions of sulfur from the ocean, which could lead to a rise in the quantity of solar energy that reaches the Earth's surface, thus increasing warming.

Explain one other way that ocean acidification affects weather and climate.

4.4 Deforestation

Deforestation is defined as the cutting down or clearance of a forest or strand of trees. The land is converted for other uses, including farming. It is stipulated that high rates of deforestation happen within tropical rainforests. Many nations' deforestation happens naturally or can be human-induced. It has numerous impacts, but this section will only explore its effects on cycles, ecosystems, weather, and climate.

4.4.1 Climate Change and Biogeochemical Cycles

Deforestation is among the major contributors to global climate change. The major problem brought about by deforestation is the effect on the carbon cycle. Greenhouse gases (GHGs) play a role in the absorption of thermal infrared radiation, and when in large quantities, it can bring about climate change. Whereas oxygen remains the second most abundant gas within the atmosphere, it fails to absorb thermal infrared radiation in the same way as GHGs. CO_2 is the most prevalent GHG and accounts for 82.2% of all the GHGs in the United States (Longobardi et al. 2016).

Deforestation mainly affects the water and carbon cycle. Provide three reasons for this tendency.

Deforestation reduces the quantity of carbon stored and releases CO_2 into the atmosphere. It is because as trees die, they tend to release stored carbon. It is heralded as the second primary anthropogenic source of CO_2 into the atmosphere. Apart from the carbon cycle, the water cycle is also affected by deforestation. It leads to lower global vapor flows from land. The slight change in vapor flows could lead to the disruption of natural weather patterns. Trees are also crucial for the water cycle as they absorb rain and produce water vapor that is emitted into the atmosphere. They also reduce water pollution by halting polluted runoff. For example, more than 50% of the water within the Amazon ecosystem is held in plants (Longobardi et al. 2016). Furthermore, the loss of forests is a loss for complex ecosystems that impact many species across the planet. When degraded, it could lead to the devastation of ecosystems and biodiversity loss. Seventy percent of the plants and animals that live within forests lose their habitats to deforestation (Longobardi et al. 2016). The loss of habitat may result in the extinction of species and has negative repercussions for populations that rely on services provided by forests.

4.5 Genetic Modification

Genetic modification or engineering refers to the use of biotechnology to manipulate the genes of an organism. They are technologies used for changing the genetic composition of cells, such as genetic transfer in and across the boundaries of species.

Genetic modification allows the introduction of new traits that would not naturally happen or the removal of undesirable traits. These changes can be done by silencing or inserting a gene into an organism's genome, which enables humans to ascertain the genetic constitution of living creatures.

Paul Berg is one of the major contributors to genetic modification. He successfully carried out genetic modification through advancements in biochemistry through the discovery of new enzymes. Through the creation of hybrid DNA molecules, he used the intensively researched SV40 monkey virus and the lambda bacteriophage to create the recombinant deoxyribose nucleic acid . In other words, Berg assembled the first DNA molecules by combining genes from various organisms. The original concept he had was to pen and splice the SV40 DNA into genes. The contribution of Paul Berg to genetic modification can further be found at the end of the chapter under Individual Matter.

Regarding the ecosystem, genetic modification arises from increased deployment of herbicides as well as the contamination of aquatic ecosystems. Transgenic crops could have indirect environmental impacts because of the changing environmental practices related to new varieties. Such indirect impacts could be harmful or beneficial based on the involved changes. Toxicity is a significant issue that surrounds chemical herbicides and pesticides (Taleb et al. 2014). Toxicity negatively affects ecosystems and animals such as bees and monarch butterflies are at risk of this effect. According to scientists, the utilization of agricultural herbicides and pesticides has led to the destruction of farmland birds, insects as well as wild plants. Ecosystem changes are the significant effects of genetic modification, but it also has minimal impact on the water cycle, weather, and climate.

Case Study 4.2: Monarch butterflies

Fig. 4.1 Monarch butterflies, *Danaus Plexippus*

(continued)

Monarch butterflies taste nectar in every September prior to heading to Mexico from the US and Canada. It is among the most extraordinary migrations found within the animal kingdom and has the aim of reaching the Mexican mountain range. According to various calculations, monarchs move approximately 75 km every day, occasionally reaching 2500 m in height. Because of the weather vagaries, they are at times compelled to cover 7000 km prior to reaching their resting place. The monarch's life cycle, similar to other butterflies, tends to revolve around the phases of larva–caterpillar–pupae–butterfly–larva. More importantly, monarchs usually eat milkweed plants.

The major threat of monarch butterflies is the low prevalence of milkweed. This has been due to the increased utilization of genetically modified organism (GMO) crops that can cope with a high dosage of herbicides. Milkweed often grows around crops like soybeans and corn in Midwest US. Milkweed acts as the primary food source of the species. However, the main problem is that contrary to GMO crops that are found close by, milkweed cannot cope against herbicides. Despite doubts raised about GMO plants causing the decline of milkweeds, there is a high correlation between higher utilization of herbicide-tolerant plants and the decline of the monarch population.

Regarding weather and climate, biotech solutions can positively affect the environment through mitigation of the effects of climate change within agriculture through GHGs. In fact, agricultural biotechnology could offer solutions, including conventional breeding, genetic engineering, and tissue culture. The advances in breeding can aid agriculture to attain higher yields as well as meet the expanding populations' needs with limited water and land resources. For example, it is estimated that due to enhanced breeding techniques, productivity increases within the global production of primary crops, such as rice and maize, has risen by about 21%. The use of fewer resources enhances sustainability and contributes less to climate change, unlike many conventional agricultural forms that contribute to the emission of GHGs.

Similarly, it positively affects the water cycle by reducing evaporation, surface runoff, evapotranspiration, and transpiration. One of the major examples of genetic modification is the Water Efficient Maize for Africa (WEMA) varieties which are based in Kenya and organized by the Bill and Melinda Gates Foundation (Beyene et al. 2015). Water Efficient Maize for Africa was launched in 2008 to assist farmers in managing the drought risk through the development and deployment of maize varieties that produce between 24 and 35% more grain. The project focuses on Sub-Saharan Africa through a combination of conventional breeding, biotechnology as well as marker-assisted breeding to initiate drought tolerance. The main countries that the project operates in are Tanzania, Uganda, Kenya, Mozambique, and South Africa. Higher and more reliable harvests can aid farmers to feed families and raise their incomes. Biotechnology plays a crucial role in ensuring that the developed hybrid maize seed utilizes water more efficiently and becomes resistant to insect pests. Farmers who use Drought tEG hybrids report that they have harvested between 20 and 35% more grain (Beyene et al. 2015). The overall effect of the WEMA project could be the availability of drought-resistant maize hybrids for smallholder farmers within Sub-Saharan Africa.

Water Efficient Maize for Africa (WEMA) was developed as a result of marker-assisted breeding. Ideally, marker-assisted breeding refers to indirect selection process whereby a trait is chosen based on a marker (could be biochemical or morphological) linked to a trait of interest(including diseases resistance, quality and disease resistance).

Regarding the carbon cycle, appropriate management practices when they are adopted in farming can be crucial in enhancing carbon sequestration. Crops that are developed using modern agricultural biotechnology minimize the importance of tillage, enabling farmers to adopt no-tilling farming practices. Consequently, the quality of the soil increases and becomes rich in carbon. Additionally, less carbon within the soil may be oxidized by being exposed to the air because the soil is not disrupted by plowing, thus leading to less emission of CO_2 into the atmosphere.

4.6 Acid Rain

Acid rain entails any form of precipitation that has acidic components, including nitric or sulfuric acid, which falls on the ground. It could include hail, dust, snow, or rain. Acid rain results from sulfur dioxide (SO_2) and nitric oxide (NO) being emitted into the atmosphere. They react with oxygen, water, and other chemicals to form nitric and sulfuric acid. They combine with water along with other materials prior to falling to the ground. The main sources of these gases in the atmosphere are the combustion of fossil fuels for the generation of electricity, vehicles, and industries.

4.6.1 Ecosystems

Acid rain affects the aquatic environment through its accumulation in water. Marine animals and plants require a specific pH to survive and become unable to cope if it gets beyond their survival level. At low pH levels, many fish eggs are unable to hatch, and it can also lead to the death of adult fish (Warren et al. 2017). Acid rain that flows from the catchment areas into water bodies reduces biodiversity because of the acidification of lakes and rivers. Species such as plant and fish have reduced, while others have been eliminated due to excess acid rain. Acid rain makes trees vulnerable to extreme weather and diseases and damages their growth. For example, the destruction of forests because of acid rain is evident in

Eastern Europe, particularly Poland, Switzerland, as well as Germany. The ravaging impacts of acid rain on soil have consistently led to the damage of altitude forests and vegetation cover because of being encircled by clouds and fogs. In addition to this, the widespread impacts of acid rain have resulted in stunted growth as well as the demise of some of the forests and vegetation cover.

Acid rain makes precipitation more acidic, thus negatively affects buildings through corrosion. Also, acid rain leads to weathering as water flows from high to the low ground. When it rains, the water pH is often around 6.0, implying that it is quite acidic (Warren et al. 2017). When gases start dissolving within the condensed water, it leads to the creation of sulfuric and nitric acid before falling as acid rain. Finally, regarding weather and climate, acid rain does not have a substantial effect on weather and climate. It only affects the content of water vapor and surface runoff.

Explain why acid rain is quite lethal for the environment and its ecosystems.

4.7 Pollution

Pollution refers to the addition of substances or any forms of energy to the environment at a pace that is faster than it is decomposed, diluted, or converted to a harmless form. The main types of pollution are water, air, and land pollution. The modern society also deals with other forms of pollution, including light, noise, and plastic pollution. Whereas environmental pollution can be attributed to natural events, including active volcanoes and forest fires, the utilization of the terms pollution mainly means that contaminants have a human source or human activities that cause pollution.

4.7.1 Ecosystems

Air pollution impacts ecosystems in numerous ways, changing basic ecosystem functions, including primary production as well as biogeochemical cycling, thus affecting ecosystem services and the benefits that humans accrue from them.

Eutrophication is one of the major impacts of pollution on the ecosystem (Langston 2017). Conceptually, eutrophication is defined as the enrichment of water using nutrient salts that lead to structural ecosystem changes, including higher algal production, the decline in fish species, and water quality deterioration.

Identify the relationship between pollution and eutrophication.

4.7.2 Biogeochemical Cycles

Pollution of the ecosystem disrupts its natural balance, thus affecting organisms in diverse ways. The introduction of pollutants, like toxic waste, can endanger various species as well as plants in the area. Also, pollutants such as nitrogen, phosphate, and oil can immensely impact the ecosystem. These changes can be done by silencing or inserting a gene into an organism's genome, which enables humans to ascertain the genetic constitution of living creatures. It can result in the stoppage of the oxygen cycle, while polluted water may also affect animals that depend on water in the lake.

Regarding the water cycle, air pollution leads to the formation of acid rain that affects lakes, rivers, and the sea. These harm living organisms due to the altered chemical compositions as well as water quality (Tian et al. 2015). Some of the factors that contribute to water pollution are organic waste, inorganic chemicals, substance organic chemicals, and radioactive substances. Contaminants that lead to water pollution also lead to disruption of the cycle, which is vital for the maintenance of the planet's temperature. Water bodies that suffer more damage as a result of pollution are rivers. Typically, rivers can easily break down and withstand contaminants, especially when it is degradable. Contrastingly, when the contaminant level is high, they cannot be eliminated, thus making the water unusable.

Pollution also leads to a reduction in the quantity of oxygen within the oxygen cycle because of factors such as logging or low vegetation cover. In the absence of CO_2 in the atmosphere, pollution could also contribute to lower potential for absorption of ultraviolet (UV) rays and the capacity to withstand ambient temperature (Tian et al. 2015). High numbers of forest fires lead to higher quantities of CO_2 within the atmosphere. Regarding the nitrogen cycle, nitrogen being the most abundant element in the air can combine with oxygen to cause acid rain.

Where do radioactive pollutants emanate from? Identify two sources.

4.7.3 Weather and Climate

Air pollution leads to significant changes in the climate depending on its type. Some air pollutants could contribute to more global warming. It includes greenhouse gases (GHGs) like carbon dioxide that causes global warming. Greenhouse gases have increased in our atmosphere today, thus leading to more global warming. Contrarily, some air pollutants may slow down the process of global warming and include tiny particles released by trucks and cars. Such tiny particles are usually called aerosols and can be made of various things including sulfates, carbon, or mineral dust. Some of the particles naturally enter the atmosphere. However, air pollution caused by the combustion of fossil fuels can also add them to the atmosphere. Aerosols influence climate by creating a cooling effect. Various computer models suggest that tiny aerosols globally lead to about 50% of the cooling, in the same way, GHGs cause warming (Zanobetti and Peters 2015). Car pollution can also lead to the creation of smog, which is a dangerous weather condition that minimizes visibility within cities. Smog develops when large quantities of air pollutants are released into the air, usually due to wind patterns.

4.8 Seasonal Changes

The change of seasons leads to the alteration of chemical and biological functions of a whole ecosystem. It is because energy consumption and resources are drastically added depending on the season. Many ecosystems function in a cause–effect correlation. The loss or addition of one resource leads to a net effect in the entire ecosystem. Seasonal shifts only occur for a short period and may not necessarily affect the ecosystem. During winter, the death of a plant may not lead to mass extinction or adverse environmental damage. This is because the ecosystem is usually prepared for the winter, thus enabling all ecosystem aspects to cope despite having limited resources. However, climate change causes ecosystems to become negatively affected by seasonal changes. For instance, a prolonged winter would severely affect the environment since other ecosystem services have not yet adapted to such a change (Zeppel et al. 2014). Consequently, ecosystems adapt to seasonal cycles, but changes in the climate drastically shift these cycles, thus harming all ecosystem parts.

Seasonal changes also affect weather since it is a measure of the weather conditions for a short period. For example, they may affect the quantity of precipitation and transpiration (Zeppel et al. 2014). However, it does not have a significant effect on the climate. Regarding the biogeochemical cycles, there are minimal impacts posed to them as a result of seasonal changes. As such, seasonal changes mainly affect ecosystems and biodiversity.

Discuss two more ways in which seasonal changes affect weather.

4.9 High Rate of Disasters

There has been a significant increase in the number of disasters that afflict the planet. Volcanic eruptions, natural bushfires, and earthquakes all impact the various ecosystems on Earth (Fig. 4.1 shows the number of reported natural disaster events

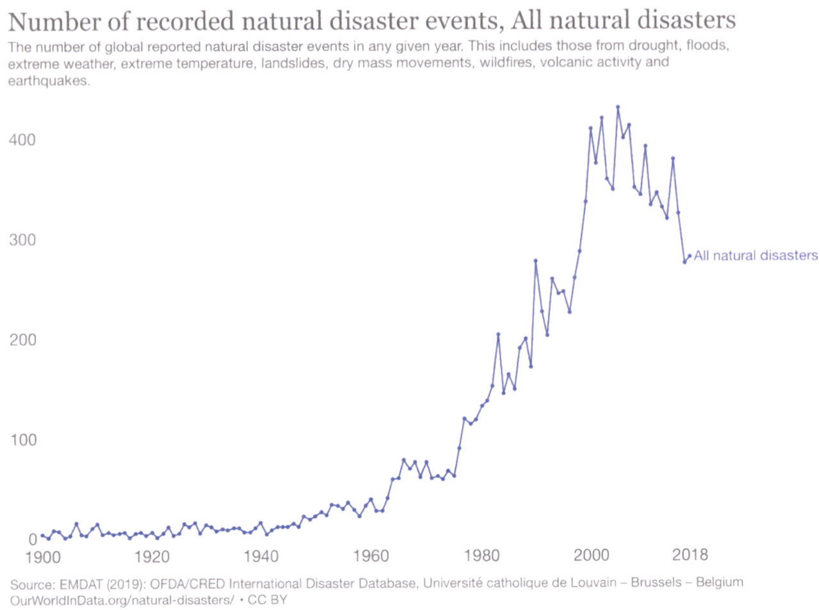

Fig. 4.2 Natural disaster overtime. *Source:* EMDAT (2017)

over time, while Fig. 4.2 reveals by type the number of reported weather and non-weather related natural disasters which can affect the Earth's ecological balance). Initially, they affect wetlands, coastal systems, and forests by spreading invasive species and causing habitat loss. Ecosystem degradation in the short-term minimizes the capacity of the forest to carry out carbon sequestration, thus aggravating climate change effectively. However, over time, many forms of natural disasters play a crucial role in the rejuvenation of the same ecosystem that they destroyed. Climate change has a broader impact on disasters, while disasters have a lesser impact on climate change. Notwithstanding, natural disasters can have a substantial impact on climate change and, simultaneously, gradual climate change could affect natural disasters (Woodward et al. 2016). The gradual rise in temperature could increase volcanic activity and tsunamis. The increasing patterns of El Niños significantly affect climate change and are poised to be a dominant influence in the next few decades. The significant impact of prolonged disasters on the biogeochemical cycle is that they will affect the various processes of the water cycle. For example, increased flooding will increase the quantity of runoff flowing into the water bodies. Also, El Niños will increase the potential of flooding and carbon sequestration because of the availability of the high quantity of water that can act as carbon sinks (Fig. 4.3).

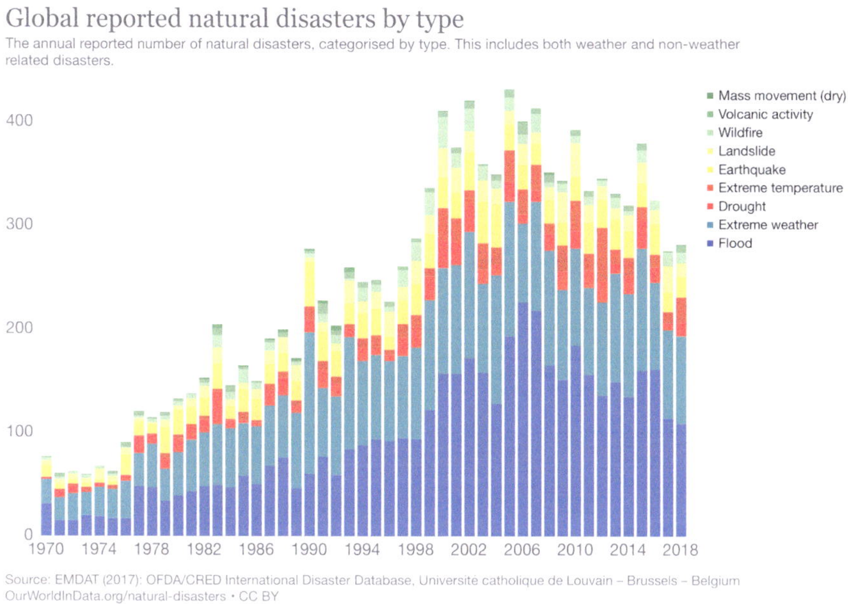

Fig. 4.3 Natural disaster by type. *Source:* EMDAT (2019), Ritchie and Roser (2019)

4.10 Extinction

From a biological perspective, extinction is depicted as the termination of an organism or various organisms. The extinction moment is mainly perceived as the demise of the last individuals of a specific species, but the ability to breed, as well as recover, can be affected. Determining extinction is quite tricky since the potential range of a species could be quite large and is often carried out retrospectively. This difficulty results in phenomena like Lazarus taxa. Ideally, this is a situation whereby a species that may have been assumed as extinct abruptly "remerges" after an absent period.

Extinction can make other species alter their behavior as a way of compensating the vacant role that extinct species leave behind. It could result in negative or positive impacts on ecosystems, thus improve or worsen outcomes. Currently, scientists believe that any contribution is entirely lost at the extinction point, resulting in a decline of the performance of the ecosystem. Biodiversity reduction negatively impacts the ecosystem. A crucial finding is that the response of the surviving species could determine the changes in the ecosystem (Goswami et al. 2017). Extinction could also affect the water and carbon cycles, but its impact on these cycles remains uncertain. This is because another section of the same cycle could offset impacts posed on one section of the cycle. Regarding weather and climate, extinction could result in the reduction in the quantity of forest cover, water bodies, and some animals that play a crucial role in maintaining the balance of gases. These could have a crucial effect on the weather and climate. However, similar to the biogeochemical cycles, the impact of extinction on weather and climate remains unknown due to limited research.

Case Study 4.3: Giant panda

Fig. 4.4 Giant panda

Giant Pandas are among the most popular animals in the world and are native to China's mountain forests (see Fig. 4.4). They have black and white body parts. The muzzle, legs, shoulders, and ears are black, while the other parts of the body are white. It is estimated that about 2060 pandas are available today in the world (Li et al. 2010). The International Union for Conservation of Nature (IUCN) considers the giant panda susceptible to extinction as they infrequently reproduce, making the population difficult to recover. Habitat destruction is a major contributor to the decline of panda populations. With the growing human population within China, the habitat of pandas is overtaken by development, thus forcing them into less livable areas. The destruction of habitats also results in a shortage of food. They also feed on various bamboo varieties that bloom at varying periods of the year. In case one kind of bamboo gets destroyed due to development, it could leave pandas without anything to eat, thus raising the vulnerability of starvation. The Chinese government has been actively restoring and protecting bamboo habitat, thus increasing the chances of survival. The measures have yielded positive results. Surveys by State Forestry Administration suggest that the

(continued)

actions have contributed to a rise of the panda population. Consequently, the IUCN upgraded the status of the giant panda in 2016 from endangered to vulnerable. Despite the higher population being good news today, predictions suggest that climate change will lead to the extinction of 35% of the bamboo habitat occupied by panda within the next 75 years (Li et al. 2010).

If you were given a chance to help the giant panda, what would you do?

4.11 Individual Matter: Paul Berg

Paul Berg was born on June 30, 1926, in New York (see Fig. 4.5). He was the first scientist to show the likelihood of making recombinant DNA. He also assisted in providing pioneer guidelines for limiting the probable harm attributed to genetic engineering. He went to Abraham Lincoln High School in Brooklyn. In 1953, he enrolled to study chemical engineering but decided not to in order to support the war effort. He became a flyer for the navy, but because of his age, he decided to study biochemistry at Penn State University (Science History.org, 2018). In 1948, he joined Western Reserve University where he did his doctorate and demonstrated the way vitamin B-12, as well as folic acid, allows animals to manufacture amino acid methionine.

Berg's first achievement was the development of techniques for splicing and joining DNA molecules which set the stage for recombinant DNA technology that resulted in the modern biotechnology industry. He is also acknowledged as a role model for raising doubts about the ethical implications associated with genetic

Fig. 4.5 Paul Berg. *Source:* U.S. National Library of Medicine/National Institutes of Health via Encyclopædia Britannica (n.d.)

engineering. Berg also played a crucial role in organizing the Asilomar Conference on Recombinant DNA in 1975 (Science History Institute 2017). This brought together 140 professionals to talk about the ramifications of the technology and guidelines for regulating its use. The Nobel Prize in Chemistry was awarded to him in 1980 in recognition for his fundamental studies about nucleic acid biochemistry.

4.12 Future Trends

4.12.1 Africa

Africa will likely face new disease challenges because of climate change. Also, rising temperatures will affect the habitability of some of the areas. The other trend is that ocean acidification will alter ocean chemistry as well as marine organisms' physiology. It could potentially have an adverse impact on coral reefs, fisheries, and aquaculture, particularly for the numerous African countries where marine and coastal ecosystems support livelihoods and act as sources of food. More critically, various organisms will continue disappearing or being vulnerable to extinction, including the African wild dog, Ethiopian wolf, Pangolin, and rhino.

4.12.2 Asia

The vast population and frequent natural disasters will ensure that climate change severely affects the region. Asia will continue contributing to global warming because of China, Japan, and India (three of the top five polluters globally). Also, transition towards non-emitting energy sources will remain uneven with some of the nations including India and Qatar continuing to champion for renewables. Extreme weather events will be experienced, including prolonged precipitation.

4.12.3 Australia

Slezak (2018) predicts that 3 million hectares of Australian forests could be lost within 15 years because of the booming livestock industry and failure by the government to initiate restrictions against deforestation.

4.12.4 Antarctica

King penguins could become extinct by the end of the century. Overfishing and climate change could force the king penguin to the edge of extinction.

4.12.5 Europe

Intensive farming practices will contribute to a sharp decline in the *Andrena transitoria* bee. Climate change will continue being a key driver of extinction risk for bee species, and specifically bumblebees. Heavy rainfalls, droughts, as well as increased temperatures could change habitats, resulting in population decline.

4.12.6 North America

As the North American continent is pushed towards further change, there is a higher risk of unanticipated impacts, with some being large and irreversible. Sea levels could continue rising if the current trends continue.

4.12.7 South America

In South America, experts predict that severe weather events may likely become a common phenomenon. In addition, record-breaking temperatures will also be experienced along with heatwaves and erratic rainfall patterns. This could lead to stunted growth and poor harvests.

4.13 Summary

Climate Change

- Climate change leads to the migration of species from habitats, while other species may flourish depending on the changing conditions.
- This will result in the extinction of numerous species and minimization in ecosystem diversity.
- Higher global temperatures could also lead to the death of numerous coral reefs.
- Climate change increases precipitation and evaporation stages of the water cycle.
- It leads to the increase of fossil fuels and the rise of carbon content in the atmosphere.

- The major impacts on weather and climate are droughts, heatwaves, heavy downpours, flooding, and hurricanes.
- Water Efficient Maize for Africa (WEMA) project is evidence of biotechnology used today.
- The giant panda is evidence for the extinction and habitat destruction today.

Global Warming

- On ecosystems, global warming leads to forest destruction, wildfires, death of fish and aquatic species, and inhabitable ecosystems. It may also change the chemical composition of water and oxygen decline.
- Massive losses of mangroves, higher salinity, increased migrations of species, and prevalence of disease-causing pathogens.
- On cycles, there will be an increase or reduction of precipitation depending on its period. River runoff will also continue increasing.
- On weather and climate, there will be a rise in global temperatures that could lead to extreme weather events, including hurricanes, heatwaves, flash floods, and lightning.

Ocean Acidification

- Higher acidity harms marine organisms and leads to coral bleaching. Some species could also find it challenging to make their shells/skeletons.
- Ocean acidification accelerates the carbon cycle, increases water temperatures, and leads to a rise in sea levels.
- It makes marine ecosystems less habitable.

Deforestation

- Deforestation leads to a disastrous impact on the carbon cycle, thus minimizing the amount of carbon stored.
- It results in a lower quantity of vapor that flows from the land.
- Land degradation can lead to biodiversity loss and could result in the extinction of various species.

Genetic Modification

- Transgenic crops may be beneficial or disastrous, based on resulting changes.
- Toxicity negatively impacts the ecosystems as well as animals.
- Biotech solutions can positively impact the environment by mitigating the impacts of climate change. They can also promote the use of fewer resources, thus enhancing sustainability.
- Appropriate management practices in genetic modification can be crucial for improving carbon sequestration.
- Water Efficient Maize for Africa (WEMA) is among the leading projects in Africa that shows the significance of biotechnology.
- Paul Berg, Nobel Peace Prize winner, is among the major contributors to the genetic modification debate because of his feats in creating hybrid DNA molecules.

Acid Rain

- Acid rain affects the aquatic environment by accumulating in water.
- It alters the pH of aquatic environments, thus affecting the ability of species to cope.
- Acid rain makes trees unable to cope with extreme weather events and diseases.
- It makes precipitation more acidic, thus leading to corrosion.
- Does not have a significant effect on weather and climate as it only affects the content of water vapor as well as surface runoff.

Pollution

- Air pollution changes the functioning of basic ecosystems.
- Leads to eutrophication in marine ecosystems, thus affecting water quality. Disrupts the natural balance of the planet, thus impacting various cycles.
- Reduces the amount of oxygen in the oxygen cycle.
- Air pollution influences the climate in various ways, including global warming or cooling of the atmosphere.

Seasonal Changes

- Lead to changes in biological and chemical functions of an ecosystem.
- Combines with climate change to cause changes within the ecosystem.
- Affect the weather conditions of a particular place, particularly rainfall patterns.

High Rate of Disasters

- Affect wetlands, coastal systems, and forests by spreading invasive species and habitat loss.
- Leads to degradation and reduces the ability of forests to act as a carbon sink effectively.
- Has a minimal impact on climate, but affects processes of the water cycle.

Extinction

- Leads to changes in species' behavior and declined the performance of ecosystems.
- Has unquantifiable effects on weather, climate, and cycles of the world.

Future Trends

- Climate change will make Africa vulnerable to new disease challenges.
- Ocean acidification will impact Africa's coral reefs and fisheries negatively.
- Climate change will adversely affect Asia because of the vast population, while intensive farming practices in Europe will lead to the decline of certain species.
- In North and South America, sea levels could continue rising; severe weather events may likely become a common phenomenon and temperatures will continue rising.
- Deforestation will result in further loss of vast hectares of Australian forests.
- Extinction of species will continue worsening in Antarctica.

4.14 Discussion Questions

1. Global warming could be making severe weather events more frequent. What are the reasons that disasters are increasingly becoming expensive and frequent?
2. Explain the way higher temperatures can affect the water cycle and result in drought in Africa.
3. Explain how the effects of drought or flooding on the water cycle could affect people and their property.
4. After learning about the way global warming could affect the water cycle, explain ways the water cycle can influence global warming and reduce temperature change.
5. How can we reverse the trend of extinction that we are currently experiencing?
6. Suggest four mechanisms on how sustainable development goals can be aligned to address the increasing number of disasters.
7. Discuss the importance of eutrophication on organisms. How can eutrophication be used as a positive avenue for livelihoods?
8. Discuss six measures that can be undertaken to save the giant panda.
9. How can the WEMA project be replicated across Africa successfully?

References

Andrews, J. A., & Schlesinger, W. H. (2001). Soil CO2 dynamics, acidification, and chemical weathering in a temperate forest with experimental CO2 enrichment. *Global Biogeochemical Cycles, 15*(1), 149–162.

Archer, S. R., & Predick, K. I. (2008). Climate change and ecosystems of the southwestern United States. *Rangelands, 30*(3), 23–28.

Beyene, Y., Semagn, K., Mugo, S., Tarekegne, A., Babu, R., Meisel, B., et al. (2015). Genetic gains in grain yield through genomic selection in eight bi-parental maize populations under drought stress. *Crop Science, 55*(1), 154–163.

Bonan, G. (2015). *Ecological climatology: Concepts and applications*. Cambridge: Cambridge University Press.

Cahill, A. E., Aiello-Lammens, M. E., Fisher-Reid, M. C., Hua, X., Karanewsky, C. J., Ryu, H. Y., et al. (2012). How does climate change cause extinction? In *Proceedings of the Royal Society B.*. https://doi.org/10.1098/rspb.2012.1890.

Cao, L., Caldeira, K., & Jain, A. K. (2007). Effects of carbon dioxide and climate change on ocean acidification and carbonate mineral saturation. *Geophysical Research Letters, 34*(5), 1–5. https://doi.org/10.1029/2006GL028605.

Doney, S. C., Fabry, V. J., Feely, R. A., & Kleypas, J. A. (2009). Ocean acidification: The other CO_2 problem. *Annual Reviews, 1*(1), 169–192.

Doney, S. C., Mahowald, N., Lima, I., Feely, R. A., Mackenzie, F. T., Lamarque, J. F., et al. (2007). Impact of anthropogenic atmospheric nitrogen and sulfur deposition on ocean acidification and the inorganic carbon system. *Proceedings of the National Academy of Sciences, 104*(37), 14580–14585.

EMDAT. (2017). Global reported natural disasters by type. Retrieved November 04, 2019, from http://www.emdat.be/.

EMDAT. (2019). Number of recorded natural disaster events, all natural disasters. Retrieved November 11, 2019, from http://www.emdat.be/.

Encyclopædia Britannica. (n.d.). Berg Paul. Photo. Retrieved September 10, 2018, from https://www.britannica.com/biography/Paul-Berg/images-videos-/media/-1/61698/151266/.

Goswami, M., Bhattacharyya, P., & Tribedi, P. (2017). Allee effect: The story behind the stabilization or extinction of microbial ecosystem. *Archives of Microbiology, 199*(2), 185–190.

Hoegh-Guldberg, O., & Bruno, J. F. (2010). The impact of climate change on the world's marine ecosystems. *Science, 328*(5985), 1523–1528.

Joos, F., Prentice, I. C., Sitch, S., Meyer, R., Hooss, G., Plattner, G. K., et al. (2001). Global warming feedbacks on terrestrial carbon uptake under the Intergovernmental Panel on Climate Change (IPCC) emission scenarios. *Global Biogeochemical Cycles, 15*(4), 891–907.

Langston, W. J. (2017). Toxic effects of metals and the incidence of metal pollution in marine ecosystems. In *Heavy metals in the marine environment* (pp. 101–120). Boca Raton: CRC Press.

Li, R., Fan, W., Tian, G., Zhu, H., He, L., Cai, J., et al. (2010). The sequence and de novo assembly of the giant panda genome. *Nature, 463*(7279), 311.

Longobardi, P., Montenegro, A., Beltrami, H., & Eby, M. (2016). Deforestation induced climate change: Effects of spatial scale. *PLoS One, 11*(4), e0153357.

Mori, N., Kato, M., Kim, S., Mase, H., Shibutani, Y., Takemi, T., et al. (2014). Local amplification of storm surge by Super Typhoon Haiyan in Leyte Gulf. *Geophysical Research Letters, 41*(14), 5106–5113.

Nerem, R. S., Beckley, B. D., Fasullo, J. T., Hamlington, B. D., Masters, D., & Mitchum, G. T. (2018). Climate-change–driven accelerated sea-level rise detected in the altimeter era. *Proceedings of the National Academy of Sciences, 115*, 2022–2025.

Pataki, D. E., Carreiro, M. M., Cherrier, J., Grulke, N. E., Jennings, V., Pincetl, S., et al. (2011). Coupling biogeochemical cycles in urban environments: Ecosystem services, green solutions, and misconceptions. *Frontiers in Ecology and the Environment, 9*(1), 27–36.

Ritchie, H., & Roser, M. (2019). Natural disasters. Retrieved November 03, 2019, from https://ourworldindata.org/natural-disasters/.

Schmittner, A., Oschlies, A., Matthews, H. D., & Galbraith, E. D. (2008). Future changes in climate, ocean circulation, ecosystems, and biogeochemical cycling simulated for a business-as-usual CO2 emission scenario until year 4000 AD. *Global Biogeochemical Cycles, 22*(1), 1–21. https://doi.org/10.1029/2007GB002953.

Science History Institute. (2017). Paul Berg. Retrieved July 18 from https://www.sciencehistory.org/historical-profile/paulberg/.

Slezak, M. (2018). Global deforestation hot-spot: 3m hectares of Australian forest to be lost in 15 years. *The Guardian*. Retrieved December 15, 2018 from, https://www.theguardian.com/environment/2018/mar/05/global-deforestation-hot-spot-3m-hectares-of-australian-forest-to-be-lost-in-15-years/.

Taleb, N. N., Read, R., Douady, R., Norman, J., & Bar-Yam, Y. (2014). The precautionary principle (with application to the genetic modification of organisms). preprint arXiv:1410.5787.

Tian, H., Chen, G., Lu, C., Xu, X., Ren, W., Zhang, B., et al. (2015). Global methane and nitrous oxide emissions from terrestrial ecosystems due to multiple environmental changes. *Ecosystem Health and Sustainability, 1*(1), 1–20.

Verisk Maplecroft. (2016). Climate change vulnerability index 2017. Retrieved December 15, 2018, from https://reliefweb.int/sites/reliefweb.int/files/resources/ve-risk%20index.pdf/.

Warren, D. R., Kraft, C. E., Josephson, D. C., & Driscoll, C. T. (2017). Acid rain recovery may help to mitigate the impacts of climate change on thermally sensitive fish in lakes across eastern North America. *Global Change Biology, 23*(6), 2149–2153.

Woodward, G., Bonada, N., Brown, L. E., Death, R. G., Durance, I., Gray, C., et al. (2016). The effects of climatic fluctuations and extreme events on running water ecosystems. *Philosophical Transactions of the Royal Society B, 371*(1694), 20150274.

Zanobetti, A., & Peters, A. (2015). Disentangling interactions between atmospheric pollution and weather. *Journal of Epidemiological Community Health, 69*, 613–615.

Zeppel, M. J., Wilks, J. V., & Lewis, J. D. (2014). Impacts of extreme precipitation and seasonal changes in precipitation on plants. *Biogeosciences, 11*, 3083–3093.

Part II
Green Practices and Environmental Responsibility

Chapter 5
Health and Diseases

Abstract Everyday activities have a significant impact on the spread of diseases. This is particularly attributed to climate change. Thus, understanding the role played by climate change in the spread of diseases is crucial to understand the most viable ways of managing them. With increased negative impacts on climate change on disease spread, it is crucial also to understand the green ways of not only managing but also preventing diseases. This chapter discusses the communicable and non-communicable diseases (NCDs) and the role of sustainability in managing them. Disease prevention and control mechanisms, particularly the sustainable mechanisms, are also deeply explored. To enhance understanding of substance and drug abuse, the role of sustainability in drug abuse is further examined. Sustainable methods of preventing drug use are then explored. In this chapter, real-life examples, case studies, an experiment, and the individual matter of Salome Karwah have been used for enhancing understanding of the concepts and aspects presented.

Keywords Communicable diseases · Non-communicable diseases · Drug abuse · Salome Karwah · Green health

5.1 Diseases

The term disease refers to any condition that prevents the body from functioning normally. Consequently, diseases impair the health of the body by affecting its normal homeostatic processes. This term has been used for referring to infectious diseases that are attributed to pathogenic microbial agents, such as fungi, aberrant proteins, bacteria, and protozoa. Any infection that may not lead to clinically evident impairment of the normal functioning of the body is not perceived as a disease. In this section, the focus will mainly be on diseases brought about by climate change. Prior to discussing this, it is worth noting that the major types of diseases are infectious, non-infectious, acquired, chronic, acute, congenital, genetic, hereditary, idiopathic, primary, terminal, and secondary disease. In this chapter, the discussion will mainly be limited to diseases spread by climate change.

A. A. Akinsemolu, *The Principles of Green and Sustainability Science*,
https://doi.org/10.1007/978-981-15-2493-6_5

The warming climate plays an ever-increasing role in influencing the global emergence, resurgence as well as redistribution of infectious diseases. Many of the most prevalent infectious diseases and specifically those transmitted by the insects are quite sensitive to climate variation. The availability of vaccines has significantly reduced the prevalence of infectious diseases. However, there is evidence to suggest that climate change may potentially result in a resurgence of some deadly illnesses on the planet. Many of the common infectious illnesses are often affected by the variation of climate, but the ones transmitted by insects are usually sensitive. One of the diseases attributed to climate change is malaria. Malaria is a bacterial infection that leads to the death of about 429,000 individuals, and 80% of these people are children under five years (Onyango et al. 2018). Despite this number decreasing, there are fears that climate change could play a crucial role in helping in the spread of malaria. Malaria is often transmitted by mosquitoes belonging to the genus Anopheles that are sensitive to variations in climatic conditions. It mainly requires moist conditions and temperatures of about 25–28 °C for breeding. The rising global temperatures may alter the range of Anopheles mosquitoes, and thus facilitate the spread of malaria.

What makes it conducive for Anopheles mosquitoes to spread in higher temperatures?

Africa is among the riskiest places, and the continent has been identified as one of the major hotspots of malaria in the world. It is estimated that 92% of the world's deaths attributed to malaria occur in Africa (see Fig. 5.1). However, an increasing number of scientific studies suggest that climate change could aggravate malaria issues in Africa. In East Africa, it is predicted that climate change may substantially expand the numbers of the Anopheles mosquito by the end of the twenty-first century, thus exacerbating the vulnerability of the disease within the region. This resonates with findings made by various studies that by 2080, climate change would result in a substantial increase in malaria transmission in both southern and eastern Africa (Beier et al. 2008; Altizer et al. 2013; Onyango et al. 2018). Other parts of the world that will also experience this are Central America, some parts of India, Nepal, and southern Brazil.

Apart from malaria, dengue fever could also spread further because of climate change. Studies have also established that climate variability may also impact the susceptibility of dengue fever and concerns exist about the same being true for yellow fever, hantavirus as well as Ebola (Kakmeni et al. 2018). There is evidence showing that climate change may already be changing the distribution of some of these diseases. Regarding the Ebola virus disease, it is severe and usually fatal. The virus spreads from wild animals to people and spreads within the population via a human-to-human transmission (Banwell et al. 2018). The first outbreaks took place within the grassroots areas of Central Africa, close to the tropical rainforests. The fatal outbreak was in 2014–2016 in West Africa that occurred in key urban

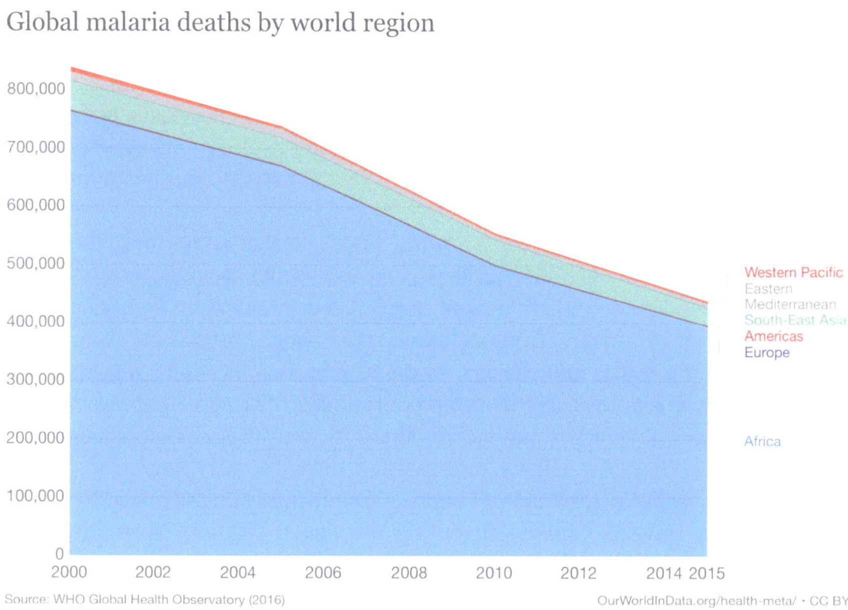

Fig. 5.1 Global malaria deaths by world region. *Source:* World Health Organisation (2016), Ritchie and Roser (2019)

centers and rural areas. Ebola can lead to severe viral hemorrhagic fever outbreaks. Viral hemorrhagic fever refers to various illnesses caused by unique virus families. The term is generally used for describing a severe multisystem syndrome, meaning that many organ systems are infected in the body. The four major types of the Ebola virus identified in Africa are Zaire, Ivory Coast, Sudan, and Bundibugyo (Banwell et al. 2018). The main symptoms are sudden fever, sore throat, headache, intense weakness, as well as muscle pain. It is usually followed by a rash, diarrhea, vomiting, impaired liver and kidney function, and bleeding. The primary mode of transmission is direct contact with the secretions or blood of the infected person. Family members usually have an elevated risk when taking care of the infected individual through close contact. To date, there are no known treatments for the disease, but few drug therapies are under evaluation. Because of the severity of the disease and the way its spreads, many medics and people who put their life on the line to save lives have been identified as heroes and martyrs.

Kwei Quartey's (The Ebola Plague: African Heroes and Martyrs) provides a comprehensive account of the heroes and martyrs who have been beneficial in ensuring that affected populations get utmost care. Nurse Kacie Hickox provided care for the Sierra Leone Ebola victims and was forcibly quarantined when she went back to New Jersey. Stella Ameyo Adadevoh, a West African physician, died while caring for victims of the virus in Nigeria (Nwafor and Nwafor 2017). She gave her life at such a difficult period when the world was not concerned about the severity of the epidemic. She played a crucial role in preventing the spread of Ebola in Nigeria that could have affected other nations. Salome Karwah is also a renowned figure because of the pivotal role she played in fighting the spread of Ebola. A

Time Magazine Person of the Year in 2014, Karwah is recognized for her efforts in combating the Ebola virus epidemic within Liberia. She was a survivor of Ebola prior to going back to assist other patients who had been affected by the disease. Her actions saved thousands of lives. The story of Karwah's heroics is highlighted in the Individual Matter section at the end of this chapter. Along with other persons, these heroes and martyrs have played a crucial role in helping affected victims and preventing the spread of Ebola.

Apart from Ebola, various recent studies highlight that some mosquito-borne diseases, such as yellow and dengue fever, and tick-borne diseases, such as Lyme disease and tick-borne encephalitis, have commenced spreading to higher latitudes, possibly because of increased temperatures (Caldwell 2018). For example, in Kenya, yellow fever has increasingly become common in some regions. It is caused by male mosquitoes, and its symptoms are dire. The spread of insect-borne pathogens is quite severe but may not be the only way that climate change may influence human diseases. Higher precipitation intensity, resulting from climate change, increases the distribution of waterborne infections that can lead to deadly diarrheal illness as well as flourish during the heavy rainfall. Other infectious diseases, including salmonellosis, giardiasis, and cholera, could exhibit increased outbreaks because of elevated temperature and flooding. Scientists have also raised concerns that permafrost soil melting within the polar regions as a result of climate change could lead to the release of disease-carrying bacteria and viruses.

Why do you think Africa is the most affected in terms of infectious diseases? Use examples from Africa to substantiate your response.

There are five major conclusions on the relationship between climate change, health, and diseases. Firstly, some groups have a higher vulnerability to climate-sensitive health effects because of their age, gender, social marginalization, or other pre-existing health conditions, including Human Immunodeficiency Virus (HIV) (Caldwell 2018). Secondly, many infectious diseases, especially water-borne diseases, are quite sensitive to climate conditions. Thirdly, climate change elongates the transmission season and the geographical distribution of numerous diseases, including dengue and malaria. For instance, the conditions associated with dengue transmission have a likelihood of expanding globally. Fourthly, climate change will be associated with new as well as emerging health problems, including heatwaves and other extreme weather events. Heat stress has a likelihood of aggravating the working conditions and increasing the vulnerability of renal, respiratory, and cardiovascular diseases. Also, it is estimated that 22.5 million individuals are displaced yearly due to weather and climate-related disasters (Caldwell 2018). There is a likelihood that these figures will increase in the future. Climate-induced human mobility is associated with a socioeconomic cost and could impact physical and mental health. Finally, undernutrition and malnutrition have been highlighted as a concern for various developing states in Africa, Latin America, and Asia.

Despite these many risks and challenges, there are numerous solutions for solving climate-related diseases and their distribution, which will be discussed in detail under Sect. 5.4 Disease Control and Prevention.

Case Study 5.1: Nipah and Hendra virus

The Nipah and Hendra virus in South Asia and Australia are among examples of how disruption of an ecosystem could lead to the outbreak of diseases. The viruses from these diseases originate from flying foxes also called fruit bats. They eat fruit through mastication of the pulp and spitting out the seeds and juice. For an extended period, these bats have evolved with the Nipah and Hendra virus, and due to this co-evolution, they are not significantly affected (Martin et al. 2018). However, the outbreak of the virus from bats to species that have not evolved with it can lead to disastrous consequences as observed in rural Malaysia's case in 1999. Scientists believe that there is a probability that a bat dropped chewed fruit into a piggery. The infection of the pigs amplified and shifted to humans. Its lethality was immense because out of the 276 individuals infected in the country, about 106 individuals died. Many others experienced crippling and permanent neurological disorders. The disease does not have any vaccine or cure. Approximately 12 other smaller outbreaks occurred since then in South Asia.

Experts contend that it is crucial to understand that any emerging diseases during the last three or four decades have been due to encroachment into natural habitats and demographical changes (Martin et al. 2018). Emerging infectious diseases can be described as new forms of pathogens. They could also be old ones that have successfully mutated. For instance, Acquired Immune Deficiency Syndrome (AIDS) spread into humans from chimpanzees during the 1920s when hunters of wild game in Africa butchered infected animals (Reperant and Osterhaus 2017). Diseases often emanate from nature and wildlife, to find their way into humans, including malaria and plague.

5.2 Sustainable Ways of Managing and Treating Infectious Diseases

There is a need to find sustainable solutions for the prevention and management of infectious diseases, including hepatitis, malaria, neglected tropical diseases, and HIV, which need timely intervention. Through progress that has been made by the Millennium Development Goals (MDGs), the World Health Organization (WHO) has come up with a collaborative health approach for accelerating progress and meeting ambitious international targets for infectious diseases. The new approach is among the sustainable ways of providing opportunities for synergistic efforts. It emphasizes major actions that include the expansion of coverage of universal care, development of a sustainable approach for strategic information, strengthening health systems, the pursuit of sustainable financing, and encouraging research across all levels.

Finally, the other sustainable way of preventing outbreaks in people is through the One Health Initiative. This is a global program that involves more than 600 professionals and scientists that propagate the concept that ecological, animal as well as human health are connected and should be managed holistically. It is a movement geared towards forging equal yet all-inclusive collaborations between the dentists, nurses, physicians, and veterinarians to unite both veterinary and human medicine.

Experiment 5.1: Infectious diseases: simulated disease transmission

Required Materials

1.5 ml tube with 1 ml of body fluid that can detect infection, marker, and transfer pipette.

Background of the Experiment

The main objective of this learning activity is to imitate or act out the spread of an infectious disease in an entire population. The population could be a classroom or community. In the population, one individual will start the infection, and it will spread through the exchange of body fluids, including saliva on a cup that you share with friends. You will be required to act out the exchange of body fluids with three individuals within the classroom. This activity will use a clear liquid in a tube. The tube should be labeled BF to indicate Body Fluid. The clear liquid lacks body fluid, and this is the activity's simulation. *Caution:* Do not swallow or ingest any materials in this activity!

Experiment Procedure

1. Select experiment participants from the whole sample, and not just in the immediate area. Be patient to wait for the signal prior to each body fluid exchange.
2. Using a marker, write your surname on the lid of the tube with body fluids.
3. Find someone that you can share body fluids with.
4. Pipette one transfer of your liquid into their tube, while they do the same. There should be about 50% of your body within the tube. Close tube and slightly shake to ensure that they mix. The body fluids have successfully shared.
5. On your table, record the person you shared the body fluids with.
6. For the second share, repeat the steps with another person.

(continued)

7. Record the person you shared the body fluids with.
8. Repeat the same steps for the third share or person and record the person you shared your body fluids with on your table.
9. Using the board, record the persons you shared body fluids within the order of sharing. See Table 5.1.

Table 5.1 Record of sample sharing

Your name	
First person you shared with	
Second person shared with	
Third person shared with	

Analysis

1. Add the solution that can detect infection to your body fluid tube and observe.
2. What occurred when the infection detection solution was added?
3. Does it imply that you have the infection, and how does it make you feel?

Note: The solution should turn pink, or nothing should happen. For pink, it means that you have the infection, and for nothing, it means that you do not have the infection.
Add the infection detection solution to your body fluid tube. From the results on the board,

1. How many people in the population are infected?
2. Who do you think infected you?

Examine the data and ascertain who started the infection: This should be accomplished by having the person infected circling their name on the board. Working backward, you will be able to ascertain who infected others in the second swapping round and first round. You will find two people who were infected initially. Because they were the first ones to be infected, they were the first cause of the disease. In addition, since you will be unable to differentiate people who were originally infected, it is critical for the instructor to take note of the person who first took the tube having the initial infection at the start of the activity.

Discussion Questions

1. What is the importance of ascertaining the way a disease spreads?
2. Why are diseases spread by climate change deemed more dangerous compared to others?
3. How can you protect yourself from an infectious disease?
4. Can you protect yourself in the same way against all the diseases?

5.3 Disease Transmission

Communicable diseases are diseases that spread from one individual to another through water, indirect or direct modes. They include flu, typhoid, dysentery, Acquired Immune Deficiency Syndrome (AIDS), and the common cold, and are often caused by pathogens that are passed from one human to the other. Transmission methods include saliva, sexual contact, blood as well as mucus. Contaminated surfaces, like countertops and doorknobs, offer a medium of passing the disease from a person to another. Communicable diseases spread through four major ways. The first way is having physical contact with a person that has already been infected (Boutayeb 2006). For example, touch can spread staphylococcus,, while sexual intercourse can spread Human Immunodeficiency Virus (HIV) and gonorrhea. Fecal or oral transmission can lead to the spread of hepatitis A, whereas droplets lead to the spread of tuberculosis and influenza. The second mode of transmission is having contact with an object or surface that has been contaminated by the pathogen, such as blood, water, and food. Thirdly, animal or insect bites can transmit diseases, such as malaria, plague, flea, and yellow fever. The fourth mode of transmission is traveling through the air, and some of the diseases that can spread this way are measles and tuberculosis.

Non-communicable diseases (NCDs) are conditions or diseases that do not result from infectious agents. They cannot be transmitted from one person to another and last for long periods. In addition, they can be caused by prolonged illness, inheritance, malnutrition, abnormal cell proliferation, and allergy. The major causes of NCDs are low vegetable and fruit intake, high consumption of fast food, physical inactivity, and high cholesterol (Boutayeb 2006). These behaviors contribute to cardiovascular diseases and some forms of cancer. Obesity could result in metabolic changes and increase the vulnerability of NCDs, such as heart disease and type 2 diabetes. In other words, risk factors like the background of an individual, the environment, and lifestyle increase the probability of some NCDs (Boutayeb 2006). Some of these risk factors are genetics, gender, age, unhealthy diet, smoking, and exposure to air pollution. Many of these conditions are perceived as preventable since they are brought about by modifiable risk factors. According to the WHO, the five leading risk factors of NCDs are high blood pressure, consumption of alcohol, high cholesterol, use of tobacco, and being overweight. The other factors are the socioeconomic conditions of a person, also called the social determinants of health. Figure 5.2 highlights the relationship between NCDs, burden and income.

5.4 Disease Prevention and Control

The prevention and control of diseases is the best way of ensuring that people are not infected or do not contract diseases that could be avoided. Various public health strategies can be used in controlling communicable diseases. The first strategy is

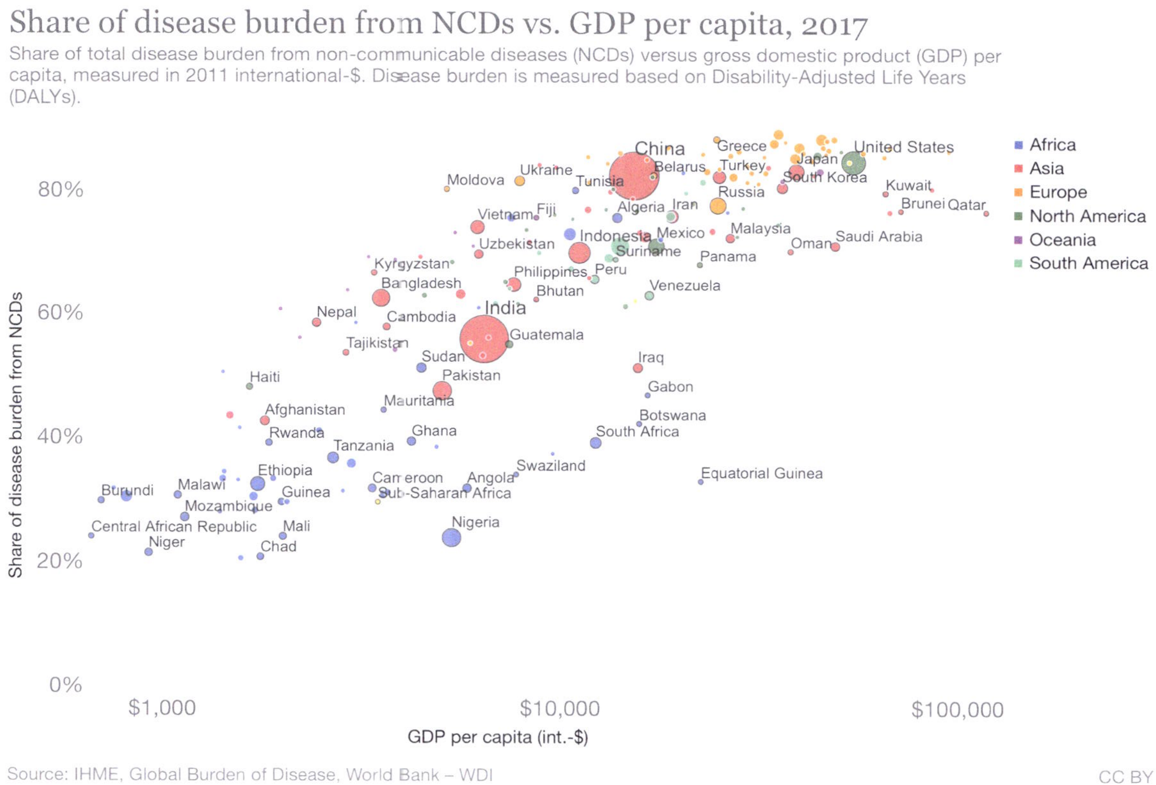

Fig. 5.2 Relationship between NCDs, burden and income. *Source:* Ritchie and Roser (2017), Institute for Health Metrics and Evaluation (IHME) (2017), World Bank (2017)

the prevention of infection through education or prophylaxis. This strategy entails immunization as a way of preventing and controlling diseases. Immunizations stimulate the immune system, the body's disease-fighting system. A healthy immune system can recognize bacteria invading the body and produce antibodies that disable or destroy them. Ideally, immunizationis vital for preparing the immune system to fight disease (Tappero et al. 2017). It is defined as the process of the body as it builds up immunity to cope with a particular disease. Vaccination, on the other hand, entails administering a weaker form of the virus to make the body immune to the specific disease. The terms immunization and vaccination are often used interchangeably and confuse people since they are related. However, one term leads to the other. A vaccine is the most convenient way of preventing an individual from infection by a specific disease. Vaccines are usually obtained from non-infective bacterial forms. For instance, a smallpox vaccine usually has dead bacteria cells obtained from the disease itself. The way of administration usually is by injections or drops (Tappero et al. 2017). The immunization process occurs after the administration of the vaccination. After the body gains the ability to fight the disease, it builds immunity. Immunization and vaccination work collaboratively to ensure that the body remains healthy and that its internal systems function appropriately. Immunizations are made effective through periodic repeat injections, also called boosters.

Apart from immunization and vaccination, the second way of disease prevention and control is the minimization of infection transmission by conducting prompt investigations and following up infectious disease incidents. Similarly, implementing protective measures can also be crucial. The third mechanism is the provision of epidemiologic data for response measures. Furthermore, continuous surveillance of disease incidences within the community is a vital mechanism.

This strategy has been labeled as unsuccessful. Why do you think this is the case?

Finally, regular physical activity, intake of a balanced diet, and healthy living are vital for preventing and controlling NCDs.

Identify diseases that can be prevented through regular exercise and a balanced diet.

Immunization, vaccination, regular exercise, and a balanced diet are sustainable mechanisms that can help alleviate communicable and non-communicable diseases. However, for infectious diseases attributed to climate change, specific adaptation solutions are required. A surveillance network can be essential for monitoring the movement of mosquitoes. An example of a successful strategy deployed is the Tiger Mosquito Surveillance Network in France (Calba et al. 2017).

Do you know of any in Africa?

Countries should embed health into their national adaptation plans and programs. This solution has been adopted in Macedonia to ensure that health becomes an integral part of its plans. The third strategy is training and awareness-raising activities. States can develop learning courses that cover climate change and health. The training should also aim to raise awareness and enhance knowledge on the health impacts of climate change among the general public. For example, the Nairobi work program has developed a five-step process that aims to address problems associated with climate change. Despite not being overly successful, the process has played a crucial role in improving healthcare. In addition, long-term collaborations could be instrumental for the development of early warning systems for infectious diseases through the consideration of climate change.

Identify other African nations that have been successful in coming up with strategies aimed at addressing tropical diseases.

5.5 Sustainable Solutions: Green Health

Green Health is a sustainable way of reducing the environmental effect of healthcare. The healthcare sector leads to the generation of large volumes of pollutants, including mercury and dioxins during the process of treatment. The disposals, including syringes that have been used, could pollute the environment because of the remnants of certain drugs. Additionally, healthcare makes use of resources, such as plastics, lumber, and others, that contribute to habitat destruction, biodiversity loss, and the impairment of an ecosystem's health. As advised by the World Medical Association (2018), there is a need for the global health community to embrace environmentally responsible approaches. Achieving change requires raising awareness as well as sustained advocacy, including making health practice response to the environment. Despite climate change making a crucial contribution to the spread of diseases, it has been established that the healthcare community itself also plays a pertinent role in harming the environment, thus increasing the likelihood of some of these diseases spreading. Towards this end, there are several ways of making clinics and hospitals respond to the environment. Some of the methods are prudent waste management, including limiting waste incineration and minimizing paper use by replacing with electronic medical records. Non-hazardous wastes also need to be recycled responsibly. The design and construction of hospitals can also be in such a way that they make efficient use of clean energy as well as optimizing recycling. In day-to-day medical practice, the same principles can be applied in various ways, including but not limited to:

1. Efficient and economical use of water heaters.
2. Usage of tapped water rather than bottled water.
3. Availing recycle bins and using recycled paper.
4. Reducing the number of mercury-containing equipment, including thermostats and thermometers.
5. Switching to electric hand dryers.
6. Use of eco-friendly office cleaning chemicals.
7. Maintenance of consistent disposal strategies in the pharmacy.
8. Carpooling, bike riding, and telecommunication.

Case Study 5.2: Green health

Green Nephrology Programme Green Nephrology is a program in the United Kingdom that seeks to enhance the sustainability of kidney care.Established by the Campaign for Greener Healthcare, it has the aim of developing climate change mitigation strategies on issues associated with kidney care (Dunbar-Reid and Buikstra 2017). The program also seeks to align sustainable healthcare provision with better quality care at affordable costs. Such programs can be beneficial for the medical community to ensure that costs, the environment, and provision of healthcare to society are integrated sustainably.

Korea Society for Green Hospitals Launched in June 2013, Korea Society for Green Hospitals is an organization geared towards low-carbon growth within medical practices to minimize the ecological effects of healthcare institutions in South Korea (Liu et al. 2017). The program also focuses on encouraging Korean hospitals to collaborate with global Green hospital networks. Heralded as an implementer of Green strategies, the program plays a crucial role in environmental stewardship. Such programs can be rolled out in various nations to foster Green health.

5.6 Drug and Substance Abuse

5.6.1 Drugs

A drug is defined as any substance that when consumed in any way or form, leads to a temporary physiological change within the body. It can be taken into the body through inhalation, injection, smoking, absorption, or dissolving under the tongue.

5.6.2 Classes of Drugs

Drugs may be categorized in various ways. In pharmacology and medicine, a drug can be categorized by the chemical activity or the condition that it treats. For instance, anticonvulsant medications are often used for preventing seizures, whereas mucolytic drugs relieve congestion.

Drugs are commonly classified into five major categories: narcotics, stimulants, anabolic steroids, depressants, and hallucinogens. All these drug categories, apart from anabolic steroids, are psychoactive, which means that they affect more than one mental faculty, including thoughts, feelings, cognition, behavior, and perception. In addition, the use of these drugs is related to various mental health as well as personal complications.

The chemical classification of drugs includes depressants, stimulants, hallucinogens, opioids, inhalants, cannabis, and new psychoactive substances (Neubig et al. 2003). Depressants are the drugs that slow brain activity as well as nerves, while stimulants accelerate brain activity. Hallucinogens influence the central nervous system to change one's perception of space, time, and reality. Inhalants are usually abused by children and include glue, pen ink, and gasoline. Furthermore, the legal classification of drugs yields five classifications that are regulated in the United States by law: Schedule I, Schedule II, Schedule III, Schedule IV, and Schedule V. These classifications are categorized based on their potential to be abused and in case they have a legitimate medical use.

5.6.3 Addiction and Dependence on Drugs

The terms addiction and dependence are often used interchangeably but have stark differences. From a medical perspective, dependence entails a physical condition whereby the body adapts to a drug. In case a person with drug dependence suddenly stops taking that drug, they experience measurable and predictable symptoms called withdrawal syndrome. Despite dependence being part of addiction, it is common to find non-addictive drugs also yield dependence among patients. Furthermore, dependence is attributed to changes in the body that result from consistent drug exposure.

The National Institute of Drug Abuse (NIDA 2018) defines addiction as a "chronic, relapsing disease that is characterized by compulsive drug seeking and use, despite harmful consequences." This means that addiction is the overwhelming need to utilize a drug. The compulsion takes an extended period and can come back unexpectedly after some period of recovery (Volkow 2014). Addiction is a psychological condition. It is the feeling one has when taking a certain drug. People may develop an addiction to illegal street drugs or even prescription medications. The two primary characteristics associated with addiction are the inability to cease substance use and the irresistible urge that one has to continue taking drugs despite experiencing negative consequences.

What do you think makes people addicted to the drugs they take? Do you know of friends who are addicts, and what do they say about their addiction?

Irrational persistence differentiates addiction from typical physical dependence. Addiction is a complex interplay of various psychological, social, and biological factors, including drug accessibility, genetic makeup, family environment, and socioeconomic status.

5.6.4 Planning for Sustainability of Drug and Substance Abuse

Sustainability in the case of preventing substance abuse and misuse entails the development of systems that promote and facilitate the delivery of efficacious prevention strategies to prevent and reduce substance use disorders. Generally, sustainability entails the maintenance of positive results within these populations. To strategically plan and achieve reductions in intake and consequences of drugs and substances, there is a need for the community to develop and reinforce a prevention system. Conceptually, prevention systems consist of the community coalition, institutions as well as organizations that implement various strategies for effectively accomplishing these changes (Reichert 2017). Stakeholders and community leaders play a crucial supporting role. There is a need for the system and stakeholders to develop the required capacities for planning strategically and sustaining the prevention interventions with sufficient intensity and effectiveness.

The sustenance of capacities needs thoughtful planning by the community, leaders, and coalition agencies. Intentional, collaborative efforts could also enable coalitions to expand the scope of their work. Three major sustainability capacities should be prioritized. Communities need to develop capacity at the three levels in order to strengthen a prevention effort. Among community agencies, communities should strengthen and expand their capacity to cope, with an objective of prevention to enable agencies to comprehend and leverage their resources for effectively addressing factors among populations that result in substance use problems and specific consequences. At the coalition level, communities should strengthen and expand the capacity of a coalition to identify and prioritize populations whose substance abuse contributes to community issues. Among the community members, the ability of the community to comprehend the role and effect of substance use is crucial for preventing planning efforts. Community members should support coalition efforts for strengthening the prevention system to enable it to deploy strategies that influence population-level consumption patterns.

5.6.5 *Methods of Drug Use and Social Risk Factors*

5.6.5.1 Methods of Drug Use

There are four primary methods of drug use: injecting, snorting, swallowing, and smoking. Methods of drug use are also called route of drug administration. Smoking is used for drugs such as tobacco, cocaine, opium, marijuana, as well as heroin. Smoking takes the drug to the lungs before it is absorbed into the bloodstream.

Why is smoking among the fastest ways for an individual to experience a high?

Smoking poses various health risks, including heart disease, hypertension, and other lung infections. Snorting or inhalation is mainly practiced by users of heroin, amphetamines, tobacco, and cocaine. One becomes high after about 15 min after the intake of the drug. Some of the negative impacts associated with snorting are the destruction of the septum and the respiratory system. Injections entail using a syringe to inject an illicit substance into one's bloodstream. It can be carried out in three major ways: intramuscular, intravenous, and subcutaneous injections. The major impacts are damage to arteries, thus leading to thrombosis and hemorrhaging, veins' scarring, and increased the likelihood of infection because of contamination (Elliott et al. 2017).

Why do you think injection is the most lethal way of drug administration?

Finally, oral administration or swallowing is among the easiest ways of using drugs. The drug moves through the mouth and into the stomach where the stomach lining absorbs it into the bloodstream. Some of the drugs that are swallowed are opium, alcohol, ecstasy, magic mushrooms, marijuana, and opium. It is among the safest methods of taking drugs for various reasons. First, the substance is slowly absorbed, thus leading to effects that are less dangerous and extreme. Second, the digestive system automatically induces vomiting in case the drug user ingests anything that could be risky. Notwithstanding, there have been reports of deaths attributed to the swallowing of cocaine, thus suggesting that this route of administration still poses certain dangers (Pramanik and Vidua 2016).

5.6.5.2 Social Risk Factors

Ideally, not all adolescents with addiction may have a history of abuse. Some other factors, including social influences that exist in adolescence or childhood, could also affect vulnerability to addiction. Some of the social risk factors are bullying, negative peer pressure, and gang affiliation. Bullying is a risk factor as individuals of both gender are victims. They have more likelihood to use hard drugs, inhalants, and marijuana. Negative peer influence also plays a crucial role in enhancing addiction. Association with peers who are addicted to drugs is a significant risk factor for substance abuse. Peer pressure also contributes to higher proportions of alcohol and cigarette use. For example, scholars report that adolescents who use alcohol and other drugs before the age of 13 stated that their peers' opinion was important to them. They rank these opinions more influential than the opinions of their mothers. Finally, gang affiliation is an antisocial behavior that entails engaging in risky behavior. Whereas not all gang-involved individuals may use drugs, there is the likelihood that gang affiliation encourages the use of drugs.

What other social factors promote drug abuse?

5.6.6 *Prevention of Drug Abuse and Misuse*

Substance abuse prevention is a process geared at preventing the start of substance use or reducing the development of issues related to the use of substances, and most efforts mainly target minors. Some of the substances that preventive efforts target are alcohol, tobacco, marijuana, cocaine, and opioids. According to experts, an optimal combination of prevention strategies is needed for addressing substance abuse and misuse problems at the individual and community level (Galanter et al. 2014). There is also a need for early intervention to treat mental illness associated with drug use. Communication, education, and enforcement are ideal strategies that can be used for preventing substance misuse. Prevention can be carried out at the family level, community level, and individual level. Family-based programs can be instrumental in strengthening protective factors. They teach parents better communication skills, the right discipline styles, as well as family management approaches. Community-based programs can also be detrimental for empowering communities to deal with drug abuse.

How do family-based programs help in preventing substance abuse?

5.6.7 *The Placebo Effect*

Also referred to as the placebo response, it is a phenomenon where a placebo can occasionally enhance the condition because the individual expects it to be beneficial.

A placebo is a substance that appears to be an actual or real medical treatment but is fake (Kermen et al. 2010). The common aspect with all placebos is that they lack an active substance that can positively influence health. When an individual believes that they will benefit from a particular treatment, there is the likelihood that they could experience the benefit. According to various studies, placebos could affect conditions like pain, menopause, sleep disorders, depression, and irritable bowel syndrome. The placebo mainly works by focusing on the expectations of a person. Combined with the other prevention strategies, the placebo effect could act as "faith" and prevent people from abusing drugs as it enables them to tap into the power of positive thinking.

5.7 Individual Matter: Salome Karwah

Salome (see Fig. 5.3) was the daughter of a doctor, she met her husband, James Harris in 2013. A year after they started dating, the Ebola epidemic struck Liberia, Sierra Leone, and Guinea in 2014. Her family was among the first to be infected with the virus in the summer of 2014. Karwah's father was the first to die of Ebola, and it eventually killed both her parents along with seven relatives. Karwah, her sister, mother, and fiancée were also soon infected by Ebola. Along with her husband and mother, they were quarantined in a Doctors Without Borders (also known as Médecins Sans Frontières (MSF) medical unit with other patients awaiting treatment (Fairall and Bateman 2017). Karwah focused on caring for her sister because she was pregnant when she contracted Ebola. She changed her clothes and cleaned fluids that had Ebola. She also did not inform her sister about the death of their mother because of the fear that it would negatively affect her health.

By September 2014, the sisters recovered and were discharged on August 28, 2014. Her boyfriend, James Harris was also released a few days later after recovering. The staff at the unit had noticed that Karwah and her boyfriend were able to care for other victims despite it being risky to their health when still patients. A few days after they were discharged, MSF employed them to act as mental health counselors within the units. Karwah went back as a nurse and counselor, and it was presumed that as survivors, they had become immune to that specific Ebola strain (Nwafor and Nwafor 2017). She worked at the unit until the Ebola epidemic came to an end. In October of the same year, she authored an article in the Guardian that reiterated that assisting others made her make sense of her life. She also emphasized that in case one has Ebola, they should not be stigmatized since the next person to contract the virus is not known. She was featured in Time Magazine as its co-Person of the Year on December 2014.

Harris and Karwah got engaged and married in January 2016 when she was pregnant with their third child. In the summer of the same year, she became pregnant again and agreed that this would be their last child. They contemplated aborting the baby but decided to keep it. She experienced difficulties in this pregnancy and underwent cesarean section despite being diagnosed with hypertension during that

Fig. 5.3 Salome Karwah. *Source:* Time Magazine (2014)

period. After a successful surgery, they went back home to care for the child and the other children. She continued feeling unwell, showed signs of hypertension, and confided to members of her family that Eternal Love Winning Africa (ELWA) hospital staff had neglected the complications she had developed. A few hours after arriving home, she collapsed, and foam began coming out of her mouth. Upon being taken back to ELWA hospital, she was in the care of a doctor who specialized in treating Ebola survivors, while the one on duty refused to treat her. She remained in care for about 3 h while experiencing seizures and convulsions.

Consequently, Harris decided to personally go to the emergency room and get a wheelchair that enabled him to take his wife in for emergency attention. Harris narrated that the staff still refused and told him to take her to another hospital. The other doctor who came in 3 h later was unable to save her as she succumbed to the illness of childbirth at 28 years. Despite her painful death, she is a martyr who remains in the hearts of many Ebola survivors, including her sister Josephine.

5.8 Future Trends

5.8.1 *Africa*

Whereas changing climate will directly impact health, there will be an increase of disease vectors including mosquitoes and the threat of transmission of zoonotic diseases, which will adversely affect health. Africa will also face other emerging disease challenges attributed to climate change happening outside the continent. For instance, the thawing of permafrost in the northern hemisphere will free up the trapped viruses that will spread through avian migration.

5.8.2 *Asia*

Climate change will impact the infectious disease burden despite sufficient hospital infrastructure to address emerging infectious diseases. There is potential transmission risk in southern parts of Asia where dengue transmission has already been reported. The Middle East will be among the most hit in terms of climate change with summer temperatures expected to rise more than twice the world average (Saha 2019). Droughts, prolonged heatwaves, and desertification will make parts of Middle East uninhabitable. In areas where people could live, climate change could spur violent competition over diminishing resources.

5.8.3 *Australia*

Warmer temperatures could result in a rise in the spread of diseases through food and water like gastroenteritis. Australia will remain malaria-free during the next few decades (La Canna 2019). Despite this, other vector-borne diseases like Ross River Virus, dengue fever, and Barmah Forest Virus may expand their range, basing on the lifestyle and socioeconomic factors that are related to travel frequency, destinations, and hygiene.

5.8.4 *Europe*

Climate change and health experts are warning of the growing threat to public health in Europe from global warming as rising temperatures help potentially lethal diseases spread quickly across the continent. There could be a further rise in West Nile Virus cases (a viral infection spread by mosquitos). Global warming could exacerbate the vulnerability to tick-borne diseases that can transmit diseases such as Zika and dengue.

5.8.5 *North America*

Future climate change will likely affect the length of the transmission season as well as the geographical range of a substantial proportion of the infectious diseases. The direct effect of climate change on habitat will adversely affect biodiversity, further affecting the emergence and spread of infectious diseases.

5.8.6 *South America*

Melting of glaciers will adversely lead to the reduction of water supplies. It is also anticipated that climate change will aggravate the water problem and increase chances of water contamination. This is because the transmission and reproduction cycle of microorganisms like *Campylobacter spp.* and *Escherichia coli* benefits from global warming. The increase in vector borne diseases in future will also have a detrimental impact on the emergence of diseases. Climate change will influence the frequency of the outbreaks of diseases associated with the microorganisms.

5.9 Summary

- The major types of diseases are infectious, non-infectious, acquired, chronic, acute, congenital, genetic, hereditary, idiopathic, primary, terminal, and secondary disease.
- Some modes of transmission are saliva, blood, mucus, and sexual contact.
- Communicable diseases spread in four main ways: physical contact, contact with an object or surface, animal or insect bites, and by air.
- Non-communicable diseases (NCDs) can be hereditary but are mainly attributed to one's lifestyle.
- The environment and background of individuals are risk factors for NCDs.
- The five major leading risk factors of NCDs are high blood pressure, consumption of alcohol, higher cholesterol, use of tobacco, and being overweight. Other factors are the social determinants of health.
- Immunization and vaccination are the primary methods used for disease prevention and control.
- The major classes of drugs are narcotics, stimulants, anabolic steroids, depressants, and hallucinogens.
- The chemical classification of drugs results in various types including hallucinogens, opioids, stimulants, depressants, cannabis, inhalants, as well as psychoactive substances.
- Addiction and dependence are interchangeable terms but have substantial differences as one leads to the other.

- There are four primary methods of drug use: snorting, smoking, oral transmission, and injection.
- The major social risk factors for drug abuse, especially among adolescents, are peer influence, bullying, and gang affiliation.
- An optimal combination of various prevention strategies is beneficial for tackling substance abuse.
- Family, community, and individual-based strategies can be instrumental in addressing substance abuse.
- The placebo effect could be used in conjunction with other preventive and control strategies to deter individuals from abusing drugs.
- Regarding future trends, Africa will experience increased incidences of disease vectors and zoonotic diseases.
- Europe will experience an increased spread of diseases, including the rise in West Nile Virus cases while North and South America's length of transmission season will prolong.
- Warmer temperatures in Australia could lead to the spread of diseases, and various vector-borne diseases will spread depending on socioeconomic factors

5.10 Discussion Questions

1. How does the placebo effect work to prevent drug abuse?
2. What makes some methods of drug use lethal compared to others? Compare and contrast each method.
3. How does drug abuse prevent cohesion in society?
4. Explain two sustainable ways of preventing drug abuse.
5. Identify other Ebola heroes and heroines that you know.
6. Explain the relationship between drug abuse and poverty.
7. How can sustainable development goals be adapted to solve drug abuse?
8. Discuss the effects of opioids on the human body.
9. Explain the difference between addiction and dependence.
10. Identify and discuss three major differences between immunization and vaccination.
11. What are the major challenges experienced during vaccinations?
12. Explain sustainable and innovative methods that are geared towards reducing the spread of infectious diseases.

References

Altizer, S., Ostfeld, R. S., Johnson, P. T., Kutz, S., & Harvell, C. D. (2013). Climate change and infectious diseases: From evidence to a predictive framework. *Science, 341*(6145), 514–519.

Banwell, N., Rutherford, S., Mackey, B., Street, R., & Chu, C. (2018). Commonalities between disaster and climate change risks for health: A theoretical framework. *International Journal of Environmental Research and Public Health, 15*(3), 538.

Beier, J. C., Keating, J., Githure, J. I., Macdonald, M. B., Impoinvil, D. E., & Novak, R. J. (2008). Integrated vector management for malaria control. *Malaria Journal, 7*(1), S4.

Boutayeb, A. (2006). The double burden of communicable and non-communicable diseases in developing countries. *Transactions of the Royal Society of Tropical Medicine and Hygiene, 100*(3), 191–199.

Calba, C., Guerbois-Galla, M., Franke, F., Jeannin, C., Auzet-Caillaud, M., Grard, G., et al. (2017). Preliminary report of an autochthonous chikungunya outbreak in France, July to September 2017. *Eurosurveillance, 22*(39), 17-00647. https://doi.org/10.2807/1560-7917.ES.2017.22.39.17-00647.

Caldwell, J. (2018). How climate, species interactions, and global change drive infectious disease dynamics. *International Journal of Infectious Diseases, 73*, 60.

Dunbar-Reid, K., & Buikstra, E. (2017). The environmental impact of healthcare and haemodialysis: The Jekyll and Hyde dilemma. *Renal Society of Australasia Journal, 13*(2), 38.

Elliott, A., Zulich, A., Cottler, L., & Striley, C. W. (2017). Drug users and non-drug users both favor uncommon informed consent practices. *Drug & Alcohol Dependence, 171*, e59.

Fairall, L., & Bateman, E. (2017). Health workers are vital to sustainable development goals and universal health coverage. *BMJ: British Medical Journal, 356*, j1357.

Galanter, M., Kleber, H. D., & Brady, K. (Eds.). (2014). *The American psychiatric publishing textbook of substance abuse treatment*. Washington: American Psychiatric Publishing.

Institute for Health Metrics and Evaluation & Global Burden of Disease Study GBD. (2017). *Disease burden from non-communicable diseases vs. GDP per capita, 2017*. Retrieved November 08, 2019, from http://ghdx.healthdata.org/gbd-results-tool/.

Kakmeni, F. M. M., Guimapi, R. Y., Ndjomatchoua, F. T., Pedro, S. A., Mutunga, J., & Tonnang, H. E. (2018). Spatial panorama of malaria prevalence in Africa under climate change and interventions scenarios. *International Journal of Health Geographics, 17*(1), 2.

Kermen, R., Hickner, J., Brody, H., & Hasham, I. (2010). Family physicians believe the placebo effect is therapeutic but often use real drugs as placebos. *Family Medicine, 42*(9), 636.

La Canna. (2019). *Curious Darwin: Why is there no malaria in Australia's northern capital*? Retrieved November 11, 2018, from https://www.abc.net.au/news/2017-09-26/curious-darwin-why-is-there-no-malaria-northern-capital/8964984/.

Liu, Z., Liu, D., Ma, B., Zhang, X., Su, P., Chen, L., & Zeng, Q. (2017). History and practice of thyroid fine-needle aspiration in China, based on retrospective study of the practice in Shandong University Qilu Hospital. *Journal of Pathology and Translational Medicine, 51*(6), 528.

Martin, G., Yanez-Arenas, C., Chen, C., Plowright, R. K., Webb, R. J., & Skerratt, L. F. (2018). Climate change could Increase the geographic extent of Hendra virus spillover risk. *EcoHealth, 15*, 509–525.

National Institute of Drug Abuse (NIDA). (2018). *Drugs, brains, and behavior: The science of addiction*. Retrieved September 2, 2019 from https://www.drugabuse.-gov/publications/drugs-brains-behavior-science-addiction/drug-misuse-addiction/.

Neubig, R. R., Spedding, M., Kenakin, T., & Christopoulos, A. (2003). International union of Pharmacology committee on receptor nomenclature and drug classification. XXXVIII. Update on terms and symbols in quantitative pharmacology. *Pharmacological Reviews, 55*(4), 597–606.

Nwafor, G. C., & Nwafor, A. O. (2017). Experiences of the Ebola victims in the West African nations: A human rights imperative. *International Journal of Public Law and Policy, 6*(1), 78–98.

Onyango, E. A., Sahin, O., & Mackey, B. (2018). Climate change and malaria risk in East Africa: Using structural analysis to rank influencing variables and identify suitable adaptation pathways. In *Proceedings of the 4th Practical Responses to Climate Change Conference:* (p. 164). Barton: Engineers Australia.

Pramanik, P., & Vidua, R. K. (2016). Sudden cardiac death of a body packer due to cocaine cardiotoxicity. *Clinical Medicine Insights: Pathology, 9*, CPath-S41070.

Reichert, J. (2017). *Fighting the opioid crisis through substance use disorder treatment: A study of a police program model in Illinois*. Retrieved November 6, 2018, from http://www.icjia.state.il.us/articles/fighting-the-opioid-crisis-through-substance-use-disorder-treatment-a-study-of-a-police-program-model-in-illinois/.

Reperant, L. A., & Osterhaus, A. D. (2017). AIDS, Avian flu, SARS, MERS, Ebola, Zika... what next? *Vaccine, 35*(35), 4470–4474.

Ritchie, H., & Roser, M. (2017). *Burden of disease*. Retrieved November 09, 2019, from https://ourworldindata.org/burden-of-disease/.

Ritchie, H., & Roser, M. (2019). *Malaria*. Retrieved November 05, 2019, from https://ourworldindata.org/malaria/.

Saha, S. (2019). *How climate change could exacerbate conflict in the Middle East*. Atlantic Council. Retrieved November 4, 2019, from https://www.atlanticcouncil.-org/blogs/menasource/how-climate-change-could-exacerbate-conflict-in-the-mid-dle-east/.

Tappero, J. W., Cassell, C. H., Bunnell, R. E., Angulo, F. J., Craig, A., Pesik, N., et al. (2017). US centers for disease control and prevention and its partners' contributions to global health security. *Emerging Infectious Diseases, 23*(Suppl 1), S5.

Time Magazine. (2014). *Salome Karwah*. Retrieved May 24, 2019, from https://-www.newser.com/story/238996/ebola-fighter-who-nabbed-time-person-of-the--year-honor-dies.html/.

Volkow, N. D. (2014). America's addiction to opioids: Heroin and prescription drug abuse. *Senate Caucus on International Narcotics Control, 14*, 1–16.

World Bank. (2017). *Disease burden from non-communicable diseases vs. GDP per capita, 2017*. Retrieved November 08, 2019, from http://data.worldbank.org/data-catalog/world-development-indicators/.

World Health Organisation. (2016). *Global Malaria death estimates from World Health Organization (WHO)*. Retrieved November 03, 2019, from http:-//apps.who.-int/gho/data/view.main.14117?lang=en/.

World Medical Association (WMA). (2018). *Green health: Reducing the environmental impact of healthcare*. Retrieved November 4, 2018, from https://www.-wma.net/what-we-do/public-health/green-health/.

Chapter 6
Renewable Energy

Abstract Everyday activities at the local and international level are crucial for tackling sustainability and climate change issues, and countries are distinctly placed to advance initiatives in renewable energy. Energy is at the core of the sustainability challenge because it is a pertinent ingredient of the current economic life. Despite this, society's long-term reliance on fossil fuels implies a future of higher energy costs and environmental damage. This chapter argues that to tackle the issue of climate change and achieve a green world, there is a need to substitute fossil fuels with other energy forms that are environmentally sustainable, including solar and wind. The various types of renewable energy sources are explored to understand their significance for the environment. Also, this chapter explores the emerging renewable energy technologies. Bunker Roy's contribution to the renewable energy sector, the case studies of the humpback whale, and the microbial fuel cell along with an experiment provide further solid evidence of the roles and importance of renewable energy sources for the environment and green society.

Keywords Microbial fuel cell · Solar · Wind · Geothermal · Bunker Roy · Solar glass · Thorium nuclear reactors

6.1 Overview

Renewable energy is produced from natural processes that continuously replenish, including geothermal heat, water, biomass forms, wind, and tides. This form of energy cannot be depleted and is consistently renewed. Other sources define renewable energy sources as those that obtain their energy from ongoing natural processes, including water, sunshine, wind, as well as geothermal heat flows (Hussain et al. 2017). The most common definition is that it is obtained from a resource that is substituted by a natural process like that generated from wind or the sun. Apart from tidal and geothermal power, many forms of renewable energy emanate from the sun.

A. A. Akinsemolu, *The Principles of Green and Sustainability Science*,
https://doi.org/10.1007/978-981-15-2493-6_6

6.2 Types of Renewable Energy

The main types of renewable energy are solar, wind, geothermal, hydroelectric power, biomass, and biofuels. As earlier stated, many of these forms rely on sunlight. Hydroelectric power and wind directly result from differential heating of the surface of the Earth, which results in wind and precipitation formation as the air gets lifted. Solar energy directly results from sunlight conversion using solar collectors or panels (Muruganantham et al. 2017). Geothermal energy does not rely on sunlight and results from radioactive decay within the crust when combining with the initial heat and tidal energy. Out of these types of renewable sources of energy, hydroelectric power remains the predominantly consumed globally, accounting for 25% of total renewable consumption (see Fig. 6.1).

Solar energy depends on nuclear fusion. After being collected, it can be converted in various ways. The sun will continue being a source of energy for an extended period. Some of the current technologies used for capturing this energy are solar collectors, thin-film sheeting, and photovoltaic cells. Ideally, solar cells convert sunlight into electricity. The main strength of this source of energy is that it is clean and does not cause any pollution. Challenges associated with this are cost and the low quantity of energy generated by these cells. Also, the battery must be advanced to hold electricity for an extended period and cover the days that have little or no sunshine (Muruganantham et al. 2017). The main issue is that solar

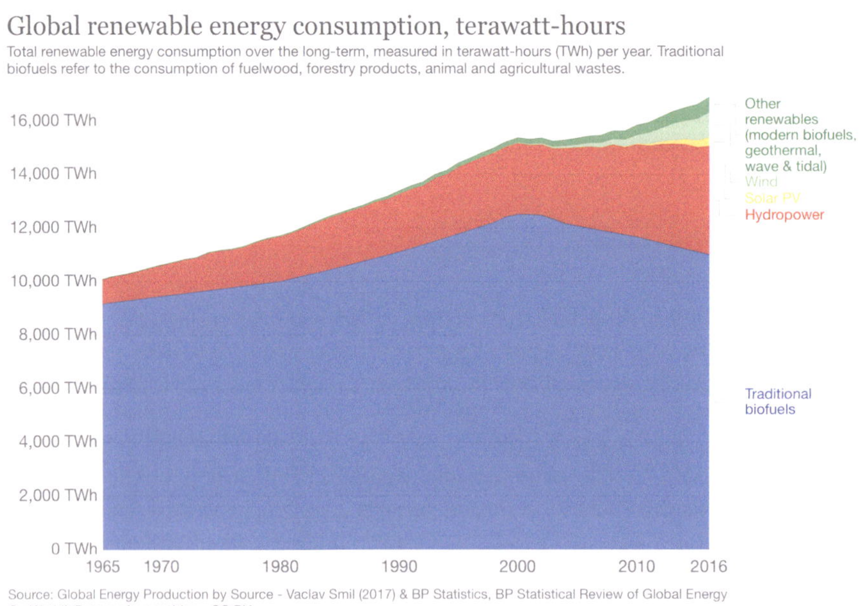

Fig. 6.1 Global hydroelectric power consumption. *Source:* Ritchie and Roser (2019); Smil (2017)

power is not sufficient to fully power modern society. One of the major contributors to solar energy is Bunker Roy, who established Barefoot College in Tilonia. The college spreads over eight acres and is powered by solar energy (Barefoot College 2017). Having studied at Delhi University, he argues that his elitist education almost destroyed him. He also believes that the formal educational system devalues and demeans traditional knowledge valued by the poor. Despite the college beginning with the objective of offering solutions to water issues of India, the mission still reverted to sustainable development and empowering the marginalized. The courses provided at the institute are grounded in the Gandhian philosophy of ensuring that villages are self-reliant. More information will be provided under the Individual Matter section at the end of the chapter.

Experiment 6.1: Solar power: photovoltaic powered motor

Materials Required Voltmeter, variable resistor, thick unshielded wire, alligator clips, solar panel wire, and magnetic wire

Background of the Experiment The photovoltaic powered motor experiment is designed to enhance learning on solar cells, current vs. voltage curves, and calculation of maximum resistance and power. It will be critical in enhancing learning on solar power and the functioning of a direct current (DC) motor (DC motor converts direct current electrical energy into mechanical energy) that is fuelled by solar cells.

Experiment Procedure

1. Cut a standard length of wire, about 10 cm.
2. Remove 2 cm from both ends of the magnetic wire.
3. Coil the wire into 3 cm loops.
4. Bend unshielded wire into a stand to get hold of the coiled magnetic wire. The stand should be taller than a magnet's thickness and the radius of the coil.
5. Connect solar cells to leads.
6. Place coil into a stand.
7. Position magnet under the coil and observe the coil's movement.

Analysis

- What do you observe?
- What from the experiment could you deduce as a determinant of the sustainability of solar power?

Participants should show an improvement in learning and understanding renewable energy source concepts by complementing theory with laboratory experiments.

(continued)

Experiment on Solar Distillation

Materials Required Water, clear plastic wrap, two (2) large plastic containers, two (2) small glasses, masking tape, two (2) tablespoons of dirt, two (2) small rocks, and two (2) tablespoons of salt.

Background of the Experiment Hydropower is perceived as a renewable source of energy since the water on the planet continually goes through the water cycle due to the energy of the sun. This experiment seeks to demonstrate how the sun could be used for making water more useful. Distill in this experiment means turning something into a gas through vaporization and converting it back into a liquid through condensation in order to purify it.

Pre-Questions

- Are you able to distill clean water from the muddy water?
- Are you able to distill clean from salty water?

Experiment Procedure

1. Fill the plastic containers with water to 2.5 cm level. Mix the dirt into one of the containers with water and salt into the other.
2. Place an empty glass into the middle of the plastic containers and ensure it remains empty.
3. Cover the plastic containers using plastic wrap and seal them using the masking tape. Place a rock in the center of the plastic wrap; it should be directly over each glass but not touch it.
4. Position the stills in a place with sunlight for 2 h. Observe the water forming in the glass and record what you observe.

Analysis
Did the stills make clean water?

Critical Thinking
Explain the way the experiments worked. Imagine a situation where knowledge could save your life. Estimates suggest that more than one billion individuals globally drink unhealthy water. How can your project assist this population?

Regarding wind power, airflow can be utilized for running wind turbines. Wind farms are mainly located in areas where winds are more robust and consistent, including high altitude sites and offshore. In 2015, it was established that wind-generated electricity accounted for 4% of the electricity demand globally

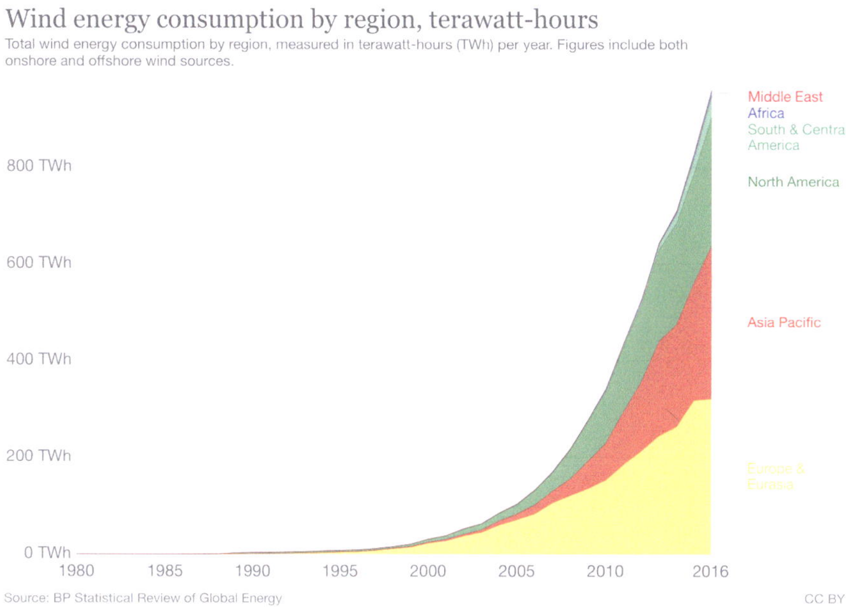

Fig. 6.2 Global wind energy consumption. *Source:* Ritchie and Roser (2019), Smil (2017)

(Hussain et al. 2017). In other states such as Denmark, wind energy accounted for 40% of the electricity demand. Figure 6.2 shows the total wind energy consumption by region for both onshore and offshore. Wind energy may be used for pumping water or generating electricity (Ehsani et al. 2018). However, the major challenge is that it needs extensive areal coverage to produce large quantities of energy. Despite being among the least costly renewable source of fuel available, opposition to towers and concerns regarding the long-term effects of turbines and towers on the habitat impede the popularity of wind energy.

What do you think should be done to increase the use of wind power as a source of renewable energy?

Case Study 6.1: Humpback whale

Fig. 6.3 Humpback whale

Lift enables an airplane to remain afloat. Scientists used the insight gained from these human-made technologies to enhance maneuverability within the air through the humpback whale idea (see Fig. 6.3). Contrary to many of other whale species, humpback whales do not survive on krill, but maneuver to catch fish. To achieve this, they have to make sharp turns. While doing so, they have to prevent their flippers from stalling in the same way; whales need more lift from higher angles of attack (Brown et al. 2018). Humpbacks have the ability to maneuver their flippers prior to beginning a stall which allows them to have more lift and make fish-catching turns. This is credited to tubercles, or bumps, on the scalloped edges on their flippers' leading side. A humpback whale often swims in circles that are sufficiently tight to yield nets of bubbles while catching prey. While large volumes of water that flow over smooth flippers split into many turbulent vortices upon crossing the flipper, the water passing through the tubercles of humpback maintain channels of water that moves at high speed. This facilitates humpbacks to maintain their grip on the water and turn the tighter corners.

Wind tunnel tests involving humpback fins highlight that aerodynamic improvements make an 8% improvement in lift as well as 32% drag reduction (Panwar et al. 2011). A practical example is WhalePower Company that applies the lessons provided by humpback whales to design wind turbines.

Table 6.1 Examples of hydroelectric power plants globally

HEP plant	Location	Gigawatts, GW, produced
Krasnoyarsk	Russia	6.0
Sayanogorsk	Russia	6.4
Longtan	China	6.43
Xiangjiaba Dam	China	6.45
Grand Coulee Dam	US	6.81
Tucurui Dam	Brazil	8.37
Guri Dam	Venezuela	10.2
Xiluodu Dam	China	13.9

Source: Conca (2017)

Geothermal energy is a form of energy that is produced and stored in the Earth. Ideally, thermal energy ascertains the temperature of matter and uses heated water as well as steam from the surface of the Earth to fuel power stations that convert steam into electricity. Despite geothermal energy being a crucial source of Green energy, some of the challenges that are associated with it are the high costs involved in building and the limited number of places for active geothermal areas. Contrarily, hydroelectric power (HEP) makes use of flowing water to move turbines that generate electricity (Pollet et al. 2012). There are numerous hydroelectric power plants globally, but the major examples in terms of size are shown in Table 6.1.

The US Geological Survey (2018) estimates that hydropower provides 20% of the energy needs of the world. There are various issues about the continued utilization of hydropower that prevent it from being fully exploited. First, rivers should be dammed, which can pose significant effects on wildlife and the land. Second, the impact also influences communities relying on fish. Finally, the costs associated with building such dams are quite high while the return takes an extended period.

From the sources discussed above, what clean energy source do you believe has the best potential for mass adoption?

Biomass energy is the other renewable energy and is obtained from plants. It is a common form of energy globally. However, the most popular form is the use of trees as sources of fuel for warmth and cooking. The process leads to the release of large quantities of carbon dioxide into the air and significantly contributes to unhealthy air in numerous areas. Some modern biomass energy forms are methane generation as well as alcohol production for fueling electric power plants (Toklu 2017). Energy obtained from oceans, tides, as well as hot hydrogen fusion, could also be utilized for generating electricity.

From the above discussion, do you believe that a nation can attain 100% renewable energy? Several countries have adopted ambitious plans for obtaining power from renewable energy but still face numerous challenges. How can countries increase their potential for 100% renewable energy?

6.3 Environmental Benefits of Renewable Energy

Renewable energy is known to be an affordable, clean, and sustainable way of enabling people to generate electricity. Whereas many individuals switch to solar and other forms of energy because of its financial benefits, the significant environmental benefits are also important. First, renewable energy minimizes air pollution. Renewable energy does not produce significant harmful gaseous emissions (greenhouse gasses) like fossil fuels which make up the majority of traditional energy sources (Panwar et al. 2011). Similarly, many renewable sources of energy provide a way for powering homes without the use of fossil fuels. This implies that renewable energy sources could be instrumental in minimizing the quantity of greenhouse gas emissions released into the atmosphere.

What is the benefit of low or zero-emission of gases to the environment?

Apart from minimizing air pollution, renewable energy also reduces water pollution and only requires a small amount of water. Solar cells do not require any energy for generating electricity. It is among the major benefits of solar that is not talked about.

How do geothermal power plants promote water pollution?

Unlike geothermal sources, solar energy does not lead to the contamination of local water resources. Also, the operation of solar cells and panels do not strain local supplies. Third, renewable energy minimizes the demand for finite resources. This contrasts with non-renewables like fossil fuels that are continually leading to the decline of finite resources. In other words, going renewable not only helps us to save money but also benefits the environment.

6.4 Overall Benefits of Renewable Energy

First, renewable energy is mainly perceived as eco-friendly. As a clean source of energy, it has zero greenhouse emissions. Fossil fuels release large proportions of greenhouse gas which cause climate change, global warming, and degradation of air quality (Panwar et al. 2011). Fossil fuels also tend to contribute to the emission of sulfur in the atmosphere, which causes acid rain. As such, the use of renewable sources of energy is mainly detrimental to the reduction of air pollution.

Second, renewable energy sources are renewable, thus implying that they do not get depleted. Energy sources like fossil fuels are perceived as limited resources, and there is a high likelihood that they may deplete in the future. As such, renewable energy could aid developing states from over-dependence on fossil fuels (Gámez et al. 2017). Sunshine and robust winds can guarantee a steady and sufficient energy supply to a country for numerous years.

Third, renewable energy is a reliable source. There has been an increase in the use of fossil fuels, which has resulted in a threat to human sustainability and security. It is because fossil fuels are vulnerable to political instabilities, unnecessary war, trade disputes, and inflation of energy prices (Gámez et al. 2017). These factors negatively influence a country and can substantially drain its economy. Whereas some contend that wind and solar energy are not reliable, a solid infrastructure can solve this. If generation plants are distributed over a wide geographical area, there could be a minimal interruption in electricity generation as weather disruptions in one place are not similar to other locations.

The fourth advantage is that it promotes job creation. Some sources of renewable energy are economically viable since they are cheaper compared to many traditional energy sources. Studies indicate that since the initiation of renewable energy, stable jobs have increased in many economies (for example, in the United Kingdom and the Netherlands) (Bulavskaya and Reynès 2018). According to experts, the campaigns geared at embracing renewable energy will lead to the creation of thousands of stable jobs.

The other benefits are stabilization of global energy prices, less maintenance of facilities, improvement of public health, and empowerment of people in the countryside. Regarding empowerment, the generation of renewable energy mainly occurs in remote environments. It implies that local towns can get a fair proportion of the generated power, thus catalyzing the regeneration of these areas in socioeconomical ways (Bulavskaya and Reynès 2018). The electrification of such areas can open up untapped development opportunities. It is achieved by advancing greenhouses through the use of geothermal power and heating of towns using hot water.

6.5 Emerging Renewable Energy Technologies

Recent headlines about the energy industry have predominantly focused on the drop of the oil and gas prices. Whereas low-cost energy is newsworthy, the over-focus on fossil fuels could be deterring innovative developments that have been taking place within the industry. The industry is among the most vibrant and transformative sectors of the economy. Declining costs, technological improvements as well as the influence of financing structures have made the sector a major driver of economic growth in the US and the world.

The renewable energy industry continues to be accessible and is poised to replace cheap yet disastrous fossil fuels. As a more significant number of people decide to understand the concept of energy, there is a critical effort to comprehend renewable energy (Hussain et al. 2017). Some of the innovations and emerging technologies in renewable energy that show the increased interest in renewable energy are explored below.

6.5.1 Solar Glass

Solar glass is a transparent glass panel with the ability to harvest solar energy. Contrary to other photovoltaic cells that absorb photons, leading to only semi-transparent glass, solar glass utilizes organic salts for absorbing ultraviolet and infrared light. This allows the spectrum of light to pass through completely. After absorption of non-visible wavelengths, they produce infrared light, which is pushed to the edges of panels. Some of the applications of solar glass include cell phones and window panels found in the high-rise constructions. The expected efficiency rates are usually as high as 10% (Hussain et al. 2017). While this may not wholly charge a phone, it can prolong a charge's length or enhance the overall efficiency of a building.

6.5.2 Thorium Nuclear Reactors

Thorium nuclear reactors make use of thorium rather than plutonium for nuclear fission, providing various benefits. Thorium reactors are often safer compared to the plutonium reactors because their designs lead to important steps for preventing a meltdown. The reactors can cut off the supply of neutrons needed for producing energy, thus halting fission whenever required. Contrastingly, plutonium and uranium yield their neutrons, thus making safeguards impossible (Humphrey and Khandaker 2018). It is easier to obtain thorium and also leads to less nuclear waste. In addition, the production of a bomb using the thorium reactors could become quite impractical, thus minimizing the likelihood of a nuclear breakdown. Some countries that actively research and prototype thorium nuclear reactors are the United Kingdom (UK), Israel, India, China, Germany as well as India. The objective of India is to establish a working prototype, whereas China seeks to ensure that it has a functioning plant online in the next 10 years. Similarly, Norway is testing thorium use in the current nuclear reactors.

6.5.3 Fuel Cell Electric Vehicles (FCEV)

Despite Honda's FCX Clarity being available commercially since 2008, hydrogen fuel cells took off in 2016. Mercedes-Benz and Toyota are providing fuel cell models, and Daimler AG has partnered with Ford and Nissan to develop fuel cells. Hydrogen fuels are made of fuel cell stacks, and each has an anode, cathode as well as a proton exchange membrane (Ehsani et al. 2018). For electricity production, compressed hydrogen is usually pumped through the anode. When it gets into contact with the proton exchange membrane, it separates into protons and electrons. Electrons are utilized for powering the car's motor as well as several other components. The proton and oxygen combine to release water, leading to substantially minimized greenhouse gas emissions.

Fuel Cell Electric Vehicles (FCEVs) are one of hydrogen fuel cell's applications. Honda built a home energy station that makes use of hydrogen fuel cell technology for powering both homes and cars (Ehsani et al. 2018). There are various limitations associated with this, including shorter recharging times and the possibility of it not being as green as originally thought.

6.5.4 Cellulosic Ethanol

It is a second-generation biofuel manufactured through the conversion of vegetation that is not suitable for human consumption into ethanol. While first-generation biofuels often make use of edible feedstock, including corn, cellulosic ethanol can be produced through the use of raw materials, including grass and wood. Biofuels are basically renewable, but cellulosic ethanol has a lesser effect on the food chain compared to first-generation since it can be produced from crops that are marginally beneficial for the production of food (Hussain et al. 2017). Conversely, the raw materials' conversion rate is significantly lower for cellulosic ethanol, and in the absence of manufacturing technology, cellulosic acid's future could be used as a fuel additive instead of a petroleum replacement.

Cellulosic ethanol remains popular in numerous parts of the world, including the US, Brazil as well as the European Union (EU). It is also touted as a probable solution for energy independence and a method for the reuse of waste materials. Some nations like the US have developed legislation that mandates fuel manufacturers for adding a certain biofuel percentage into their products. Such legislation is among the major market drivers for growth, leaving the technology susceptible to economic decline. Cellulosic ethanol could provide excellent opportunities for African nations by contributing to better rural incomes, economic growth, and employment (Hussain et al. 2017). Some African states could rely on them as a crucial export and offer low-cost labor. However, the major challenges for the African context are that biofuels could result in an increase in environmental risks and food security problems. Furthermore, biofuel is significantly subsidized and cannot compete within the current market in the absence of government intervention. In this regard, the viability of cellulosic ethanol remains in doubt. However, it could be a great source of sustainable energy in the current paradigm that is characterized by high levels of climate change attributed to human activities.

6.5.5 Space-Based Solar Power (SBSP)

This is geared towards the collection of solar power within outer space. The potential benefits of this technology are at a higher rate and a more prolonged period of collection. A significant proportion of incoming solar energy is lost as it goes through the Earth's atmosphere (Green and Newman 2017). These systems are instrumental for converting sunlight into microwaves. It is lucrative for those who are looking for large-scale solutions to human-induced climate change or the depletion of fossil fuels.

Space-Based Solar Power (SBSP) has been actively pursued by countries like Russia, China, and Japan. As highlighted by Mankins in his 2014 book "The Case for Space Solar Power," orbiting plants would supply abundant free fuel. After being launched, the only waste stream could be heat that poses severe technical issues (Mankins 2014). However, the advantage is that it is eco-friendly and requires less routine maintenance. From these discussions, it is imperative to note that renewable energy sources will be crucial for sustaining life and ensuring that energy production is sustainable.

Case Study 6.2: Microbial fuel cell (MFC)

This is a bio-electrochemical device that is detrimental for harnessing respiring microbes for the conversion of organic substrates into electrical energy (see Fig. 6.4). At the core, it transforms chemical energy into electricity through the process of oxidation-reduction reactions. They have been quite crucial in wastewater treatment. The application in wastewater treatment is premised on the background that microbes have an affinity for sewage and the conditions in such plants are ideal for such bacterial types. Bacteria consume

(continued)

large quantities of the sludge that is usually present in the wastewater. An example is the Emefcy Company in Israel that reduces the proportion of sludge by 80% in its processes, thus saving them money and time (Jadhav et al. 2018).

In 2017, two organizations partnered to establish Fluence Corporation to become the dominant provider of treatment solutions. Fluence offers local and sustainable treatment solutions for community and business empowerment. Fluence provides a combined range of services across the water cycle. It has its major operations in South America, North America, Europe, and the Middle East (Sevda et al. 2018). Its core objective going forward is to expand the vast Chinese market for wastewater treatment in rural areas.

Another organization takes the microbial fuel cell importance for wastewater a step further through the production of essential hydrocarbons. Ecovolt, a flagship product of Cambrian Innovation, uses an MFC for the conversion of carbon-rich wastewater streams. It has been necessary for the production and provision of clean energy as well as heat.

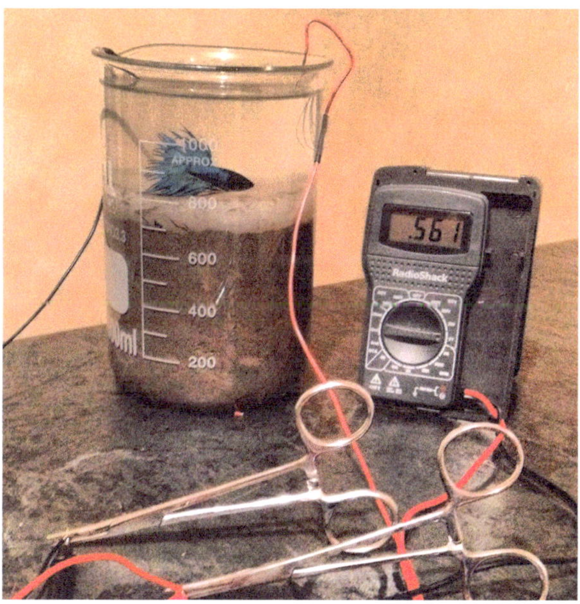

Fig. 6.4 Microbial fuel. *Source* Dan (n.d). via www.instrutables.com

6.6 Individual Matter: Bunker Roy

Bunker Roy (see Fig. 6.5) was born on August 2, 1945. He is an Indian educator and social activist; famous for establishing the Barefoot College. Roy was chosen as one of the 100 most influential personalities by Time 100 in 2010 for the work he did in teaching semi-literate Indians in rural areas. After completing Doon School between 1956 and 1962, he graduated from St. Stephen's College in 1967. He is also credited for the establishment of the Social Work and Research in 1972 whose mission changed to sustainability and empowerment (Latchem 2018). The program aimed to plant water pumps and train the population in maintaining them without relying on outside mechanics. His work earned him recognition because of the college's programs that have successfully trained at least three million individuals in skills, including solar teachers, architects, doctors, engineers as well as weavers.

At Barefoot, they also harness the Sun's power. By recognizing that solar energy offers electricity and minimizes carbon emissions, the college observed that it is also a catalyst for employment creation, boosting income, and providing self-reliant solutions for these communities (Latchem 2018). A similar program called Women Barefoot Solar Engineers of Africa has played a critical role in improving the lives of rural populations who earn less than \$1 a day. Arguably, the college has played an impactful role in protecting the environment and empowering local communities.

Fig. 6.5 Bunker Roy. *Source:* Shankbone (2010)

6.7 Future Trends

6.7.1 Africa

Africa will use renewable energy sources for fueling a larger proportion of its future growth. This would be economically competitive, unlock economies of scale and would provide substantial benefits through local value creation, equitable development, energy security, and environmental sustainability. Africa could also leapfrog to modern renewable energy since such technologies are the most economical solution for mini and off-grid electrification within remote areas and grid extension within some cases of the centralized grid supply with appropriate renewable energy sources.

6.7.2 Asia

The use of renewable energy will continue expanding at the highest rate. This is the continent with the highest potential of renewables growing, apart from spurring development. Renewables will also have the highest potential of curbing carbon emissions and mitigating climate change.

6.7.3 Australia

Even in the absence of subsidies, solar and wind energy will remain more cost-competitive compared to other energy resources. Solar could reach price parity in all the markets.

6.7.4 Europe

European renewable energy will significantly continue growing in the coming decades. European Union (EU) member state policies will also remain the major drivers that will shape the power market. The lower demand for renewable energy raises concerns for the energy producers regarding the long-term prospects. There is also a potential for rapid offshore wind expansion that would lead to a rise in the exposure for not only thermal but also renewables generation.

6.7.5 North America

The capacity of renewable energy will continue growing and becoming more cost-competitive. The core fundamentals that will continue driving growth are reducing costs of solar and wind generation, advanced battery storage technology, and the growing expertise of grid operators as well as the more toolsets for the integration of intermittent renewable energy into the grid. There will also be more demand for green (renewable) energy from both individuals and businesses.

6.7.6 South America

Investment in renewable energy will increase due to the frequency of natural disasters within the region. It is also poised that countries in this region will continue experiencing storms, droughts, and catastrophic floods. In nations such as Columbia and Peru, the number of deaths and fatalities related to climate change will be higher. Other South American nations including Uruguay will increasingly deploy wind and solar energy. More importantly, Mexico and other South American nations will lower their emissions.

6.8 Summary

- Renewable energy sources are sources that continually provide energy without being depleted because of their ability to replenish themselves.
- Primary renewable sources are geothermal heat, water, biomass forms, wind, and tidal.
- Main types of renewable energy are wind, biomass, hydroelectric power, biofuels, geothermal, and solar.
- Solar energy is clean and does not lead to any pollution. Challenges associated with it are high costs and the low amount of energy supplied.
- Wind energy is also an enormous supply of electricity, but the major challenge is the need for wide-area coverage to produce a large quantity of energy.
- Geothermal energy also suffers from high costs and the limited number of places that are considered active.
- The main challenges facing hydroelectric power generation are the need for dams, the impact on communities and ecosystems, and high development costs.
- Biomass energy is unviable as it contributes to environmental degradation.
- Renewable energy benefits the environment by minimizing water and air pollution, reducing the amount of greenhouse gas emissions, and reducing the demand for non-renewable resources.

- The overall benefits of renewable energy are that it is eco-friendly, renewable, reliable, promotes job creation, stabilizes the prices of energy, requires less maintenance of facilities, empowers people living in the countryside, and enhances public health.
- Renewable energy in Africa will continue becoming popular, and its use will increasingly be preferred because of the role it plays in sustainability while in North and South America, its capacity will continue becoming more cost-competitive.
- In Europe, renewable energy will continue growing in the coming years.
- In Asia, renewable energy use will continue growing at a great rate, while Australia could experience price parity.

6.9 Discussion Questions

1. Explain how renewable energy is essential for achieving sustainable development goals.
2. In reference to the five principles of Going Green identified in Chap. 1, discuss how renewable energy is vital for promoting the "Going Green" initiative.
3. Using your local community, critically explore the benefits of solar energy.
4. Has this chapter increased your understanding of Going Green? Explain the information that you will use in promoting the Going Green initiative.
5. Why do we not utilize renewable energy despite its benefits?
6. Explain how wind turbines function.
7. Which African country derives the most energy from solar power?
8. What are the two main challenges that are preventing countries in Africa from exploring or tapping on renewable energy sources?
9. How can we start the journey towards a 100% renewable energy world?
10. Explain how renewable energy can be beneficial for the youth.
11. What roles should schools play in the adoption of renewable energy sources? Identify and explain four roles.
12. What is the major contribution of Bunker Roy in sustainable energy development?
13. Cellulosic ethanol remains a controversial technology. Provide three reasons for these controversies.
14. Discuss the future of renewable energy in the next three decades.

References

Barefoot College. (2017). Barefoot college founder Sanjit 'Bunker' Roy receives Princeton doctor of law degree honoris causa. Retrieved November 4, 2018, from https://www.barefootcollege.org/barefoot-college-founder-sanjit-bunker-roy-receives-princeton-doctor-of-law-degree-honoris-causa/.

Brown, D. M., Robbins, J., Sieswerda, P. L., Schoelkopf, R., & Parsons, E. C. M. (2018). Humpback whale (Megaptera novaeangliae) sightings in the New York-New Jersey Harbor Estuary. *Marine Mammal Science, 34*(1), 250–257.

Bulavskaya, T., & Reynès, F. (2018). Job creation and economic impact of renewable energy in the Netherlands. *Renewable Energy, 119*, 528–538.

Conca, J. (2017). The biggest power plants in the world – Hydro and Nuclear. *Forbes*. Retrieved November 4, 2018, from https://www.forbes.com-/sites/jamesconca/2017/08/10/the-biggest-power-plants-in-the-world-hydro-and-nuclear/#3be6adcc2c88/.

Dan, S. (n.d.). *Microbial fuel cell of science*. Retrieved November 05, 2019, from https://ourworldindata.org/grapher/renewable-energy-consumption/.

Ehsani, M., Gao, Y., Longo, S., & Ebrahimi, K. (2018). *Modern electric, hybrid electric, and fuel cell vehicles*. Boca Raton: CRC Press.

Gámez, M. R., Pérez, A. V., Será, A. S., & Ronquillo, Z. M. (2017). Renewable energy sources and local development. *International Journal of Social Sciences and Humanities, 1*(2), 10–19.

Green, J., & Newman, P. (2017). Citizen utilities: The emerging power paradigm. *Energy Policy, 105*, 283–293.

Humphrey, U. E., and Khandaker, M. U. (2018). Viability of thorium-based nuclear fuel cycle for the next generation nuclear reactor: Issues and prospects. *Renewable and Sustainable Energy Reviews, 97*, 259–275.

Hussain, A., Arif, S. M., and Aslam, M. (2017). Emerging renewable and sustainable energy technologies: State of the art. *Renewable and Sustainable Energy Reviews*, *71*, 12–28.

Jadhav, D. A., Ghangrekar, M. M., & Duteanu, N. (2018). Recent progress towards scaling up of MFCs. In *Microbial Fuel Cell* (pp. 443–457). Cham: Springer.

Latchem, C. (2018). Education for sustainable development. In *Open and distance non-formal education in developing countries* (pp. 155–165). Singapore: Springer.

Mankins, J. (2014). *The case for space solar power*. Virginia: Virginia Edition Publishing.

Muruganantham, B., Gnanadass, R., & Padhy, N. P. (2017). Challenges with renewable energy sources and storage in practical distribution systems. *Renewable and Sustainable Energy Reviews, 73*, 125–134.

Panwar, N. L., Kaushik, S. C., & Kothari, S. (2011). Role of renewable energy sources in environmental protection: A review. *Renewable and Sustainable Energy Reviews, 15*(3), 1513–1524.

Pollet, B. G., Staffell, I., & Shang, J. L. (2012). Current status of hybrid, battery and fuel cell electric vehicles: From electrochemistry to market prospects. *Electrochimica Acta, 84*, 235–249.

Ritchie, H., & Roser, M. (2019). *Renewable Energy*. Retrieved September 24, 2019, from https://ourworldindata.org/renewable-energy/.

Sevda, S., Sarma, P. J., Mohanty, K., Sreekrishnan, T. R., & Pant, D. (2018). Microbial fuel cell technology for bioelectricity generation from wastewaters. In *Waste to wealth* (pp. 237–258). Singapore: Springer.

Shankbone, D. (2010). Bunker Roy. Retrieved March 06, 2018, from https://en.-wikipedia.org/wiki/Bunker_Roy#/media/File:Sanjit_Bunker_Roy_at_Ti-me_20-10.jpg/.

Smil, V. (2017). *Energy Transitions: Global and national perspectives*. Retrieved November 09, 2019, from http://vaclavsmil.com/2016/12/14/energy-transitions-global-and-national-perspectives-second-expanded-and-updated-edition/; https://www.bp.com/en/global/corporate/energy-economics/statistical-review-of-world-energy.html/.

Toklu, E. (2017). Biomass energy potential and utilization in Turkey. *Renewable Energy, 107*, 235–244.
United States Geological Survey (USGS). (2018). *Hydroelectric power water use*. Retrieved November 4, 2018, from https://water.usgs.gov/edu/wuhy.html/.

Chapter 7
Water and Wastewater Treatment

Abstract Water pollution remains a crucial problem in the quest for sustainability today. Safe drinking water is pivotal for good health, but a significant proportion of the water used for drinking today remains contaminated. This chapter seeks to shed light on the purpose of water and wastewater treatment. It also seeks to enhance understanding of preliminary, primary, secondary, and quaternary treatment, as well as discuss wastewater treatment plants. The aforementioned processes are crucial for enhancing the quality of water and ensuring that it is safe and acceptable for end-use. The arguments in the chapter are reinforced by the case study of the curse of oil in Nigeria, experiment on bioremediation, and individual matter of Ken Saro-Wiwa. Future trends in the various continents are also explored to determine the future of water and wastewater treatment.

Keywords Oil · Primary treatment · Secondary treatment · Tertiary treatment · Quaternary treatment · Ken Saro-Wiwa

7.1 Overview

Water treatment is defined as a process that enhances the quality of water in order for it to become more acceptable for end-use. This can be industrial water supply, river flow maintenance, drinking, water recreation, or safely returning it to the environment. On the other hand, wastewater treatment is defined as a process used for converting wastewater into an effluent that can be returned to the water cycle with less effect on the environment (Qasim 2017).

Water reclamation enables wastewater to be utilized for other purposes. Treatment occurs in a wastewater treatment plant, usually called a water resource recovery facility. Wastewater treatment forms a crucial part of the sanitation field. It is worth noting that sanitation also entails human and solid waste management. Sewage sludge, screenings as well as grit are some of the by-products from plants that can be treated in treatment plants.

A. A. Akinsemolu, *The Principles of Green and Sustainability Science*,
https://doi.org/10.1007/978-981-15-2493-6_7

What is the significance of screening? How does water treatment contribute to better water quality?

7.2 Preliminary Treatment

Many urban wastewater sewers have various ranges of matter, usually bulky, particularly in combined systems. Preliminary treatment is crucial for protecting systems and pipelines against blockages. Ideally, preliminary treatment involves treating wastewater using industrial and commercial facilities to eliminate harmful pollutants before discharging the water to a sewer system. For any waste to be perceived as pre-treated, it should comply with the basic definition of treatment, which entails meeting a three-point test (Neyens and Baeyens 2003). First, the water should go through a specific process that involves chemical, biological, thermal, or physical treatment. Second, the process should alter the waste's characteristics. Third, it should do so to minimize the volume of the waste, enhance its management, minimize its hazardous nature, and improve its recovery. This implies that there is a need to separate a part of each waste stream to facilitate recycling. Furthermore, this process could be more complicated when technologies like anaerobic digestion and mechanical biological treatment are used. There are two significant trends in today's modern plants. The first one is the minimization of by-products that must be discharged outside the plant resulting in residual grease and grit treatment systems (Qasim 2017). The second entails the need to establish an effective preliminary treatment and part of a compact system. Ideally, pre-treatment is crucial for removing all materials that can be collected easily from raw sewage before they damage or clog primary treatment clarifiers. The major objects removed in this phase are large objects, such as branches, leaves, and trash. The sewage material goes through a bar screen to facilitate the removal of all the large objects that are contained in the sewage stream (Qasim 2017). It is commonly carried out with a mechanically raked bar screen in processing plants that serve large populations. The design of the mechanical bar screen is such that it aligns with the flow rate. The solids collected during this phase are often disposed of in landfills or occasionally incinerated. If large solids remain in the waste stream, they can become entrained within the treatment plant's pipes and moving parts.

Briefly discuss the anaerobic digestion treatment (refer to Neyens and Baeyens 2003).

Prior to primary treatment, preliminary treatment is carried out for grit removal, flow equalization, and fat and grease removal. Grit is made up of heavy materials, including sand, cinders, and gravel. It also consists of organic matter, including seeds, coffee grounds, eggshells as well as bone chips. Grit removal is beneficial for (1) minimizing the amount of massive deposits that may form in the aeration tanks, channels, conduits as well as aerobic digesters, (2) reducing the rate/frequency of digester cleaning attributed to excessive grit accumulation, and (3) protecting moving mechanical equipment from abnormal wear and abrasion (Jelic et al. 2011). Ideally, abrasion is defined as the process of wearing or scraping. There are three types of grit chambers: horizontal, aerated, and vortex. The grit removal system must operate efficiently during the regular and sustained peak flows when the water holds the highest grit volume.

At what point does grit removal fail?

Regarding flow equalization, secondary treatment and clarifiers function optimally under normal flow conditions. As such, equalization basins are utilized to store diurnal flow peaks temporarily. Basins offer a temporary place for holding incoming sewage in the course of plant maintenance. They also provide a means for dilution and distribution of batch discharges that may deter secondary biological treatment (Jelic et al. 2011). Flow equalization basins need discharge control that is variable and includes provisions for aerators, cleaning as well as a bypass. Finally, the removal of fat and grease is carried out by passing sewage via a small tank. In this tank, skimmers gather the fat floating on the surface. Also, air blowers in the tank's base are utilized for helping recover the fat. Contrastingly, many plants utilize primary clarifiers.

What materials facilitate the removal of grease? Discuss two challenges that may be experienced during the removal of grease.

7.3 Primary Treatment

In this stage, wastewater collects within primary clarifiers, which are ideally large sedimentation tanks and basins. Smaller particles settle due to gravity.

Why are sedimentation tanks made to be large?

Solid matter is then collected by mechanically driven scrapers and directed to hoppers that are connected to equipment for treating sludge. In case the plant failed to remove oil and grease previously, this is achieved at this step. Some plants utilize equipment for saponifying the collected fats (animal fat or vegetable oil) to produce glycerol and soaps by combining it with a strong alkali. In other words, during the primary sedimentation stage, the sewage moves through large tanks called pre-settling basins (Qasim 2017).

Discuss two benefits of saponification.

7.4 Secondary Treatment

This is designed to degrade organic content derived from food waste, detergent, soaps, and human waste. Many municipal plants use aerobic biological processes for treating sewage. For effectiveness, the biota needs both food and oxygen to live. Biota is defined as the plant and animal life of a specific region, geological period, or habitat. Protozoa and bacteria consume soluble organic contaminants that are biodegradable, including fats, sugars, and carbon molecules. The bacteria bind the less soluble fractions to form floc (Qasim 2017). Conceptually, secondary treatment systems are categorized as suspended-growth or fixed-film systems. The main objective of secondary treatment is to treat further the effluent emanating from primary treatment to eliminate residual organics as well as suspended solids. In many cases, secondary treatment also entails the removal of biodegradable dissolved organic matter through aerobic biological treatment by microorganisms using oxygen.

Between primary and secondary treatment, which one is more cumbersome? Provide three valid reasons.

The final phase entails treatment of biosolids and the remaining water. Because of gravity, separation of organic waste from the heavier grit occurs. Organic waste is deposited in landfills. The remaining primary sludge moves to a thickener whereby it undergoes centrifuging before being fed to digesting tanks. This process produces methane which can be used for powering the plant. Stabilized sludge is the final solid product obtained and can be used as fertilizer. The other wastewater is then treated to eliminate nitrogen, phosphorous, and other nutrients.

7.5 Tertiary Wastewater Treatment

The primary purpose of tertiary wastewater treatment is increasing water quality to industrial and domestic standards, or meeting particular requirements governing safe water discharge. If done by municipalities, this form of treatment also entails removing pathogens, ensuring that water becomes safe for drinking. Tertiary treatment involves various more steps after secondary treatment to further minimize nitrogen, pathogens, turbidity, metals, as well as organics. Many of these processes entail some form of physicochemical treatment like filtration, reverse osmosis, more disinfection, coagulation, and adsorption of organics. It is carried out when the wastewater is to be redeployed for irrigation, recreational purposes, or drinking water.

Tertiary processes can eliminate not less than 99% of the impurities found in sewage, leading to the production of an effluent that is safe for drinking (Hamoda et al. 2004). However, tertiary treatment could be quite costly, usually twice more than the secondary treatment cost. This form of treatment is usually utilized under particular circumstances. This stage also removes parasites, viruses, and bacteria which harm the public health. The process involves wastewater flowing from the biological reactor into a pumping station where they meet within the flash mixer. Alum is then used for the removal of extra phosphorous particles and grouping the remaining solids to easily remove them from the filters. Through gravity, wastewater emanating from flash mixers goes through sand filters. Due to alum being added, tiny particles cluster in masses known as floc. Sand traps the floc, whereas the clear water is often gravity-fed to the chlorine tank. Backwashing of the filters is done every 24 h to get rid of the accumulating floc. The backwash water is taken back to the primary treatment stage to traverse the full treatment (Hamoda et al. 2004). The tertiary treated wastewater is disinfected by the chlorine contact tank. Chlorine eliminates microorganisms within treated wastewater, such as parasites, bacteria, and viruses like Cryptosporidium and Giardia. Any of the remaining chlorine is then eliminated through the addition of sodium bisulfite prior to it being discharged. Chlorine is eliminated from the water because it could harm the water quality.

7.6 Quaternary Wastewater Treatment

The purpose of quaternary treatment is to lower the dissemination of pollutants whose presence within surface waters rises regularly, involving issues for people's health as well as the aquatic environments' biodiversity. The process is vital for eliminating micropollutants in sewer water. Adsorption and oxidation are used for eliminating micropollutants. The method recognizes that the increased use of industrial chemicals, hormones, and pharmaceuticals results in rising volumes of pollution in waters (Costa-Pierce 1998). Since the substances manifest in surface waters, they are called micropollutants or trace substances. Various substances are initiated into the sewer system at various points. The concentration of the substances increase to the sewage treatment plant until the optimal micropollutant concentration is reached. One of the elimination methods is the use of active carbon powder as granulate.

7.7 Wastewater Treatment Plants

Drinking water is a valuable commodity for all people in the world. Access to drinking water in our homes is not an unusual thing. Have you ever asked or wondered what happens to wastewater after you flush it down the toilets and drains? Often, we think that it does not concern us. However, it is a pertinent concern that should be addressed. As earlier defined, wastewater is water that has been used in domestic, agricultural, medical as well as industrial activities, and becomes wastewater following alterations to its quality, temperature, and composition (Qasim 2017). Wastewater plants deal with three major types of wastewater: sewage, agricultural, and industrial wastewater. Sewage wastewater refers to all the wastewater utilized within domestic dwellings, including from showers, sinks, or toilets.

What is your understanding of the term sewage wastewater?

On the other hand, industrial wastewater results from industrial, commercial, and production activities and has a unique chemical composition compared to sewage water. A crucial example in this section is the oil industry. It is crucial to identify activities that promote the generation of waste in the oil and gas industry. From a general perspective, there are six major stages in the industrial operations of oil. The first stage is a seismic survey, which is the most appropriate geophysical method utilized for locating petroleum oil as it facilitates subsurface imaging (Spellman 2013). Common wastes that emanate from seismic operations are explosive waste, equipment maintenance waste, and domestic waste. The second process is called exploratory drilling. It is carried out after the identification of a promising geological structure (Chan et al. 2009). Some common wastes attributed to exploratory drilling are drilling muds, cementing waste, production testing waste, domestic wastes, drilling chemicals, and cuttings. The third stage, **construction**, produces wastes such as lubricants, domestic waste, paints, and solvents (Gupta et al. 2012).

Furthermore, the production stage occurs after the installment of treatment facilities. Typical forms of waste during this stage are production chemicals, produced water, and sludge. The maintenance stage operations include drilling, vehicle repair, and all equipment inspections (Spellman 2013). Some of the common wastes generated during this stage are solvents, used parts, lubricants, filters, and contaminated soil. The abandonment stage involves the decommissioning of a production field. It entails abandoning wells, demolition of facilities, landfill closure, site re-vegetation as well as site monitoring. Some of the common wastes that are obtained from abandoning include plant equipment, contaminated soil, oil sludge as well as construction materials.

Define exploratory drilling and site re-vegetation. What is the benefit of site re-vegetation?

7.8 Wastewater Treatment of Oil Industries

Wastewater that is produced from oil-based industries can be managed using simple conventional treatment processes since water and oil do not mix. The process of treatment only becomes complex in the presence of soap that naturally forms when caustic soda is added. Generally, the most crucial step in oil removal is the use of a mechanical separation unit. The dissolved air flotation (DAF) system treatment also needs to reduce the chemical oxygen demand (COD) level prior to the final biological process taking place at the aeration tank.

Dissolved air flotation (DAF) is a process of water treatment that clarifies wastewaters by removing suspended matter, including solids or oil. Removal is attained through the dissolution of air within the water and then released at atmospheric pressure within a flotation tank basin (Asatekin and Mayes 2009). Released air develops tiny bubbles that adhere to suspended matter, thus making suspended matter float to the water surface, whereas removal could be facilitated by a skimming device.

Chemical oxygen demand (COD) is the measure of water capacity required for consuming oxygen during organic matter decomposition and the oxidation of inorganic chemicals, including nitrite and ammonia. Large buffer tanks play a beneficial role in containing wastewater and providing more control of the system setup since fluctuations regarding incoming water quality will not result in substantial differences that impact other processes. This could deter automatic systems, like pH adjustment devices, that occasionally are unable to respond to changes. Ideally, the first step towards oil recovery entails the discharge of water to equalization tanks. These special-made tanks consist of two unique compartments and have a plate that enables water to go through. Despite this being crude separation, it also provides a pertinent way of recovering back oil waste. Furthermore, the next step is usually the process of acidulation as the pH is reduced to between 3.0 and 4.0 (Asatekin and Mayes 2009). This can be attained through the addition of sulfuric acid to decompose large oil particles that elude from the equalization tank into smaller particles. Lighter molecules often float to the top that can be scraped off using a scraper. Wastewater will then undergo the chemical and physical treatment processes by entering the aeration tank where additional purification takes place.

In wastewater treatment plants, the first process entails draining water to the main sewer system. The first mechanical stage is often referred to as preliminary treatment or pre-treatment. Grit and large object separation occur at this stage as water flows through gravel chamber and then disposed of.

The primary treatment in these plants allows the wastewater to flow to the pre-settling basins. It is the point where pre-treatment ends while secondary wastewater treatment begins. After primary treatment, wastewater pollution level is reduced to 60% (Spellman 2013). Secondary treatment/biological stage is founded on natural processes. Wastewater treatment plants utilize bacteria which consumes organic residue then turns it into sludge. In the course of the biological stage, excess sludge is often pumped and moved to settling tanks. At this point, the sludge gets transported to digestion tanks for more treatment.

Sludge in digestion tanks is heated and combined after being pumped out of settling tanks. Biogas is generated during this process, which can be reused by wastewater treatment plants. When the digestion of sludge reaches the maximum level, the second process of digestion occurs within the storage tanks. Separation of water from the semi-solid sludge takes place before being transported for more treatment, while the residual semi-solid goes through mechanical dewatering (Jelic et al. 2011). After about 4 weeks, it is sufficiently dried out to ripen. As long as it complies with agricultural standards, it may be utilized for fertilizing industrial crops. The final wastewater treatment step entails service water inspection. The objective is to conduct an analysis of the contamination level and ensure that treated water conforms to the best standards.

Case Study 7.1: The curse of oil in Nigeria

Fig. 7.1 Crude oil adverse effect on the environment, *Source:* Adaoyichie (2018)

The production of oil in Nigeria has been associated with both fortune and misfortune, depending on the person or entity that is affected. The country considers it a significant fortune because it is the source of national wealth. Oil accounts for almost 90% of Nigeria's foreign exchange earnings (Idemudia 2012). The government, at all levels, relies on it for funding its development projects and programs. In the coastal areas where oil is produced, the much-desired technology transfer has been attained particularly within exploited and explored areas. The membership of Nigeria in important bodies, including the World Petroleum Congress, African Petroleum Producers Association, and Organization of Petroleum Exporting Countries (OPEC), has raised its profile globally and made Nigeria highly influential (Kashi and Watts 2008).

(continued)

The Niger Delta is the area that the Niger River's natural Delta covers as well as the areas found to the west and east that produce oil. The northern boundaries are found close to the Niger River of Aboh, whereas the eastern and western boundaries are located close to the Imo and Benin Rivers, respectively. It is estimated that the area covers about 25,000 km^2 (Mähler 2010). The Niger Delta is also made up of nine states: Ondo, Abia, Cross River, Imo, Akwa Ibom, Bayelsa, Delta, Edo, and Rivers. The region consists of some distinct characteristics, including having one of the sandiest coastal ridge barriers, freshwaters, seasonal swamp forests, and lowland rainforests. The region is quite heterogeneous in terms of ethnicity and culture. The five main cultural and linguistic groups are Igbo, Delta Cross, Ijaw, Edo, and Yoruba.

The Delta Cross is predominantly made up of Ogoni, Abua as well as Ogbia. The Ogoni are the most famous globally due to their agitation for autonomy and resource control. Despite the Igbo and Yoruba being the two largest ethnic groups within Nigeria, they are among the smallest in the Niger Delta. The major Yoruba groups are the Ikale and Ilaje in Ondo State borderlands and the Itsekiri in Delta State. Despite oil being beneficial to the country, oil-bearing communities found within the Niger Delta perceive oil as a curse rather than a blessing. In these communities, erosion, deforestation, land dereliction, as well as destroyed farmlands are the major signposts for the destruction left behind by oil (see Fig. 7.1). It has also contributed to the pollution of creeks and destruction of aquatic life in the coastal Niger Delta. In places where there are spillages, the losses witnessed cannot be quantified in monetary terms. The other evidence of oil being a curse is the increasing problem of acid rain that leads to the destruction of houses. Following over 50 years, exploration, environmental degradation, and pollution have continued to afflict local communities, leading to brutalizing poverty, health hazards, deaths, unemployment, and diseases among the Ilaje and Ogoni found in the Niger Delta. Niger Delta produces 100% of the country's oil and gas, but the Ilaje community survives without electricity or clean running water (Ross 2012). For instance, for one to see a doctor, they have to travel many hours using boats. Oil has occasionally been spilled into these creeks, thus disrupting the fishing communities or causing violent conflict. When the Delta region occupants try to protest, they are usually paid off, set against each other, or shot. High rates of criminality, youth unrest as well as lawlessness have also plagued the Delta. The tragedy and curse of oil in the region can be narrated well through the eyes of the Ilaje and Ogoni communities. One example of the devastation caused by oil pollution on the Ogoni community is the execution of Ken Saro-Wiwa for fighting for the rights of the Ogoni Community (Ross 2012). More information about Kenule Saro-Wiwa can be found under the individual matter section.

7.9 Bioremediation of Wastewater

Bioremediation is a technique used in waste management that entails using organisms to neutralize or eliminate pollutants from contaminated sites. The Environmental Protection Agency (EPA) (2018) defines bioremediation as a treatment that uses natural organisms to decompose hazardous substances into non-toxic or less toxic ones. Many industries across the world are exploring the need to use bioremediation as a pertinent way of cleaning up wastewater. While treating wastewater, bioremediation predominantly uses natural microorganisms from Mexico. Such initiatives provide economically and environmentally sustainable treatment techniques (Gratia et al. 2009). Bioremediation is cost-effective because of the time and planning required for successful treatment. In many developed nations, optional wastewater treatment methods are needed (Kshirsagar 2013). Lagoons can offer a viable solution for wastewater treatment in developing nations as well as in rural areas. Algae is often used for the bioremediation process and sewage treatment. It plays a crucial role in the aerobic treatment of waste within the secondary treatment process. Algae-based wastewater treatment systems are predominantly utilized for the removal of nutrients, especially phosphorus and nitrogen.

Experiment 7.1: Bioremediation for wastewater treatment: Microalgae selection

Materials Required

Batch photoreactor, cylindrical borosilicate Pyrex bottles, and cylindrical bioreactors.

Background of the Experiment

This experiment is an attempt to establish the significance of algal bioremediation as a viable technology for the treatment of wastewater within a sustainable and economical way.

Experiment Procedure

1. Perform the experiment in batch photobioreactors on a lab-scale using 2000 mL cylindrical borosilicate Pyrex bottles.
2. Drill three 1/4 inch holes into the caps of cylindrical bioreactors to get two inlets and one outlet. Secure carbon dioxide and airlines within the two inlet openings while leaving the third outlet unobstructed to deter pressure and oxygen build-up within the reactor.
3. Maintain the gas flow rate into the bioreactors using two rotameters. Subject the cultures to illumination using four fluorescent lamps placed in a horizontal and parallel way to the front as well as the back of the tubular photobioreactor.

(continued)

Findings and Discussion
The findings should demonstrate that algal bioremediation is a viable technology for the treatment of wastewater sustainably and economically compared to other conventional processes of treatment. If it is used, the process could result in huge chunks of savings that could be reused for meeting the needs of war-fighters.

7.10 Individual Matter: Ken Saro-Wiwa

Kenule (Ken) Beeson Saro-Wiwa was born on October 10, 1941 (see Fig. 7.2). He was a Nigerian writer, environmental activist, TV producer, and winner of the Goldman Environmental Prize. He was also a member of the Ogoni community whose homeland in the Niger Delta remains a target for crude oil extraction since the 1950s. The community has also been at the center of extreme environmental degradation from indiscriminate petroleum waste dumping. As a spokesperson and later as president for the Movement for the Survival of the Ogoni People (MOSOP), Ken organized a non-violent campaign to oppose environmental damage of the community waters and land brought about by petroleum industry operations, particularly the Royal Dutch Shell Oil Company. He is also renowned for openly criticizing the Nigerian government (Pegg 2000).

Ogoniland has been characterized by grave environmental damage because of the more than 100 oil wells found in the area. The non-violent MOSOP was formed in 1990 and has been met with violent resistance through the imprisonment of many of its leaders and the assassination of some of them. Notwithstanding, the Ogoni have been successful in compelling Shell to stop its activities and raising significant global awareness of the problem bedeviling the region. Led by Ken Saro-Wiwa, MOSOP was instrumental for the publication of the Ogoni Bill of Rights in 1990 (Pegg 2000). The bill marked the beginning of the non-violent struggle that opposed the oil companies and the government. He highlighted the lack of basic social

Fig. 7.2 Ken Saro-Wiwa, *Source:* Adebambo (2016) via the Guardian

services, political marginalization as well as the maltreatment they experienced from Shell. According to the bill, there was a need for Ogoniland's environment to be protected, self-determination, cultural rights, and representation of the Ogoni in Nigerian institutions. More importantly, there was a need for a fair proportion of the revenue from oil sales.

In years preceding conflict escalation in 1992–1993, Ken Saro-Wiwa tried to gain the support of global non-governmental organizations. MOSOP combined with Unrepresented Nations and Peoples Organization which enabled it to have better global recognition through meetings with the United Nations and media attention. The conflict became more entrenched after 2 years without significant progress. MOSOP refocused its energies on the three oil-producing organizations in the region: Chevron, Nigerian National Petroleum Company, and Shell (Pegg 2000). Shell became the primary target as it had the largest share. Saro-Wiwa led the group in presenting the organization with an ultimatum that demanded 10 billion dollars in royalties and damages to the Ogoni. He also demanded an end to all forms of damage caused by the three companies against the environment. The threat was that in case these conditions were not met, MOSOP would rally the Ogoni community in widespread protests to oppose the presence of these companies. To respond to these threats, the Nigerian government decided to announce that any form of disturbance of oil production would be punishable as treason. It also prohibited all public assemblies and meetings. Despite the prohibition of all public assemblies, MOSOP organized a massive peaceful protect that rallied at least 300,000 Ogonis on January 4, 1993. The protesters held green twigs to symbolize the environment. With more than 60% of the Ogonis in the area protesting, they demanded fundamental environmental rights and self-determination, referred to as the First Ogoni Day (Pegg 2000). The protests led to the withdrawal of Shell from the region, thus reducing the quantity of petroleum extracted from the region as well as reducing profits to oil organizations that continued operating within the area.

The government tried to respond to the success of the First Ogoni Day and withdrawal of Shell by forcefully suppressing Ogoni activities. In April 1993, soldiers fired live ammunition to disperse 10,000 Ogoni protestors, resulting in the death of 10 protesters. The government also commenced wiping out entire villages. In December, the government killed Ogoni people who were living in Port Harcourt. In 1994 (Easter Sunday), it raided Ogoni villages found close to Afam and killed its inhabitants. Later investigations disclosed that NATO bullets and mortar bombs were utilized in the massacre. By June 15, the government destroyed 30 villages. Despite the death and displacement of thousands, widespread protests continued in the entire region. With more repression in the country, global support and awareness of the Ogoni struggle continued growing. The repeated arrests of Ken Saro-Wiwa in 1994 made Amnesty International and Greenpeace lead an international campaign for him to be released. Contrastingly, dissent in the Movement led to divisions between the president and other movement leaders. While Saro-Wiwa left the country on May 21, 1994, four Ogoni chiefs who disagreed with him were killed. On the next day, Saro-Wiwa was arrested along with eight other activists. He was held without legal representation, medical care, or charges for several months.

Despite the government not producing any evidence against Saro-Wiwa, the men were executed on November 10, 1995, by hanging. They were later buried in Port Harcourt Cemetery. This action led to the suspensions of Nigeria from the Commonwealth. Supporters of the Ogoni organized protests globally, while many world leaders stated that there was a need for an oil embargo and economic sanctions. Arguably, Saro-Wiwa played a crucial role in illuminating the tribulations and suffering that the Ogoni community experienced. In addition, despite the Ogoni suffering substantial losses during the campaign, they succeeded in removing Shell from the region.

7.11 Future Trends

7.11.1 Africa

Africa will respond well to various solutions on wastewater, with many clients appreciating the significance of water treatment as well as environmental benefits and the welfare of the people. The future of waste treatment in the continent could be challenging in case the rules on environmental issues will be brought up. One of the defining issues in the coming decades will be access to clean water. Reverse osmosis will increasingly be used as a water treatment technology that uses a semi-permeable membrane for removing dissolved ions and larger water particles. The increasing international initiative of Going Green and embracing more energy-efficient systems imply that institutions will have to increasingly invest in research to increase water treatment plants' efficiency and integrate the renewable sources of energy like solar.

7.11.2 Asia

Large-scale seawater desalination will be the major driver of water and wastewater treatment in the region. Scientists will continue working to improve and ensure that saltwater purification is affordable. Developing regions in the Middle East will have more demand for these plants, thus providing growth opportunities for major suppliers within the water and wastewater treatment equipment market. Since seawater desalination needs large volumes of energy and chemicals, top companies will focus on developing eco-friendly and innovative desalination processes as well as machinery for futuristic use at minimized costs and environmental impact.

7.11.3 Australia

Economic development and urbanization will increase demand for freshwater, thus promote the use of water treatment chemicals. Substituting products that act as alternatives to the water treatment chemicals will likely deter market growth. More production of biogas and biosolids and changing focus to the use of green chemicals will act as market opportunities in the future. Since many natural water resources have reached allocation limits, the industry needs an economic valuation and water innovation and co-operation across sectors which could lead to more demand for water treatment chemicals for the municipal operations in coming years. Many municipal dewatering facilities (medium to large) will make the change to screw press-based dewatering.

7.11.4 Europe

Juyal (2018, p. 1) predicts that "the global water and wastewater treatment equipment market can attain a valuation of around USD 103.41 billion by 2025." There will be a higher demand for freshwater from various segments, thus driving the growth of the international industry. The industry will create substantial growth opportunities for both tertiary and secondary water and wastewater treatment equipment.

7.11.5 North America

North America will mainly be characterized by reduced groundwater pumping. Industries will reduce the use of underground aquifers. Water control bodies and governments will develop sustainable water systems that include designing water treatment models and the development of centralized systems for appropriate distribution.

7.11.6 South America

In South America, concerted efforts will be put in place to enhance preparedness for better resiliency. Industry stakeholders will develop resilience plans for mitigating frequent natural disasters. Managing high stormwater flows needs water treatment devices with high capacity. Higher budgets allocated for emergencies will create more business opportunities for instrument manufacturers. It will also encourage budgetary reforms for developing advanced technologies. In addition, there will be the proliferation of advanced wastewater reuse technologies. Despite the reuse of water not being a new thing for some end users, more efforts will revolutionize the water distribution and infrastructure. It could make water and wastewater treatment to be more sustainable and cost-effective. Natural water sources are unavailable in numerous arid and semi-arid areas. Thus, industry stakeholders will have to focus on affordable options to augment the available water sources and halt events such as groundwater salting and seawater intrusion. This could offer recycling solutions to operations.

7.12 Summary

- Wastewater treatment has various phases that are crucial for ensuring that wastewater is harmless to the environment.
- Preliminary treatment helps protect the pipes and systems against blockages.
- Waste removed can be grit, oil, grease, sludge, branches, trash, and other large objects.
- The benefits of grit removal are a reduction of heavy deposits, reducing the frequency of digester cleaning, and protection of mechanical equipment from abrasion and wear.
- The main objective of primary treatment is to remove organic and inorganic solids through sedimentation. Oil and grease are some of the products that can be separated.
- Secondary treatment's main objective is to further separate the waste in the waste stream, including the suspended solids and residual organics.
- Sludge removal acts as the final process of ensuring that materials are separated from wastewater.
- Tertiary wastewater treatment's main objective is to increase water quality to industrial and domestic standards, or meeting particular requirements governing safe water discharge.
- Quaternary treatment is used for eliminating micropollutants in sewer water.
- Oil has been both a curse and a blessing for various entities, which was highlighted by Saro-Wiwa in Nigeria.
- For Africa, future trends entail more demand for wastewater treatment and more investment in research, while for North and South America, the major trends will be reduced groundwater pumping, preparedness to boost resilience, and advanced wastewater reuse technologies.
- Europe will mainly be characterized by higher demand for freshwater that will drive the industry's growth, and Asia will have many large-scale sweater desalination plants.
- Urbanization and economic development will raise the demand for freshwater in Australia.

7.13 Discussion Questions

1. Differentiate between grit and grease.
2. Explain the significance of preliminary treatment.
3. What is a waste stream?
4. Define saponification and list the chemicals used for facilitating the process.
5. Explain what happens if large solids are entrained in pipes of a treatment plant (Hint: consider the efficiency of the process).
6. Which is the most critical step in wastewater treatment and why?
7. Why do many plants use primary clarifiers that have mechanical surface skimmers?
8. In grit chambers, why is the water velocity maintained sufficiently high?
9. Communitors are often adopted for supplementing coarse screening. What is their major function?
10. Identify wastes separated during each phase of wastewater treatment.
11. Discuss the major wastes collected during the primary treatment phase.
12. Differentiate between tertiary and quaternary wastewater treatment.
13. What is the importance of sludge in farming?
14. How does wastewater treatment contribute to sustainable development?
15. Use the 3Rs (reduce, reuse, and recycle) to discuss the phases of wastewater treatment.
16. Oil is a curse to the communities living close to the Niger Delta. Explain how gold and black diamonds have also brought about the devastation to local communities in Africa. Select two countries and identify communities affected.

References

Adaoyichie, G. (2018). More oil companies are dumping Nigeria's crude. *The Pulse*. Retrieved February 21, 2019, from https://www.pulse.ng/news/more-oil-companies-are-dumping-nigerias-crude/pkqx1xn/.

Adebambo, G. (2016). Ken Saro Wiwa. *The Guardian*. Retrieved August 22, 2018, from https://guardian.ng/wp-content/uploads/2016/08/ken-saro-wiwa-1024-x614.jpg/.

Asatekin, A., & Mayes, A. M. (2009). Oil industry wastewater treatment with fouling resistant membranes containing amphiphilic comb copolymers. *Environmental Science & Technology, 43*(12), 4487–4492.

Chan, Y. J., Chong, M. F., Law, C. L., & Hassell, D. G. (2009). A review on anaerobic-aerobic treatment of industrial and municipal wastewater. *Chemical Engineering Journal, 155*(1–2), 1–18.

Costa-Pierce, B. A. (1998). Preliminary investigation of an integrated aquaculture–wetland ecosystem using tertiary-treated municipal wastewater in Los Angeles County, California. *Ecological Engineering, 10*(4), 341–354.

EPA. (2018). *Bioremediation. Overview*. Retrieved December 15, 2018, from https://-cluin.org/techfocus/default.focus/sec/Bioremediation/cat/Overview/.

Gratia, E., Weekers, F., Margesin, R., D'Amico, S., Thonart, P., & Feller, G. (2009). Selection of a cold-adapted bacterium for bioremediation of wastewater at low temperatures. *Extremophiles, 13*(5), 763–768.

Gupta, V. K., Ali, I., Saleh, T. A., Nayak, A., & Agarwal, S. (2012). Chemical treatment technologies for waste-water recycling – An overview. *RSC Advances, 2*(16), 6380–6388.

Hamoda, M. F., Al-Ghusain, I., & Al-Mutairi, N. Z. (2004). Sand filtration of wastewater for tertiary treatment and water reuse. *Desalination, 164*(3), 203–211.

Idemudia, U. (2012). The resource curse and the decentralization of oil revenue: The case of Nigeria. *Journal of Cleaner Production, 35*, 183–193.

Jelic, A., Gros, M., Ginebreda, A., Cespedes-Sánchez, R., Ventura, F., Petrovic, M., et al. (2011). Occurrence, partition and removal of pharmaceuticals in sewage water and sludge during wastewater treatment. *Water Research, 45*(3), 1165–1176.

Juyal, V. (2018). *4 top trends in the water and wastewater treatment equipment market*. Retrieved January 4, 2019, from https://www.watertechonline.com/4-trends-water-wastewater-treatment-equipment-set-affect-market-growth/.

Kashi, E., & Watts, M. (2008). *Curse of the black gold: 50 years of oil in the Niger Delta*. New York: Powerhouse Books.

Kshirsagar, A. D. (2013). Bioremediation of wastewater by using microalgae: An experimental study. *International Journal of Life Science Biotechnology and Pharma Research, 2*(3), 339–346.

Mähler, A. (2010). Nigeria: A prime example of the resource curse? Revisiting the oil-violence link in the Niger Delta. *GIGA Working Paper No 120*. Retrieved November 20, 2018, from https://papers.ssrn.com/sol3/papers.cfm?abstract_id=1541940/.

Neyens, E., & Baeyens, J. (2003). A review of thermal sludge pre-treatment processes to improve dewaterability. *Journal of Hazardous Materials, 98*(1–3), 51–67.

Pegg, S. (2000). Feature review Ken Saro-Wiwa: Assessing the multiple legacies of a literary interventionist. *Third World Quarterly, 21*(4), 701–708.

Qasim, S. R. (2017). *Wastewater treatment plants: Planning, design, and operation*. Abingdon: Routledge.

Ross, M. (2012). *The oil curse: How petroleum wealth shapes the development of nations*. Princeton: Princeton University Press.

Spellman, F. R. (2013). *Handbook of water and wastewater treatment plant operations*. Boca Raton: CRC Press.

Chapter 8
Waste Management

Abstract The rise in the global population continues to contribute to the generation of waste, especially in urban areas. This makes the issue of waste management to become one of the priorities for government and private entities. To this end, this chapter seeks to focus on solid waste management and particularly the concepts, steps involved in managing, treating, and disposing of waste. The benefits related to waste management and challenges experienced by African nations in the quest for waste management are also explored. To further explore these concepts, the case study of how African can build inclusive and sustainable cities, a practical application, and Tom Szaky's contributions to the topic have been embedded. Future trends in the diverse continents are also provided to show a glimpse of what the future of waste management will look like.

Keywords Waste · Tom Szaky · Solid waste management · Waste hierarchy · Recycling

8.1 Overview

The global population continues to rise, along with the populations in major urban centers across the globe. The increase in people correlates to an increase in the quantity of waste generated. Consequently, waste management has increasingly become an issue of focus as people consider new strategies to treat, dispose, and minimize the waste generated. The previous chapter examined wastewater treatment; this chapter will mainly focus on solid waste management. The waste management process will form the bulk of the discussion in this chapter as well as some of the challenges that African countries grapple with in trying to manage waste.

A. A. Akinsemolu, *The Principles of Green and Sustainability Science*,
https://doi.org/10.1007/978-981-15-2493-6_8

8.2 Concepts of Waste Management: Definition of Terms

8.2.1 Waste

Waste is defined as any substance that is discarded after its primary use or is valueless, defective, and of no use (Silva et al. 2017). Some examples of waste are municipal solid waste, hazardous waste, radioactive waste, and wastewater.

8.2.2 Waste Management

Waste management is defined as the actions and activities that are needed for managing waste from the start to the final stage of disposal. Some aspects are the collection, transportation, treatment, disposal, monitoring, regulation, and the legal and regulatory framework (Silva et al. 2017).

8.2.3 Waste Hierarchy

Waste hierarchy refers to the processes used for ensuring that the volume of waste generated is reduced (see Fig. 8.1). In other words, it categorizes waste management strategies in accordance with their waste minimization desirability. Waste hierarchy acts as the foundation for many waste management strategies. The core objective of waste hierarchy is to get the optimal practical benefits from commodities while generating the minimum quantity of end waste. The main premise is that policies need to enhance measures for preventing waste generation (Tangwanichagapong et al. 2017). Subsequent actions are to find alternative uses for waste by reusing, recycling, or composting. The final action is to dispose of waste in landfills or incinerators.

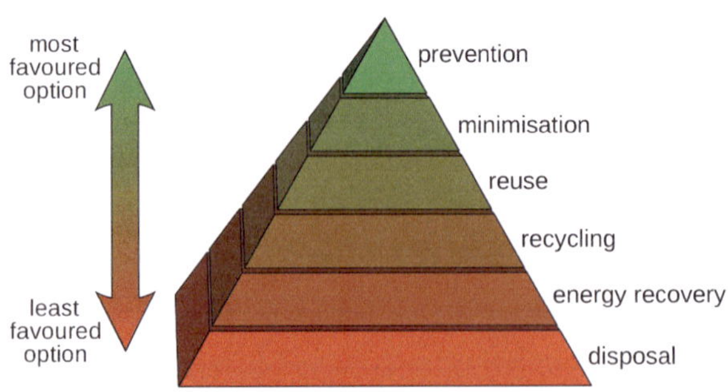

Fig. 8.1 Waste hierarchy, *Source:* Drstuey (2008)

8.2.4 Product Lifecycle

A product lifecycle is the sequence of stages that a new product undergoes during its life span. The lifecycle is as follows: design, manufacture, distribution, primary use, reduce, reuse, and recycle.

Each of these stages in the lifecycle provides opportunities for policy intervention, including rethinking the importance of the product, minimization of waste potential, and extending its use. In other words, product lifecycle is a way of maximizing the use of limited resources of the planet by ensuring that there is no unnecessary waste generation.

How does this support the theory of sustainability?

8.2.5 Resource Efficiency

Resource efficiency acknowledges that economic development and growth is unsustainable at normal production and consumption patterns. Humans have a tendency of extracting more resources for producing goods than the Earth's capacity to replenish (De Sadeleer 2014). As such, resource efficiency is the minimization of the environmental effect from goods' production and consumption to the extraction of raw materials and disposal.

How can this process of resource efficiency tackle sustainability?

8.2.6 Polluter-Pays Principle

The Polluter-Pays principle states that the polluting party should pay for the environmental effect created by their actions. Regarding waste management, the principle embodies the preference that the waste generator pays for disposing of waste appropriately. This principle is enacted in environmental law to compel the party responsible for pollution to pay for cleanup and any damages caused to the natural environment. It is perceived as a regional custom due to the robust support in many European Union (EU) nations (De Sadeleer 2014).

The Polluter-Pays principle is beneficial for greenhouse gas emissions since these emissions harm the environment. They also damage the environment through effects on the climate while contributing to air pollution. When enacted into law, the principle acts as a deterrent by mandating the polluter to incur the costs or harm caused to the environment. When pollution costs are not imposed on polluters or emitters, the costs are externalized to society. The society is compelled to bear these costs through the concept of global commons. The idea of global commons is that each person shares, and is allowed to use the resource. The principle could be applied to emitters through the carbon price that imposes a certain charge of greenhouse gas emissions that resonates with the potential cost.

Do you know of any countries that impose a carbon price on greenhouse gas emissions? How effective have they been?

One approach to limit carbon emissions has been to assure that emitters pay a cost for the pollution they release. The carbon price compels the polluter to pay using two policy instruments: a carbon tax and a quota-based system/cap-and-trade. These instruments operate by setting a limit on the optimum emission level for a specific period and distribute permits for every unit of greenhouse gas among organizations that produce emissions. In other words, the emitters pay by ensuring that they have sufficient permits for covering the number of emissions they produced for the specific year.

Does this encourage or discourage pollution?

8.3 Process of Waste Management

Waste management has increasingly become a concern in industrial, rural, and urban areas. Improper management of hazardous waste leads to environmental contamination in numerous ways (McDougall et al. 2008). Discharging hazardous substances into water bodies leads to the death of aquatic life. In addition, uncontrolled waste disposal could easily lead to the pollution of groundwater and the soil. Consequently, prudent waste management is among the most significant concerns of modern society. Ideally, waste management entails the following crucial steps:

1. Waste reduction.
2. Waste recycling.
3. Waste treatment.
4. Waste disposal.

8.3.1 Waste Reduction

Reduction is the first process of waste management and entails reducing the quantity of waste generated. It can be achieved through source reduction, concentration, and segregation. Source reduction entails the modification of industrial processes to optimize the use of raw materials and reduce the number of hazardous wastes generated (Menikpura et al. 2013).

Identify some of the ways that can be used for source reduction.

Regarding concentration, generated waste is concentrated through the use of modern methods, including decantation and precipitation techniques. This plays a crucial role in reducing the volume of wastes. Wastes are segregated by their chemical composition, physical nature as well as hazard potential, which is beneficial during their eventual disposal. Also called waste minimization, reduction is a basic waste management method that helps prevent the creation of waste material. Some methods used for avoiding waste generation are reusing second-hand products, discouraging disposable products, and designing products to remain reusable or refillable.

8.3.2 Waste Recycling

Many waste forms have various beneficial substances. Recycling is a method for enabling the waste to be reused through reprocessing. Ideally, recycling refers to two things: reuse and recycling. Reuse is the scenario where some waste can be used again. For instance, discarded containers, lubricants, and some chemicals can be used further without the need for processing. Recycling, on the other hand, is different from reuse because waste has to be rigorously treated and reprocessed prior to being used again. Recycling entails collecting and reusing waste materials, such as empty beverage containers, to reprocess these materials into new commodities (Menikpura et al. 2013). Collection vehicles and dedicated bins can be used for collecting material for recycling through the curbside collection procedure. Within some communities, the waste owner separates materials into various bins prior to its collection. Other communities use single-stream recycling in which all the recyclable materials are placed in one bin for collection and sorting is done in a central facility through the single-stream recycling. Aluminum, copper, steel, rubber tires, newspapers, and corrugated fiberboard boxes are the most common consumer products recycled. These items are made of a single material type, thus making them easily recyclable. Recycling complex products is more difficult because of the additional dismantling and separation needed. Each nation and city has a unique recycling program that can manage various forms of recyclable materials.

What is the recycling program established in your local community?

Tom Szaky's TerraCycle is evidence of how individual initiatives can play a crucial role in recycling waste. Detailed information on the initiative is found in the Individual Matter section at the end of the chapter.

Energy recovery and biological reprocessing are common ways of reusing materials. Organic recoverable materials can be recovered through digestion and composting. Recycling of organic material is then done to produce mulch or compost that can be used for agriculture and landscaping. Waste gas can also be captured and utilized for cogeneration, which is crucial for energy recovery (Menikpura et al. 2013). Cogeneration involves the conversion of waste materials that cannot be recycled into usable electricity, heat, or fuel using various processes, including gasification, anaerobic digestion, combustion as well as landfill gas recovery. The process of cogeneration leads to the production of a renewable source and could minimize carbon emissions by balancing the energy demand from fossil fuels and reducing the generation of methane from landfills.

What is the advantage of cogeneration, and how does it accelerate the decomposition of organic matter? What is pyrolysis, and how is it important for energy recovery?

8.4 Waste Treatment

Various processes can be used for the treatment of hazardous wastes. The aim is to alter the chemical and physical composition of these waste products in order to reduce their quantity and neutralize their toxicity. Treatments vary depending on the chemical nature and physical state of the wastes. There are three major types of waste treatment: physical, chemical, chemical and biological.

8.4.1 Physical Treatment

This entails mechanical segregation and separation to facilitate the process of disposal. Regarding liquid wastes, it mainly entails phase separation, which involves a sedimentation process whereby silt, sand as well as other particulates settle at the bottom (Pichtel 2005). Coagulation combines the fine colloidal particles and enables them to settle.

Identify some of the wastes treated through physical treatment. Is physical treatment effective and environmentally friendly?

8.4.2 *Chemical Treatment*

Various chemicals are used for treating hazardous wastes in order to neutralize them and eliminate their toxicity. It entails modifying the chemical properties of wastes using oxidation, precipitation, neutralization, and reduction.

What are some wastes that can be managed by chemical treatment?

8.4.3 *Biological Treatment*

Biological treatment entails the actions of microorganisms on waste products. Microorganisms such as fungi and bacteria are scavengers, which decompose organic substances in waste (Polprasert 2017). Ideally, biological treatment entails the processes of conditioning, digestion, composting, and disposal. Some microorganisms used for biological treatment belong to the genus Porphyromonas, Veillonella, Actinomyces, Bacteroides, Clostridium, Fusobacterium, Peptostreptococcus, Propionibacterium, Porphyromonas, and Prevotella.

8.5 Solid Waste Management Process

The various steps that are involved in solid waste management: waste generation, storage, collection, and disposal to dumping site are shown in Fig. 8.2.

Waste generation refers to the material discarded after serving its purpose or when it is not useful. Storage and processing refer to activities linked to the management of storage and processing of solid waste at generation points. On the other hand, collection entails the activities geared towards solid waste collection at particular locations. Transfer and transport activities entail waste transfer from points of collection to vehicles where they are transported to disposal sites (Menikpura et al. 2013). Processing and recovery are facilities and techniques used for recovering wastes for various treatments. Disposal is the final step where waste is disposed to the landfill site. Solid waste is generated within many urban centers in the world, thus making solid waste disposal an issue that affects urban centers. Various activities produce large quantities of wastes that should be appropriately

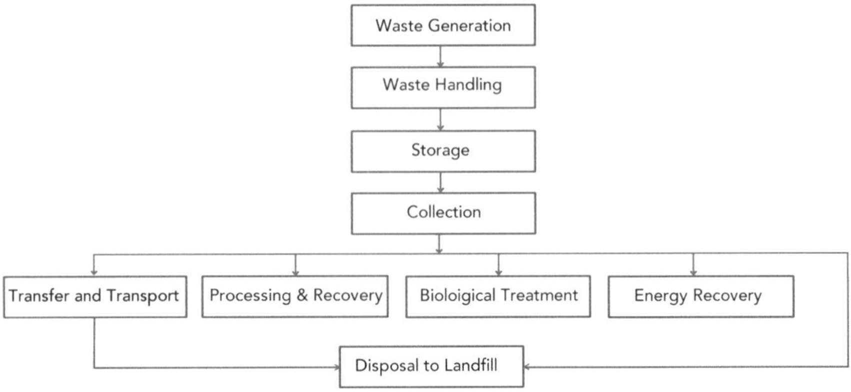

Fig. 8.2 Solid waste management process

handled. The most common methods of solid waste disposal are open dumps, drainage, sanitary landfills, ocean dumping, composting, and incineration.

8.6 Waste Disposal Methods

Solid waste management is usually associated with the dumping of garbage in landfills or incineration. Whereas these activities form an essential aspect of the process, various elements are involved in creating an optimal integrated solid waste management system.

What is integrated waste management? What is the difference with other waste management methods?

For instance, treatment techniques are crucial for reducing the toxicity and volume of solid waste. These steps could change into a more convenient disposal form. Ideally, methods of waste treatment and disposal are chosen based on the composition, form, as well as the number of waste materials (Menikpura et al. 2013). The major methods used for waste treatment and disposals are discussed below:

8.6.1 Thermal Treatment

Thermal treatment is a type of waste treatment technology that involves high temperatures in the processing of waste materials using techniques such as incineration, gasification and pyrolysis, open burning and dumps, and landfills.

8.6.1.1 Incineration

Incineration is the combustion of solid and organic waste in order to convert them into gaseous products. This method is beneficial for disposing of solid residue and municipal solid waste. It minimizes the quantity of solid waste by an average of 25% (McDougall et al. 2008). Incineration is commonly used for energy recovery to generate electricity. It also has various advantages, including reduction of waste volume, minimization of transportation costs, and lowering the emission of harmful greenhouse gases.

Incineration further converts waste materials into ash, heat, and steam. It is done on both a small and large scale by individuals and industries, respectively. Also, it is used for disposing of solid, gaseous as well as liquid waste and acknowledged as a practical way of disposing of some hazardous waste materials. However, it is a controversial way of eliminating waste because it contributes to problems such as gas emissions that pollute the environment. Concerns have also been raised on persistent organic compounds, including polycyclic aromatic hydrocarbons (PAHs), that may be created and could have serious environmental repercussions (McDougall et al. 2008).

What are the negative impacts of incineration on the environment? What alternatives could you propose for incineration?

8.6.1.2 Gasification and Pyrolysis

These are similar methods that decompose organic waste materials by exposing waste to low volumes of oxygen and extremely high temperatures. Pyrolysis does not use any oxygen, whereas gasification limits the amount of oxygen (Thomsen et al. 2017). Comparing the two, gasification is more beneficial to the Going Green initiative because it allows the process of burning to recover energy without harming the environment.

8.6.1.3 Open Burning

This is a thermal waste treatment that causes harm to the environment. Incinerators used during such processes do not have pollution control devices and have a tendency to release substances such as dioxins, carbon monoxide, volatile organic compounds, ash, hexachlorobenzene, and particulate matter (Das et al. 2018). Unfortunately, many local authorities use this method as it is an affordable solution to solid waste. Despite being the cheapest disposal method, it is esthetically displeasing, breeds microorganisms and pests, and pollutes groundwater through leaching.

8.6.1.4 Dumps and Landfills

A landfill is a site for disposing of waste materials by burying them. It is the most common and oldest form of waste treatment. Some landfills are also utilized for waste management as temporary storage. Unless there is stabilization, these areas could experience severe shaking. High winds could lift some part of the landfill, and cause a wave or tremor that shakes the ground as wind mixes with the area below the cap. Wind could carry the odor from the landfill into other areas, thus making people who may be living nearby to be uncomfortable because of the smell. However, well-managed landfills meet certain predefined conditions: confining waste to the smallest area possible, compacting waste to minimize volume, and covering waste with soil layers or other material types, including fine particles and woodchips (Orlova et al. 2017). The first advantage of landfills is that they are the most cost-effective means for waste disposal, particularly in nations that have large open spaces. Whereas incineration and resource recovery need extensive infrastructural investments, landfills require fewer costs, thus enabling them to compete favorably. Additionally, landfill gas can be upgraded to natural gas through landfill gas utilization to generate revenue. Furthermore, landfills make use of specific locations that can be monitored. However, landfills are also sources of various environmental problems. They disrupt the infrastructure, such as damaging access roads. They are also a source of environmental pollution, including the contamination of aquifers, soil, or groundwater.

How can wheel washing systems be used in mitigating the pollution of local roads and infrastructure?

Sanitary landfills are the most common solution for waste disposal. They are designed to minimize or eliminate the risk of public health hazards brought about by waste disposal. Sanitary landfills are located in places where land features act as natural buffers between the landfill and the environment (Orlova et al. 2017). For example, a landfill area could consist of clay soil that is resistant to hazardous wastes, thereby preventing susceptibility to water pollution. Using sanitary landfills poses the least environmental and health risk. However, the costs associated with the establishment of such landfills are relatively higher compared to other methods of waste disposal. Notably, various precautions have to be considered prior to operating a sanitary landfill. First, appropriate site selection is crucial. Second, they should be situated far from rivers, flood plains as well as groundwater recharge zones. The third consideration is the topographic relief and depth of the groundwater table. In addition, the landfill's base needs to have impermeable clay or plastic material lining. Appropriate drainage systems also need to be installed around landfills to check water accumulation (Orlova et al. 2017). Finally, monitoring wells for potential groundwater pollution need to be installed. The main advantages of sanitary landfills are less odor, no open burning, less costs, and accommodation for large waste quantities. The major disadvantages are more noise pollution, more dust, groundwater pollution, space issues, emission of greenhouse gases, and slow decomposition.

On the other hand, bioreactor landfills are the latest innovations obtained from recent technological research. They make use of superior microbiological processes for hastening waste decomposition (Reddy et al. 2017). Liquid is continuously added through the re-circulation of the landfill leachate to maintain moisture for microbial digestion. In the case of inadequate leachate, liquid waste such as sewage sludge is often deployed. Regarding controlled dumps, they have various similarities to sanitary landfills and have many characteristics of a sanitary landfill but may lack one or two features.

8.6.2 *Biological Waste Treatment*

Composting is a standard method of waste disposal and treatment. It is an aerobic decomposition method that uses microorganisms and invertebrates to consume and breakdown the waste product. The most common composting techniques are vermin composting, in-vessel composting, and windrow composting. The concept behind windrow composting is that compost is produced through piling of biodegradable waste or organic matter, including crop residues and animal manure in long rows or windrows. Similarly, anaerobic digestion utilizes biological processes for decomposing organic materials (Antonie 2017). However, it uses an environment that is devoid of bacteria and oxygen to decompose waste.

Expand on anaerobic digestion as a waste treatment method. Is it effective for waste treatment?

8.7 Benefits of Waste Management

Waste should not be disposed of without considering the future. Consistent and rational waste management practices are crucial for reaping various benefits. These benefits are classified as social, economic, environmental, and inter-generational equity.

8.8 Challenges of Waste Management in African Countries

Waste management in developing economies experience outdated collection services that are inadequately handled and dumpsites that are insufficiently controlled. These problems are exacerbated by governance issues. The management of wastes in these areas is a consistent challenge because of rapid urbanization, weak institutions, and under-resourcing. Along with various factors, these challenges influence waste treatment.

Poor urban planning has been one of the major challenges facing waste management in African cities such as Kampala, Nairobi, Kisumu, and Tripoli. The island of Zanzibar is one of the major examples of obstacles linked to rapid urbanization in Africa. It does not have an official strategy regarding biomedical waste management. It is estimated that about 50% of the waste generated in Zanzibar Town is dumped in the neighborhood or burnt locally (Guerrero et al. 2013). Some of these challenges are also brought about by insufficient funds that would have enabled the development of an integrated waste management system. Consequently, there is a need for African countries to re-strategize and develop means for building safe cities (see Fig. 8.3).

Case Study 8.1: How Africa can build inclusive, safe, and sustainable cities

Fig. 8.3 A busy street in Accra, Ghana, West Africa

World leaders met in New York in 2015 while committing to new sustainable development goals. It was the first time a specific urban goal was formulated to be attained by 2030 (Klopp and Pailer 2015). The objective was to "make

(continued)

cities and human settlements inclusive, safe, resilient, and sustainable." This target reflects the increasing acknowledgment that human well-being and development relies on the management of urbanization. Cities are accelerators of development but can also be sources of environmental health hazards. This is crucial for Africa, where in spite of high urbanization rates, development focus has mainly been in rural areas.

Consider Ghana; the urban population has increased from four million since 1984 to no less than 14 million. Today, more than 51% of Ghanaians live in cities (Klopp and Pailer 2015). Whereas the rates of urbanization vary across Africa, Ghana reflects the overall global trend towards an urban future and demonstrates how African cities can be highly productive. A World Bank report highlights a clear relationship between productivity, urbanization as well as poverty reduction. During the same period, annual Gross Domestic Product (GDP) growth has been growing at 5.7%. The number of service and industrial jobs has also grown by 21%, while Accra has had a 20% lower poverty rate. The Nairobi Metropolitan region also generates about 50% of the country's GDP (Klopp and Pailer 2015). Despite having many unemployed youth and substantial poverty, many rural counties in the country are usually the poorest.

As cities in Africa grow, the challenge will remain to offer adequate services as well as equitable access to opportunities. Today, there are significant gaps that exist between the required and current infrastructure and services. A result of this gap is the affordable housing crisis that yields slums and suburbs. Similar to numerous other states in Africa, the urban housing stock in Ghana is growing. However, similar to numerous other cities globally, the larger proportion of this housing is for upper and middle classes (Klopp and Pailer 2015). In addition, the housing is not growing at an expected pace. If managed effectively, the high housing demand can be a significant investment and growth opportunity. However, because of existing housing inequalities, the question remains whether the sector will develop in an inclusive, safe, resilient, and sustainable way.

The major challenge in Africa is politics. Poor planning has impeded unlocking the urbanization promise. There is no need to reinvent the wheel on matters about planning, and while various master plans have been made, none has yielded any meaningful results in cities such as Nairobi. There is a need, however, to involve communities in this planning. For projects to achieve the desired results, more affordable housing or better transit, there is a need to reshape incentives.

What are other solutions to the problems afflicting African cities?

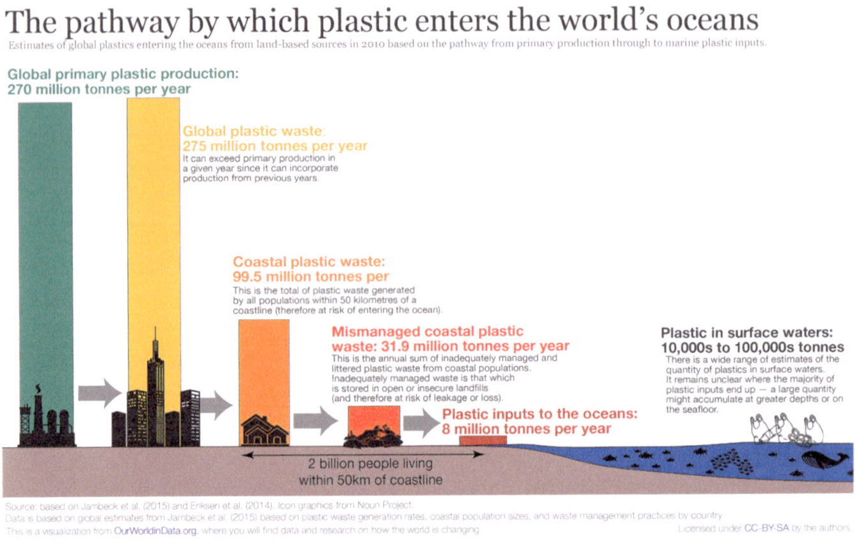

Fig. 8.4 Plastic pollution, *Source*: Jambeck et al. (2015) and Eriksen et al. (2014)

Plastic pollution remains one of the major concerns that should be addressed. Figure 8.4 highlights the pathway of plastics entering world's ocean, while Fig. 8.5 depicts the plastic waste generation by industries which ultimately poses detrimental effect to the marine ecosystem. United Nations (UN) member states should focus on four primary areas to beat plastic pollution. First, it should reduce single-use plastics. Fifty percent of all consumer plastics are designed only to be used once. Therefore, the elimination of these plastics is a crucial first step towards reducing plastic pollution. The second solution is the improvement of waste management. Thirty percent of all plastics elude from collection systems. After they get into the environment, plastics decompose into smaller pieces (microplastics) that find themselves into the food chain. There is also a need to phase out microplastics in all countries. An example is a Kenyan initiative that began last year to phase out all forms of plastics. Despite the initiative not being 100% successful, it has made significant progress towards a plastic-free environment (Chow et al. 2017). Finally, there is a need for further research to make sustainable plastic options economically viable and available. Tackling plastic waste pollution needs more holistic approaches that can focus on upstream activities. Big businesses can play a crucial role in eliminating unnecessary packaging from supply chains. Corporations need to utilize sustainable procurement strategies for encouraging the reduction of unnecessary single-use plastics.

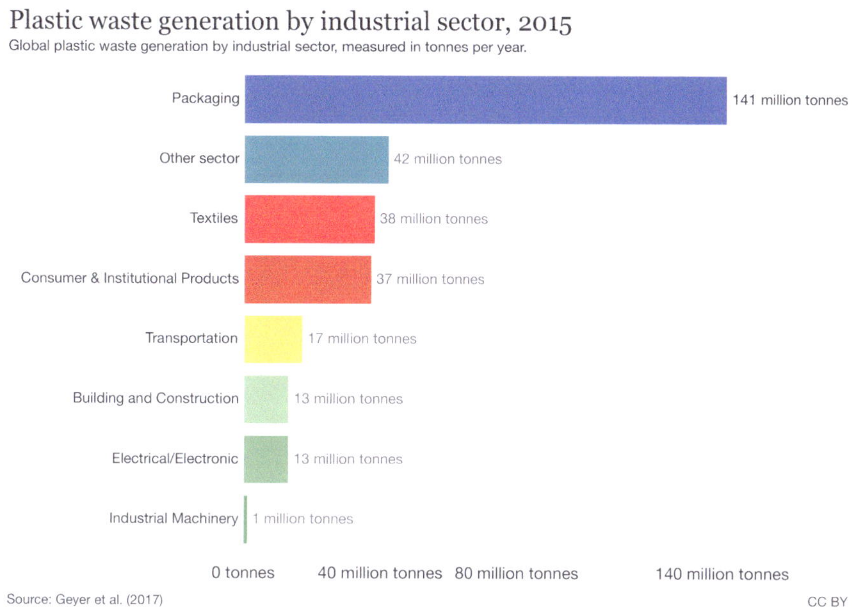

Fig. 8.5 Plastic waste generation by industrial sector, *Source*: Ritchie and Roser (2019) and Geyer et al. (2017)

What are the actionable solutions to beating plastic pollution?

8.9 Practical Application: Recycling at School

Schools usually produce large quantities of waste. It is estimated that 25% of school waste is recyclable paper, while 50% is food waste and non-recyclable paper (Lara et al. 2017). This creates an opportunity for students and teachers to collaborate on minimizing the waste produced. This application explores how to create a recycling program in a school environment. The procedure is outlined in the sub sections below:

8.9.1 Choose What to Recycle

First, students, faculty, and administrators should collaborate on recycling teams. This will enable them to decide how to approach the recycling program. Students

Fig. 8.6 Floppy disk

should analyze the amounts and types of waste that the school is producing. Sorting and recording the generated waste is a crucial step for determining the types of waste in the school. After establishing what is thrown away, the school can set recycling goals. Identify recyclable materials (see Fig. 8.6) and ascertain how to collect them. You can establish various recycling programs, but the most common recyclable materials are identified below.

8.9.1.1 Paper

Some recyclable paper products are copy paper, notebook paper, computer paper, file folders, paper grocery bags, school letterhead, and colored paper. By separating paper from the trash, the school can significantly reduce the quantity of waste sent to landfills. One of the ways that schools reduce their paper use is to keep a scrap paper tray within each classroom and ensuring that both sides of the paper are used.

8.9.1.2 Electronics

The school can establish an e-waste recycling program to recycle printer ink cartridges, monitors, and computers. The school's administration staff can be involved through the donation of old computers, monitors, and printers to corporations that accept them.

8.9.1.3 Plastic

Schools should develop new policies for reducing plastic trash created within the cafeteria. The other strategies that could be applied are recycling high-grade paper, switching to recycled paper, refilling plastic bottles, composting food waste, switching to rechargeable batteries, recycling batteries, and setting up an eco-team. An eco-team could be established by six to ten members who meet regularly to share ideas and collaborate to gain insights on how to change their lifestyles that could improve the environment.

8.9.2 Choose How to Recycle

After selecting what to recycle, the school should develop a program to document recycling objectives.

For Paper, schools should develop a program for recycling in all classrooms. All rooms used for student activities need to have trash bins for regular waste. To be efficient, the bins need to have various colors in order to differentiate them (Lara et al. 2017). Both containers need to be clearly labeled as paper or trash only. There is also a need to place paper-recycling bins in areas across the entire school.

For food waste, paper, and plastics, recycling containers within the cafeteria can be quite useful for recycling numerous kinds of paper, plastic, and materials. Bins should be positioned strategically for recycling plastic bottles, waste paper, juice boxes, and food containers.

For Metals, recycling bins need to be placed close to vending machines that sell their products in aluminum cans. These bins should also be close to trash cans.

The major ways of recycling and reducing waste at school are:

- Using electronic memos.
- Reusing of classroom supplies.
- Holding recycling contests between classrooms.
- Using copy paper for class tasks and recycling it.
- Using old magazines for any art projects.

Finally, using these tips, schools can change the way they think and approach waste. This could be beneficial for ensuring that only small quantities of waste find their way into collection systems.

8.10 Individual Matter: Tom Szaky

Tom Szaky (see Fig. 8.7) was born on January 14, 1982, and is the CEO of TerraCycle; an organization that recycles waste to make consumer products. His parents are medical doctors. The Szaky family left their home when Tom was

Fig. 8.7 Tom Szaky, *Source:* Crane (2014) via TerraCycle

four years old due to the Chernobyl disaster and moved to Toronto, Canada, in 1987 (Bhasin 2011). He attended Upper Canada College before joining Princeton University. During his sophomore year, he dropped out of school to concentrate on TerraCycle. Early in his career, he established dot-com companies and was named in 2006 as the best CEO under 30 years because of his work in TerraCycle.

TerraCycle plays a crucial role in reducing the amount of waste generated through recycling of the non-recyclable by collecting and recycling all forms of waste. It partners with individual collectors and major consumer product organizations, manufacturers, municipalities, as well as small businesses across 20 nations (Bhasin 2011). With help from people, the organization can divert large volumes of waste from incinerators and landfills every month.

Szaky's vision is that nothing is garbage. As a startup, TerraCycle has enabled Szaky to build a $24 million business with the belief that all things or waste are recyclable (Bhasin 2011). He has successfully convinced major retailers and brands, including Office Depot Inc., Colgate-Palmolive Co, and Procter & Gamble, that there is a benefit in ensuring that garbage is not in landfills. TerraCycle started in Tom's dorm room as he had a passion for solving the problem of waste. Over the years, Szaky has made a pivotal contribution by shifting TerraCycle's model onto garbage. Arguably, his initiative is one of the green initiatives that should be exploited to solve the waste menace afflicting the world.

8.11 Future Trends

8.11.1 Africa

Smart African cities will achieve zero waste, enhance services, and minimize operational costs due to smarter and more efficient systems of collecting waste. The global technology revolution will transform the way people think regarding waste management. Also, higher population growth rates and higher consumer demands will necessitate a new connected approach (Pilson 2017). The Internet of Things could help African countries to offer timeous, integrated, and sustainable solutions. More importantly, Africa will move towards a circular economy for waste management with the aid of technology (including a cradle-to-cradle approach).

8.11.2 Asia

With markets shrinking and stricter management standards resulting in more recycling costs, waste management firms that recycle will have to rethink their pricing models. Because of ongoing labor shortages, workflow enhancements in specific can potentially improve profitability with technologies such as driverless trucks set to change the industry. Smart firms could soon help organizations maintain the increasingly strict contamination requirements, minimize needless pickups, reduce costs, and make recycling easier. China will continue repositioning itself as a green power and fighting pollution through ecological civilization. However, it will continue facing new environmental dilemmas regarding what it can do with the large quantity municipal waste generated by the trend of consumerism. The Middle East will increasingly shift towards privatization of waste management. Privatization will be the key driver towards a circular economy as more countries in the region shift towards zero waste cities.

8.11.3 Australia

In spite of significant breakthroughs in waste management, the region grapples with numerous challenges that need special attention like uncontrolled open dumpsters and low rates of recovery of waste fractions. The adoption of sophisticated waste management technologies, including anaerobic digestion or incineration, will continue lagging. Adopting better technologies and increased recycling could improve waste management within the region and curb environmental effects.

8.11.4 Europe

Technology will continue being the leading solution to the issue of unmanaged and spiraling waste. Locally appropriate solutions will be used for addressing open dumping as well as other rudimentary waste management methods. A circular economy will become more common with green ways of waste management being prioritized.

8.11.5 North America

The approach to waste management will continue evolving from open dump to integrated waste management. Stopgap measures will also continually be dropped in favor of proactive mechanisms. Holistic approaches will be promoted with waste considered to be a resource instead of a liability.

8.11.6 South America

Integrated waste management will continue being used as the main way of managing waste in the cities. There will also be a proliferation of informal sector integration within the region's recycling systems. The private sector, particularly, waste collectors and informal recyclers could continue being among the major collectors of waste. Also, high technology solutions may be used for automating waste management. Labor-intensive solutions could be further deployed to increase the rates of recycling.

8.12 Summary

- Each stage in a product lifecycle is crucial for optimizing the use of limited planet resources.
- Resource efficiency helps minimize the environmental harm that can be associated with the production of goods.
- The polluter-pays principles hold the polluter responsible for any harm caused to the environment.

- Waste hierarchy includes prevention, minimization, reuse, recycling, energy recovery, and disposal (in order of importance/priority).
- Major steps in waste management are reduction, recycling, treatment, and disposal.
- Waste management and collection differ between countries, but many of the African states do not have formal systems for collecting waste.
- Waste treatment is classified into three major types: physical, biological, and chemical.
- Waste disposal methods include thermal (incineration, gasification and pyrolysis, open burning, landfills, and dumps) and biological methods.
- Landfills are the most common way of waste management. The major advantages are cost-effectiveness, ease of upgrading, and consistent monitoring.
- The disadvantages of landfills are that they cause environmental issues, including infrastructural disruption and soil contamination.
- Incineration entails burning organic waste, which minimizes waste volume.
- The 3Rs (recycling, reducing, and reusing) are crucial for waste management.
- The sustainability principle encourages companies to improve their environmental efficiencies.
- The major benefits of waste management are socioeconomic, environmental, and inter-generational equity.
- In Africa, the major challenges are outdated collection systems, poor management, ineffective governance, weak institutions, and a lack of support.
- A holistic approach to waste management is necessary for building safe, inclusive, and sustainable cities.
- Plastic pollution needs to be addressed to reduce the menace they cause on the environment.
- Various methods can be applied to recycle waste in schools.
- Tom Szaky's TerraCycle provides insight into how individual efforts can contribute towards a more sustainable planet.
- Smart African cities will attain zero waste and reduce operational costs.
- In Europe, technology will revolutionize waste management, and the techniques used will be locally appropriate.
- In Asia, smart companies could help reduce costs and make recycling easier.
- In North and South America, integrated waste management will become common, while Australia will be characterized by the adoption of better technologies and recycling techniques.

8.13 Discussion Questions

1. How effective is carbon trading and carbon tax?
2. Explain the significance of resource recovery and how it applies the life cycle analysis for managing waste.

3. Briefly elaborate on the socioeconomic, environmental, and inter-generational equity benefits of waste management
4. How can developed and developing countries use carbon trading as a way of enhancing sustainability?
5. What are the major waste management challenges?
6. What do you think is the future of waste management?
7. Distinguish between incineration and landfills. Of the two, which is the better alternative for waste management? Justify your response with real-world examples.

References

Antonie, R. L. (2017). *Fixed biological surfaces-wastewater treatment: The rotating biological contactor*. Boca Raton: CRC Press.

Bhasin, K. (2011, August 29). The incredible story of how TerraCycle CEO Tom Szaky became a garbage mogul. *Business Insider*. Retrieved November 4, 2018, from https://www.businessinsider.com/exclusive-tom-szaky-terracycle-intervie-w-2011-8?IR=T/.

Chow, C. F., So, W. M. W., Cheung, T. Y., & Yeung, S. K. D. (2017). Plastic waste problem and education for plastic waste management. In *Emerging practices in scholarship of learning and teaching in a digital era* (pp. 125–140). Singapore: Springer.

Crane, C. (2014). *Tom Szaky*. Retrieved January 21, 2019, from https://en.wikipedia.org/wiki/Tom_Szaky#/media/File:TerraCycle_CEO_Tom_Szaky.jpg/.

Das, B., Bhave, P. V., Sapkota, A., & Byanju, R. M. (2018). Estimating emissions from open burning of municipal solid waste in municipalities of Nepal. *Waste Management, 79*, 481–490.

De Sadeleer, N. (2014). Polluter pays principle. In *Essential concepts of global environmental governance* (pp. 155–156). London: Routledge.

Drstuey, S. (2008). *Waste hierarchy*. Retrieved September 06, 2018, from https://com-mons.wikimedia.org/w/index.php?curid=3518269/.

Eriksen, M., Lebreton, C. M., Carson, H. S., Thiel, M., Moore, C. J.,Borerro, J. C., et al. (2014). Plastic pollution in the world's oceans: more than 5 trillion plastic pieces weighing over 250,000 tons afloat at sea. *Plos One, 9*, e111913. Retrieved October 09, 2019, from https://doi.org/10.1371/journal.pone.0111913/.

Geyer, R., Jambeck, J. R., & Law, K. L. (2017). Production, use and fate of all plastics ever made. *Science Advances, 3*(7), e1700782. Retrieved November 10, 2019, from http://advances.sciencemag.org/content/3/7/e1700782/.

Guerrero, L. A., Maas, G., & Hogland, W. (2013). Solid waste management challenges for cities in developing countries. *Waste Management, 33*(1), 220–232.

Jambeck, J. R., Geyer, R., Wilcox, C., Siegler, T. R., Perryman, M., Andrady, A., et al. (2015). Plastic waste inputs from land into the ocean. *Science, 347*(6223), 768–771. Retrieved November 04, 2019, from http://science.scienc-emag.org/content/347/6223/768/.

Klopp, J., & Pailer, J. (2015, October 9). How Africa can build inclusive, safe and sustainable cities. *The Conversation*. Retrieved November 4, 2018, from https://theconversation.com/how-africa-can-build-inclusive-safe-and-sustainable-cities-48445/.

Lara, E. R., De la Rosa, J. R., Castillo, A. I. R., de Jesús Cerino-Córdova, F., Chuken, U. J. L., Delgadillo, et al. (2017). A comprehensive hazardous waste management program in a Chemistry School at a Mexican University. *Journal of Cleaner Production, 142*, 1486–1491.

McDougall, F. R., White, P. R., Franke, M., & Hindle, P. (2008). *Integrated solid waste management: a life cycle inventory*. Hoboken: Wiley.

Menikpura, S. N. M., Sang-Arun, J., & Bengtsson, M. (2013). Integrated solid waste management: An approach for enhancing climate co-benefits through resource recovery. *Journal of Cleaner Production, 58*, 34–42.

Orlova, T., Melnichuk, A., Klimenko, K., Vitvitskaya, V., Popovych, V., Dunaieva, I., et al. (2017, October). Reclamation of landfills and dumps of municipal solid waste in an energy efficient waste management system: Methodology and practice. In *IOP Conference Series: Earth and Environmental Science* (Vol. 90(1), p. 012110). Bristol: IOP Publishing.

Pichtel, J. (2005). *Waste management practices: Municipal, hazardous, and industrial*. Boca Raton: CRC Press.

Pilson, G. (2017). There's a strong chance a third of all people on earth will be African by 2100. *Quartz Africa*. Retrieved November 4, 2019, from https://qz.c-om/africa/1099546/population-growth-africans-will-be-a-third-of-all-people-o-n-earth-by-2100/.

Polprasert, C. (2017). *Organic waste recycling: Technology and management*. London: IWA Publishing.

Reddy, K. R., Kumar, G., & Giri, R. K. (2017). Influence of dynamic coupled hydro-bio-mechanical processes on response of municipal solid waste and liner system in bioreactor landfills. *Waste Management, 63*, 143–160.

Ritchie, H., & Roser, M. (2019). *Plastic pollution*. Retrieved November 09, 2019, from https://ourworldindata.org/plastic-pollution/.

Silva, A., Rosano, M., Stocker, L., & Gorissen, L. (2017). From waste to sustainable materials management: Three case studies of the transition journey. *Waste Management, 61*, 547–557.

Tangwanichagapong, S., Nitivattananon, V., Mohanty, B., & Visvanathan, C. (2017). Greening of a campus through waste management initiatives: Experience from a higher education institution in Thailand. *International Journal of Sustainability in Higher Education, 18*(2), 203–217.

Thomsen, T. P., Sárossy, Z., Ahrenfeldt, J., Henriksen, U. B., Frandsen, F. J., & Müller-Stöver, D. S. (2017). Changes imposed by pyrolysis, thermal gasification and incineration on composition and phosphorus fertilizer quality of municipal sewage sludge. *Journal of Environmental Management, 198*, 308–318.

Chapter 9
Environmental Ethics

Abstract Environmental conservation and sustainability require that society takes more concerted efforts apart from the contemporary ones. Environmental ethics is perhaps more important because it can ensure that individuals, corporations, and communities engage in morally sound practices. The main objective of this chapter is to explore the concept of environmental ethics. To achieve this, the chapter has explored the idea of environmental risk management, environmental policies in Nigeria, health and safety measures, and environmental impact assessment. More importantly, this chapter examines the ethics of waste and seeks to understand the seven principles of environmental management that have a correlation with the ethics of waste. Using the case studies of environmental policies in Nigeria and Apple, practical application of environmental ethics in schools, and individual matter of Irving Stowe, this chapter plays a critical contribution towards understanding the approach towards ethics in the environment. The seven principles of environmental management are particularly important as it sets a pre-cursor for understanding the ethics of waste. Future trends in the continents are also briefly described at the end.

Keywords Ethics of waste · Environmental impact assessment · Environmental risk assessment · Policies · Irving Stowe

9.1 Overview

Environmental ethics is an area of philosophy that examines the relationship between humans and the environment from an ethical and moral perspective. It tries to define our responsibilities and to ascertain appropriate actions. Environmental ethics has various perspectives, but the major focus is on the way humans value nature. Instrumental and intrinsic values are the most important values in environmental ethics. Ideally, instrumental value describes the benefit or value of something in the way it could be used in relation to human needs and wants. Intrinsic value refers to the value that something has for its own sake, irrespective of its usefulness to humans.

A. A. Akinsemolu, *The Principles of Green and Sustainability Science*,
https://doi.org/10.1007/978-981-15-2493-6_9

Ethical issues about the environment have been discussed for centuries, but environmental ethics emerged as a philosophical discipline during the 1970s. It resulted from increased awareness during the 1960s about the adverse effect of population, technology, and industrial growth on the environment. The development of this awareness was helped by the publication of two important books: Silent Spring, published by Rachel Carson in 1962, and The Population Bomb that was published by Paul Ehrlich in 1968. Silent Spring raised awareness about the widespread use of chemical pesticides and asserted that it was seriously threatening public health, as well as leading to wildlife destruction. On the other hand, the Population Bomb book warned about the devastating impacts incurred on the resources of the planet from our increasing population.

Natural resource depletion and pollution have not been the only environmental concerns in recent years. Other concerns include declining biodiversity, loss of wilderness, ecosystem degradation, as well as climate change. These are "Green issues" that have emerged in public policy and consciousness in subsequent years. Environmental ethics seeks to outline the moral obligations associated with these concerns.

Environmental ethics helps define the ethical and moral obligations of citizens towards the environment. However, human values are essential when examining environmental ethics. Conceptually, human values are things that are beneficial to people that they then utilize for evaluation of events or actions. In other words, humans assign value to some things and then use this assigned value for decision-making regarding the rightness or wrongness of something. Human values are distinct to each person since not everyone places the same significance on each life element. For instance, someone who lives in abject poverty within an undeveloped nation could find it morally acceptable to practice deforestation as a way of making room for a farm to plant food for their family. Contrarily, someone in a developed nation could find the action morally unacceptable, since deforestation leads to an increase in carbon dioxide emissions into the atmosphere that can negatively impact the environment. There are various important theories, concepts, and definitions that are vital in the field of environmental ethics.

9.2 Ecocentrism

The ecocentrism perspective focuses on interests of natural features and species of the Earth's ecosystems. It mainly focuses on the intrinsic value of the entities while appreciating their instrumental value as part of the natural process. According to ecocentrists, the ecosystem is the most fundamental source of value. This is contrary to the anthropocentric views that prioritize human needs and wants as being more valuable and beneficial compared to other natural entities. The earliest ecocentrist is Aldo Leopold because of his writings during the 1940s (Marshall 1993). For instance, environmental ethicists still cite his book, A Sand County

Almanac, today. His work is the foundation for land ethics. Leopold views "land" as the whole ecological community of natural settings or a place. However, the boundaries between species, natural communities, and ecosystems are still contested topics.

9.3 Approaches to Environmental Ethics

The three primary forms of ethical theory are deontological ethics, consequentialism, and virtue ethics (Sandler 2009). They differ in terms of their central evaluative categories. In virtue ethics, the main evaluative concepts are virtues instead of the consequences or obligation or duty. Environmental ethics is considered virtue-oriented because it evaluates our practices, relationships, and policies regarding the environment in terms of vices and virtues. It emphasizes the cultivation of environmental virtue, that is, the character traits that promote both human and non-human prosperity, apart from performing specific actions. One of the core advantages of a virtue-oriented approach is that it has a richer, more diverse, and nuanced language compared to consequentialism and duty.

Kantian ethics and utilitarianism focus on what makes an action right or wrong, while virtue ethics focuses on what makes one a good person and the meaning of living a good human life. Aristotle believed that each living thing has a specific telos or purpose that it suits. He argued that the human purpose emanates from our unique abilities to reason. Consequently, a great human life needs the development and practice of theoretical reason (Sandler 2009). Aristotle further believes that moral virtue starts with habit as we imitate moral models. Becoming fully virtuous, however, requires one to have practical wisdom that enables him or her to assess the requirements of each session and fully integrate all virtues in one's action and thought. Virtue ethics has various strengths, including its ability to focus on overall moral development instead of individual acts, thus drawing attention away from tragic moral dilemmas. Notwithstanding, virtue ethics has been criticized on various grounds. For instance, the concept of a single telos or good for humans has been questioned. The second criticism is about the universality of Aristotelian virtues. Some believe that Aristotle's ethics are largely anthropocentric since, in his hierarchical worldview, Aristotle ranks humans above other living organisms (Sandler 2009). Contemporary moral philosophers have extended approaches of virtue ethics to the environment. For instance, some believe that virtues like compassion and humility could be expressed in relation to nature and animals, while others suggest that figures like Aldo Leopold and Rachel Carson could be models for an environmentally virtuous life.

Consequentialists believe that only the consequences of action determine whether the action is moral or not. Utilitarianism is a branch or form of consequentialism that argues that a morally right action should maximize utility. According to Jeremy Bentham, a hedonistic, utilitarian, utility is the presence of pleasure and absence of pain (Attfield 2014). On the other hand, preference utilitarianism vies utility as the satisfaction of preferences. Utilitarianism is associated with various strengths, including accounting for consequences being important in ethics, being theoretically simple with a single prescription of maximizing utility, bringing all values together on one scale for them to be aggregated and weighed against each other, and easily leading to a decision-making procedure of cost–benefit analysis (CBA). This entails evaluating whether the benefits outweigh the costs to determine whether the action is morally right or wrong. However, utilitarianism faces criticism on many fronts, such as not directly factoring in distribution, suffering epistemic issues because of the difficulty in predicting all the impacts of a specific action, and issues with aggregating diverse goods on one scale (Attfield 2014).

Regarding Kantian ethics and the environment, Kant argued that humans have both a rational and animal nature, the animal nature acts as the source of our desires, but humans remain special since they possess reason and can think rationally. Kant believes in one major moral principle called Categorical Imperative. According to the Categorical Imperative, each individual should act on universalized principles or the principles that can be rationally and universally accepted. Consequently, it forbids actions such as lying, cheating, or making false promises (universal law formula) (Bowie 2017). The Categorical Imperative Humanity Formula, on the other hand, states that each individual should respect others and act in ways that appreciate other people as autonomous and rational beings. This means that humans should treat others as ends-in-themselves rather than means to our ends. The major strengths of Kantian ethics are the ability to capture the moral perception that some things should not be done irrespective of their impact on overall utility, providing a foundation for universal human rights, and not treating all preferences as equally deserving moral weight. The main critiques of Kant's ethics are that it is too strict in the categorical prohibition of some actions irrespective of the situation, it provides limited guidance in dilemmas and difficulty in interpreting formulations of Categorical Imperative to ascertain the best course of action.

Considering the strengths and limitations of hedonistic utilitarianism and preference utilitarianism, which one would you determine as better or more superior. Give two reasons.

In your opinion, what is the major disadvantage of environmental cost–benefit analysis? Can you come up with a better option?

Despite Kant's ethics being deeply anthropocentric, can you come up with a scenario where Kant's ethical theory could be beneficial in environmental protection?

Think of an environmentally virtuous role model that you look up to. What are particular values embodied by this person?

9.4 Environmental Risk Assessment

Environmental risk assessment (ERA) is an estimate of the probability of human activities leading to adverse environmental effects. It is an environmental aspect that is significant due to its socioeconomic and ecological significance to an ecosystem, potential repercussions of its loss, and the socioeconomic relevance. In this concept, the term "value" is not restricted to only the economic sense. Some examples are domestic water supplies and the old-growth dependent species. The following sections will be dedicated to the definition of terms under environmental risk assessment before further information is provided on ERA.

9.4.1 Definition of Terms in Environmental Risk Assessment

9.4.1.1 Indicator

This is a parameter that can be measured, observed, and offers information on environmental trends or patterns. Quantitative and qualitative information are some of the indicators and can include the number of endangered species, distribution changes of single species, and percentage of protected areas.

9.4.1.2 Natural Conditions

These are defined as the environmental conditions that are in the range of historical variability before European settlement. Often, natural conditions need to be described of specified conditions that have been in existent for an extended period (Kosma et al. 2014).

Identify four natural conditions that you know.

9.4.1.3 Pressure

This is a factor that influences ecosystem processes that are linked to human actions. Effects from past pressures have led to various cumulative impacts, including dam construction and decisions on land use.

9.4.1.4 Risk

This is the probability that an action can lead to an adverse environmental event or outcome.

9.4.1.5 Risk Class

This refers to the level of risk, including very high or very low with limits that can be defined for each respective class.

9.4.1.6 Risk Management

This is a process that seeks to ascertain the environmental risks associated with a particular event and ascertain the best ways for managing those risks in a way that is suited for protecting people's health and the environment. In addition, it factors in the ecological, behavioral, economic, and legal factors of each option. Some examples of risk management are determining the allowable amount of discharge that a company can release, ascertaining the extent to which a hazardous waste site can be cleaned, the establishment of national ambient air quality standards, and ascertaining the allowable contamination levels in drinking water (Kosma et al. 2014).

How does risk management promote environmental conservation?

9.4.2 Description of Environmental Risk Assessment (ERA)

As earlier defined, environmental risk assessment (ERA) is a process used for evaluating the likelihood of human activities causing environmental damage or harm. It is a formal procedure that is necessary for the identification and estimation of the risk of environmental damage. The procedure is usually triggered when an important decision can affect the environment. Despite having numerous stages, it can be categorized into three major stages: preparation, carrying out the assessment and interpretation, and application of assessment's results.

Preparation entails the collection and review of relevant background information as well as establishing the assessment's focus. Environmental risk assessment (ERA) can be applied in various situations, including ones with minimal data to ones with complex systems modeling, and detailed inventories. There are six major stages involved in environmental risk assessment, see Fig. 9.1. The first step entails establishing a context for the assessment, while the second step involves the identification of hazards or anything that could cause harm. The third step entails identifying the persons, objects, or places that could be affected and the way that

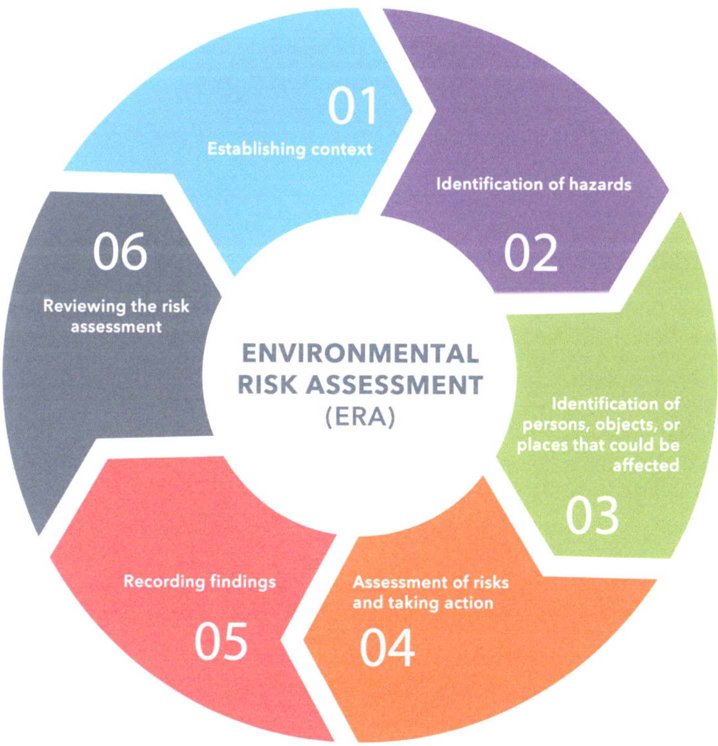

Fig. 9.1 Environmental risk assessment stages

they could be affected. The fourth step involves an assessment of risks and taking action. The fifth step involves recording the findings, while the sixth step entails reviewing the risk assessment (Hernando et al. 2006).

Environmental risk assessment (ERA) is important because it highlights the environmental repercussions of decisions. By doing so, it shifts the focus from only the merits of specific action to also identify the probable outcomes and their desirability. The awareness of risk encourages accountability and collective responsibility for risk management, thus increasing the likelihood of decisions that minimize risk. It is also flexible and can be applied to complex analyses or situations that require rapid responses. Notwithstanding, it is also associated with various limitations.

Despite having the capacity to clarify risk to the environment, it cannot set an acceptable risk threshold. The first limitation is that risk tolerance is a relative phenomenon. Institutions and people have varying viewpoints, risk acceptance, and tolerance. The second limitation is the challenge associated with isolating risks that are associated with a decision. This is difficult because of the wide range of natural variability in ecosystems. The third limitation is the wide tolerance to stress and various recovery rates (Ahuja 2013).

To what extent can ERA be beneficial for enabling managers to communicate the forecasted risks? At what point is ERA useful for businesses? What other limitations do you know?

9.5 Environmental Risk Identification

In this section, land management will be used as a reference for explaining the concept of environmental risk identification. It is crucial to identify the risks of environmental damage that could arise from various actions and activities in order to determine the ones that are significant enough to come up with solutions for avoiding, minimizing, or mitigating them. For example, for fertilizer use, there could be various approaches for the assessment of environmental risks. Many land managers may use a simple way of assessing risks independently, but some could require more formal risk assessment in case their industry organization needs this (Cooper et al. 2005). In some situations, one needs a higher standard for proving that they have scientifically assessed the risks. For example, regulatory authorities may require a more detailed assessment in case activities geared towards protecting nutrients in the soil do not align with the "allowed activity" category. For every land management unit, one is expected to assess the inherent environmental risks related to the major nutrient management activities.

Conceptually, inherent risks mean the risks that could arise due to the activity and location. For example, when dealing with land management, the use of proper management does not eliminate all inherent risks. Some typical risks that arise from nutrient management activities are contamination of the waters, undesired changes within the soil nutrient status, application of fertilizer to the non-target land as well as the accumulation of impurities within the soil profile (Cooper et al. 2005). Site factors and the number of applied nutrients govern inherent risk factors. This implies that risks increase with increased levels of nutrient inputs.

From the explanations provided above, briefly differentiate environmental risk assessment and environmental risk management. Do this using the objectives of each, processes, and stakeholders involved.

9.6 Environmental Risk Management

Environmental risk management entails the identification, evaluation as well as prioritization of risks followed by the application of resources to reduce and control the likelihood of unfortunate events. The principal objective of environmental risk management is to ensure that uncertainty does not affect the realization of business goals. Risks could emanate from different sources, including financial markets uncertainty, project failure threats, legal liabilities, accidents, natural disasters, and natural causes. Event can be classified as negative or positive with the adverse events being perceived as risks, while the positive events were categorized as opportunities. Various environmental risk management standards are used today, including the Project Management Institute, National Institute of Standards and Technology (NIST), and the International Organization for Standardization (ISO) (Pritchard 2014). Some strategies used for management of threats are avoidance, reduction of negative impact, transfer of risk to another party, or retention of all actual/potential repercussions of a specific threat.

Briefly explain these strategies of environmental risk management.

Five major risk management steps combine to deliver an effective risk management process. These are an identification of the risk, analysis of the risk, evaluation or ranking of the risk, treatment of the risk, and monitoring and reviewing the risk. Before clearly understanding risk and dealing with it, there is a need to understand the context. The context could be established through consideration of the organizational or strategic context. The strategic context refers to the environment in which the organization operates, while the organizational context refers to the core activities, objectives, and operations. Establishing the context entails identifying risks within a domain of interest, planning the other part of the process, mapping out the social scope, identifying stakeholder objectives, and the basis on which risks would be evaluated (Pritchard 2014). The identification of risks could begin with the source of the existing problems and the ones faced by competitors. It is also crucial to conduct a source analysis to determine whether the source of the risk is internal or external to the system. Some of the risk sources are project stakeholders or company employees. Also, there is a need to conduct problem analysis to determine existing threats. Environmental threats could exist with several entities, including the risk associated with losing money or human error. The method selected for identifying the risk can be determined by the culture, compliance, and industry practice. Some standard environmental risk identification methods include objective-based risk identification, scenario-based, taxonomy-based, and risk-charting identification.

Identify environmental problems that can be solved using risk-mapping in your community.

Identification of the risks seeks to identify the likelihood of bad things happening and the consequences of these occurrences. Risks may be legal, ethical, financial, or physical. Physical risks are the ones that involve personal injuries, weather, and environmental conditions. Financial risks are the ones that entail the organization's assets and include license fees, insurance costs, fraud lease payments, and loans. Ethical risks entail potential or actual harm that negatively affects the reputation of the organization, whereas legal risks are made up of responsibilities that are imposed on providers and consumers (Pritchard 2014).

Risk analysis and evaluation entails analysis of the probability and repercussions of every identified risk and making decisions on the risk factors that will have the most impact. Risk evaluation entails a comparison of risk levels found in the course of the analysis process with previously established risk criteria and making decisions on whether these risks could be accepted. In case the risk falls into acceptable or low categories, they could be accepted with less treatment. These risks need to be monitored and reviewed periodically in order to ensure that they remain acceptable. After the identification of risks, an assessment is done in accordance with various risk options. The first crucial aspect is designing a new business process that has sufficient built-in risk containment measures from the beginning, periodically re-assessing risks that can be accepted in the ongoing processes, and the transfer of risks to an external company or agency, like an insurance company, and avoiding the risks entirely.

Risk treatment entails identification of the range of alternatives for treating risks, evaluating alternatives, preparation of risk treatment plans, and implementation of the plans. It is mainly about consideration of treatment options and selection of the most appropriate method for achieving the desired outcome (Mihelcic et al. 2014). Treatment options should be proportionate to the significance of the risk, while the treatment cost needs to correspond with potential benefits. Treatment options include accepting the risk, avoiding the risk, reducing the risk, transferring the risk, retaining the risk, and financing the risk.

Risk avoidance entails avoiding an activity that could pose a risk. Despite avoidance appearing as the solution to all the risks, it also implies losing out on potential gains that could be gained if one would have undertaken the risk. However, for the sake of environmental sustainability, it is vital for businesses and entities need to adopt avoidance as a way of minimizing risks.

Avoidance has been heralded as a challenging risk assessment measure. Identify three reasons why avoidance is a challenge to businesses and society.

Contrarily, risk reduction entails the minimization of the severity of the likelihood of loss from happening. For instance, the design of sprinklers is to extinguish a fire and minimize the loss that would have been caused by a fire. The method could result in greater loss through damage of the water, thus making it highly unsuitable. For example, halon systems for suppressing fire could mitigate that risk, but the cost could be prohibitive. Halon systems are made of liquefied and compressed gas that prevents fire from spreading through a chemical means or way that disrupts combustion. Reducing risks implies striking a balance between the activity's benefit and the negative risk. Outsourcing could be a great example of risk reduction in case the outsourcer could demonstrate a higher ability of managing and minimizing risks. For instance, a city could outsource garbage collection to concentrate on city development.

Identify other examples of outsourcing in real life, especially in business. Is outsourcing effective in environmental risk assessment? Briefly explain.

Risk-sharing entails sharing the benefit of gain or burden of loss with another party. Risk transfer usually refers to how one can transfer risk to another party through outsourcing or insurance. Contrarily, risk retention entails the acceptance of loss from a risk after it happens. It is an effective strategy for the small risks, whereby the costs associated with insuring against the risk may be higher over time compared to the cumulative losses sustained. Furthermore, the risk management plan is crucial for the development of the appropriate countermeasures for mitigating every risk. It requires adequate security controls for risk management.

Financing the risk implies that the organization or entity funds the consequences associated with the risk (Giannakis and Papadopoulos 2016). One would have to carefully consider the appropriate procedures, policies, and strategies for treating the risk. This includes determining what is required for treating the risk, who is the responsible person or institution, the time frame for addressing the issue, and determining what entails managing the risk successfully. The final steps for environmental risk management are implementation, review, and evaluation of the management plan. Risk management is critical for ensuring that wastes are managed appropriately.

Identify four benefits of evaluation in environmental risk assessment. What challenges are experienced during the implementation phase?

Monitoring and reviewing are ongoing processes of risk management that are integral for every step. In addition, it is critical for ensuring that relevant information generated by the process is captured, used, and maintained.

9.7 Ethics of Waste

The morality of waste is an area that has received little attention in contemporary research. However, Gay Hawkins' book, The Ethics of Waste, provides a detailed account of the matter and puts forth arguments raised in consideration of ethics of waste. Waste is provocative, but people and industries still find it convenient to generate and dump in unethical ways, meaning that they are dumped in streams, rivers, oceans, and open grounds. Hawkins's (2006) argument is premised on the notion that if humans would become ethically responsible for the waste they generate, then we would have fewer incidences of pollution and the spread of diseases or other issues associated with the waste. Hawkins believes that waste is constructed as bad in moral terms. Despite this, our consumption economy tends to encourage redundancy, excess, and disposal. It appears that our life is influenced or governed by two major competing imperatives: a "don't care attitude" and "just produce waste." Environmental responses to the two imperatives have been dominated, as Hawkins (2006) states in Chap. 1, by "disenchantment stories" whereby nature is perceived as the victim of contamination and exploitation. At the same time, humans are perceived as being alienated from the natural world. Politically, this results in various dominant responses. First, the emphasis by deep ecology to eliminate a sense of separateness regarding nature and the person in which waste is made to become redundant by adopting an ethos of constant renewal. The second response is the incremental focus on basically altering waste practices by developing new values and behaviors at a personal level. Hawkins does not dismiss the two perspectives outright but implies that they easily slide into moralism and idealism modes, where nature becomes sacred and environmentalism a duty. They also espouse voluntarism whereby "environmentally conscious" behavior becomes a voluntary decision. Hawkins argues that such political and interpretative strategies could result in resentment, whereby people feel that an ecological duty is being imposed on them, and immobility, whereby people develop a sense of resigned despair that nature is destroyed beyond repair.

Contrastingly, Hawkins (2006) seeks to explore the various ways of analysis and understanding rubbish at the daily level of material and social practice. She explores varying ways of constructing and forging a waste politics that is non-reliant on transcendence ideas on guilt-tripping, moralism, or conscious raising. As she states, "My concern is with our most quotidian relations with waste, what they mean and how they might change" (Hawkins 2006, p. 3). This is the basis to her approach as she seeks to examine waste in a phenomenological rather than ecological way; despite doing so for politically confronting the environmental repercussions of overconsumption. Consequently, she deals with the pressing ecological realities of waste. Hawkins believes that the way we interact with waste often expresses specific ethics. The chapter on "Plastic Bags" suggests that capitalism, through commodity culture, has produced an ethos of disposability and waste habits.

There are two core ethical concerns regarding waste management. The first concern is about the attainment of "intergenerational equity" by selecting strategies and technologies that reduce the risk burdens as well as resources passed to the future generations by today's generation that produce waste. It is factual that every generation leaves burdens and benefits, and that the current decisions we make could foreclose alternatives or open better horizons for our future. This cannot be avoided, but our behaviors and decisions could become more acceptable so that appropriate equity levels are respected and we do not restrict future generations' freedom of choice. For example, in the case of producing nuclear energy and managing radioactive wastes, there is a need to carefully scrutinize the balance between benefits enjoyed by generations and the liabilities that could be imposed by current generations on future generations. The objective is to manage wastes in a manner that impacts the future at an acceptable level both in terms of safety and ethics. The second concern is the attainment of "intragenerational equity" and especially an ethical approach to the management of resource allocation as well as public participation in the process of decision-making (Okereke 2006). Intragenerational equity mainly deals with equity between people of the same generation.

Intragenerational equity is in tandem with the Brundtland's Commission that played a critical role in popularizing sustainable development by defining it as "the development that meets the needs of the present without compromising the ability of future generations to meet their own needs." Subsequently, the report produced by the Commission in 1987 about Our Common Future was endorsed by the United Nation and its definition adopted by countries globally (Okereke 2006). Equity is derived from social justice and represents a notion that there are basic needs that need to be fulfilled, and policies need to be directed with justice, fairness, and impartiality towards these ends. Waste is among the environmental inequities that exist today. Consideration of the intergenerational and intragenerational equity concerns results in various principles that should act as the foundation for ethical choices on waste management strategies. These are liabilities of waste management that need to be considered when doing new projects, people who generate waste need to take responsibility, and offer resources needed for managing these materials in a manner that does not impose excess burden on future generations, and waste needs to be managed at an acceptable protection level for the health of the environment and humans alike. From this, it is imperative that the development of waste management plans and policy needs to be discussed openly with representatives from all affected societal sections. The three principles described above are part of the seven major principles of environmental management that can be used for waste management. The principles are founded on the ethics of waste concepts discussed. In the previous chapter, environmental management was defined as the activities that promote beneficial links and reduce adverse impacts on the environment and resources (Okereke 2006). The principles of environmental management are beneficial for making decisions on environmental issues, especially waste.

9.7.1 *Polluter Pays Principle (PPP)*

Many economists have suggested that organizations that discharge contaminants in the environment need to be compelled to pay a price that resonates with the harm they cause the environment. If one produces waste, the principle assumes that the ethical thing to do is to incur the costs for the waste generated. The principle is used for the allocation of costs of pollution prevention as well as control measures to promote the rational utilization of scarce environmental resources and avoid damage to international investment and trade. The fundamental concern of this principle is that polluters need to bear abatement costs without subsidy (Khan 2015). Abatement refers to the act of reducing or ending pollution (Khan 2015). The Supreme Court of India weighed in on the Polluter Pays Principle by stating that absolute liability for environmental harm extends to compensate pollution victims and cost associated with the restoration of environmental degradation. Consequently, it includes direct and environmental costs to property or people. The damaged environment's remediation is a crucial part of the sustainable development process, hence making the polluter liable for paying the costs to individual sufferers and costs associated with reversing the damaged environment or ecology.

9.7.2 *The User Pays Principle (UPP)*

The User Pays Principle is considered part of the Polluter Pays Principle. It states that all users of resources should pay for the full long-term marginal cost of using a resource as well as related services, including any related treatment costs. Its application is mainly when resources are being consumed and used.

9.7.3 The Precautionary Principle (PP)

This principle holds the assumption that when an activity threatens the environment or human health, precautionary measures need to be taken even if no cause and effect relationships have been established scientifically. The core objective of this principle is to ensure that an activity or substance that poses a threat to life or the environment does not cause harm, even when there is lack of conclusive scientific proof (Hussen 2018). The Rio Declaration is a non-binding instrument that presents principles important for governing international environmental law, particularly sustainable development. Principle 15 of the Rio Declaration emphasizes this principle by stating that in case of threats of severe or irreversible damage, lack of full scientific proof or certainty should not be used as the reason for the postponement of cost-effective measures for preventing environmental damage. Consequently, the principle is beneficial for the protection of human health and the environment, including waste management.

9.7.4 Principle of Effectiveness and Efficiency

This principle states that resource efficiency must be accomplished using policy instruments that develop incentives for minimizing wasteful use. It is also applicable to various environmental governance issues by streaming procedures and processes for minimization of environmental costs. In this regard, the use of resources should be done in an efficient way to prevent and reduce the generation of waste.

9.7.5 The Principle of Responsibility

This principle is premised on the notion that all individuals, organizations, and states should maintain ecological processes. Access to ecological resources carries responsibilities for using them in ecologically sustainable, socially fair, and economically efficient ways (Hussen 2018). The consideration of ecological and social dimensions seeks to ensure that people are responsible for resource management as well as conservation, thus minimizing waste generation.

9.7.6 The Principle of Participation

It is the role of all individuals to take part in environmental decision-making activities. For example, there is a need for participation in issues about solid waste, including garbage and chemically hazardous waste. Activities that generate pollution must involve public participation to ensure that they are informed and participate in making recommendations on how waste can be abated.

9.7.7 The Principle of Proportionality

This principle is based on the idea of balance and maintaining economic development while promoting environmental protection. Every development has some adverse impacts on the environment, and it is crucial to adjust the people's interests to maintain the environment. Comparative hardships also have to be balanced while maintaining benefits to a larger population.

The above principles are critical for the ethical management of waste. Ideally, the ethics of waste provides insight that management should begin at the personal level (Hussen 2018). Every individual is responsible for the way they use resources, the way we dump our waste, and for considering the impact of our activities, not only to the environment but also to other people and future generations.

Differentiate between intragenerational and intergenerational equity. What do you understand by the term "ethics of waste?" What is our role in promoting equity, and what is the significance of equity in environmental resource management?

Case Study 9.1: Environmental policies in Nigeria

The foundation of environmental policies in Nigeria is found in the 1999 Constitution. Section 20 of the Constitution empowers the state to protect and enhance the environment, including land, water, air, wildlife, and forest. Section 2 of the Environmental Impact Assessment (EIA) Act of 1992 also states that the public or private sector should not engage in projects without considering their effects on the environment. Consequently, the Nigerian government has promulgated several laws to safeguard the environment. The Federal Environmental Protection Agency (FEPA) Act of 1988 is one of the policies that have led to the development of various regulations

(continued)

(Ayanlade and Proske 2016). They include the National Environmental Protection Regulations that govern effluent limitations, pollution abatement within industries, and management of hazardous and solid wastes. The Environmental Impact Assessment Act of 1992 and the Harmful Wastes Act of 1988 are other significant environmental policies. Apart from these acts, the National Environmental Standards Regulations Agency (NESREA) is responsible for protecting the environment, conserving biodiversity, and sustainable development of its natural resources.They also allow each local and state government to create its agency to protect and improve the environment. Each of the states has also been empowered to make laws for protecting the environment in their jurisdictions.

Identify and explain three more environmental policies in Nigeria.

9.8 Health and Safety Measures

Health and safety measures are otherwise called occupational health and safety or workplace health and safety. They are concerned with the safety, welfare, and health of people at work. They also seek to protect co-workers, customers, and employers. Occupational Safety and Health Administration (OSHA) can also protect family members, clients, employers, and colleagues influenced by the work environment (Rosner and Markowitz 2016). In many jurisdictions, employers are charged with the responsibility of taking care of the safety of their employees. Health and safety measures are crucial in the modern workplace that is characterized by many environmental concerns and hazards. Some of these concerns are health hazards and risks, chemical exposure, poor air quality, electric hazards, and faulty equipment. In other words, occupational health and safety seek to achieve three major objectives, maintaining and promoting the health and working capacity of workers, improving the work environment and work to ensure that it supports health and safety, and developing work organizations as well as cultures in a way that supports safety and health at work.

What are some of the achievements of OSHA?

Various health and safety measures can be adapted to eliminate and reduce the likelihood of injuries, fatalities, and environmental harm. The first measure can be to appoint someone to monitor safety and health issues. The person should be competent and possess adequate training, knowledge, and expertise. The second measure can entail writing a safety and health policy for the business. It is crucial to inform the employees regarding the implementation of a safety and health strategy to provide them information allowing them to contribute to their safety. This could be achieved through the creation of a safety and health policy that the management can place in an area where all employees can easily access (Rosner and Markowitz 2016). The third measure is the environmental risk assessment by examining the aspects that could harm the employees. Involving the employees in the course of the risk assessment will provide more knowledge of what should be changed. As earlier discussed, the five easy steps for conducting risk assessment are identification of hazards, determination of who may be harmed, evaluation of risk and decision of precautions, recording of the findings, and review of the assessment.

How can organizations develop training measures that can help them reduce the likelihood of injuries?

The fourth measure can be training and information. Useful information like the risks and hazards that employees could face, measures in place for dealing with specific hazards or risks, and emergency procedures are essential for any form of safety and health training. Employees should also be asked for their input by getting feedback from the training they require (Rosner and Markowitz 2016). Furthermore, workplace facilities are also quite crucial. Personal protective equipment (PPE) is particularly necessary for protecting employees from adverse risks. Other facilities that are required include, welfare facilities, including hand basins, toilets, drinking water, and a communal area. Other facilities that should be considered are those that can address health issues and safety issues. Fire safety and first aid are also crucial measures that can be used for enhancing health and safety.

How can fire safety be enhanced at the organization?

9.9 Importance of Environmental Impact Assessment (EIA)

An EIA is the categorization and estimation of the environmental effects of a planned project. These effects could be direct or indirect and may include both social and environmental consequences. The primary aim of an EIA is to protect the environment by making sure that a local planning authority makes the right decision. The major activities that occur in the course of an EIA are scoping, stakeholder involvement, environmental assessments, and related legal aspects. The main methods used for performing environmental assessments are checklists, laboratory testing, landscape evaluation, expert opinions and systems, and interaction matrices, and impact tree networks.

What is the importance of scoping in EIA? Explain three major challenges experienced during scoping.

The first benefit is that EIA provides opportunities for raising environmental issues. Its public nature implies that interested groups can raise concerns. It further implies that people who have specialized knowledge of potential effects can contribute to expert analyses. The second benefit is the creation of alternatives and mitigation of impacts. Upon the identification of an impact, EIAs can protect the environment through the development of alternatives that mitigate or avoid the impact. Finally, EIA is beneficial for monitoring and collecting data (Biswas and Agarwal 2013). Baseline monitoring considers the current environmental state to inform impact analyses. Impact monitoring could be used for better prediction of the impacts of future projects.

Explain two other benefits of EIA. Environmental impact assessment has been successful in many European and developed nations. However, in Africa, it remains largely ineffective. Explain three factors that impede the success of EIA in Africa.

9.10 Alan Marshall's Ethical Approach

It is prudent to discuss the approach undertaken by Alan Marshall in categorizing the ways valuation of the natural environment is done. Marshall believes that three major ethical approaches have been in existence during the last four decades: Libertarian Extension, Ecologic Extension, and Conservation Ethics (Marshall 1993).

Briefly discuss the contribution by Michael Smith in this category (Use Peter Vardy's The Puzzle of Ethics).

9.10.1 Libertarian Extension

The Libertarian Extension resonates with the civil liberty approach. Ideally, the civil liberty approach is a commitment for extending equal rights to all community members, made up of humans and non-humans in environmentalism. One of the major proponents of ecological humanism was Andrew Brennan. Eco-humanism contends that ontological entities, whether inanimate or animate, can be valued from an ethical standpoint purely on the premise that they exist. In addition, the work carried out by Arne Naess is also under this category, though they preferred the concept "deep ecology." Conceptually, deep ecology supports the idea of the intrinsic value of the environment. This argues that the environment is valuable in itself or has inherent worth. Peter Singer is a proponent of humanist ethics (Singer 2017). The work done by Singer, including the argument about how the world can solve poverty through donations and animal rights, could also be classified under the Libertarian Extension. Peter Singer believes that moral values should be expanded to include the rights of animals.

9.10.2 Ecologic Extension

According to Marshall, environmental ethics should prioritize the recognition of interdependence regarding biological entities and their diversity. While Libertarian Extension could be perceived as having a political depiction of the natural world, ecologic extension is a scientific depiction of the natural world. Ecological Extension supports the intrinsic value found in ecological entities, including ecosystems. This approach has been taken by Holmes Rolston and James Lovelock with their Gaia hypothesis. The Gaia hypothesis contends that Earth changes its geophysical structure over a given period to ensure a constant equilibrium of inorganic and organic matter (Bowen and Roman 2005). It characterizes the planet as a holistic entity with ethical value. The foundation of this argument is that the human race does not have any specific significance in the long-term. The alternative definition of the Gaia hypothesis is as an ecological hypothesis proposing that the physical components of the Earth and the biosphere closely integrate to form a complex interacting system that maintains the biogeochemical and climatic conditions in a preferred homeostasis.

9.10.3 Conservation Ethics

Conservation ethics mainly focuses on the value of the environment regarding its usefulness/benefit to humans. It is a contradiction to the intrinsic value concepts of deep ecology, thus called shallow ecology. Generally, it supports environmental preservation on the foundation that it is of extrinsic value. Ideally, extrinsic value implies that it is beneficial for people's welfare.

Consequently, this approach assumes that conservation is a means to an end. It should, therefore, deal with humankind and intergenerational considerations. One could argue that this ethic formed the basis for the Kyoto Summit arguments in 1997 as well as three agreements contained in the 1992 Rio Summit. The 1997 Kyoto Summit was held in Japan and aimed to minimize the emission of gases that contribute to global warming (Brechin 2003). The Kyoto Protocol acted as the first agreement between states to foster the reduction of greenhouse gas emission. Kyoto emanated from the United Nations Framework Convention on Climate Change of 1992 that was signed by almost all countries in a meeting popularly called the Rio or Earth Summit. The framework aims to stabilize greenhouse gas levels or concentrations at a level that would deter dangerous human interference with the climate system. The Kyoto treaty sought to actualize the pledge and became effective in 2005. Today, almost all nations have ratified the treaty apart from the US. Developing nations, like India and China, were not forced to minimize emissions because they had contributed to a comparatively small proportion of the current accumulation of carbon dioxide.

Case Study 9.2: Apple

Fig. 9.2 Apple logo

Environmental Responsibility Report (2018) suggests that Apple's (See Fig. 9.2) key mission is to change the world. The company strives to minimize its effect on the planet by focusing on three major priorities that can make a stark difference: minimizing the effect on climate change, conservation of finite resources, and use of safer materials in its processes and products. Apple's 2018 Environmental Responsibility Report suggests that Apple has attained its target of 100% renewable energy used in all its operations,

(continued)

43 nations globally. This target was set more than 10 years ago, and the organization has since established many solar and wind farms for sustaining existing data centers, co-location facilities, retail stores as well as corporate offices. Perhaps, Apple Park is among the current projects that best illustrates an ethically responsible company. Apple Park has a 17-megawatt solar installation on the rooftop. Apple has been successful in ensuring that all its in-house activities and operations run on renewable energy, but many of its manufacturing procedures are outsourced to third party suppliers. Apple has further invested in Elysis; an organization that will help it produce carbon-free aluminum (Baker 2018). The company also initiated a program dubbed Supplier Clean Energy program, which by the end of May 2018, has enabled 23 manufacturers to commit to a 100% renewable energy target, with many aiming to achieve this target by the end of 2018 (Baker 2018). The company is also on course to reach its target of introducing 4 GW of clean energy into its supply chain by 2020, further minimizing the carbon footprint of its products. Efforts like minimizing the average energy that Apple products consume as well as investing in carbon-free aluminum research will be beneficial for minimizing Apple's effect on climate change. Whereas the company has various limitations, it has generally been responsible for its programs to ensure that it protects the environment and conserves finite resources.

9.11 Practical Application: Environmental Ethics in School

The carbon footprint of many schools is at an alarming level. This necessitates the need for green schools by ensuring that all schools activities strive to reduce the harm posed on the environment. The following key areas have been identified as critical in ensuring that schools join the Environmental Ethics movement.

9.11.1 School Food

Schools should establish a partnership program to transform their food culture. This should entail the creation of websites that have information on how students can participate, events, and Food Tube short films. Schools should also establish a social media page where dissatisfied parents/guardians can post about their experiences.

9.11.2 *Energy*

Schools should develop Carbon Detectives Kit to be able to calculate their carbon footprint and make comparisons to other schools. They should also develop sustainable learning programs to encourage staff and learners to make energy reductions of about 10% and to have a better understanding of the way water and energy are used in their buildings (Yang 2006). They should also develop a movement that can be called "Walk to School" managed by a charity to encourage learners and parents to leave their vehicles at home.

9.11.3 *Other School Activities and Initiatives*

There are ten recommendations that teachers can use to ensure that schools are environmentally responsible.

1. **Connect the dots:** They need to teach students about global issues. This should entail connecting the problems to the community.
2. **Calculation of the carbon footprint:** They should particularly use footprint calculators to see the impact they have on their environment. This should entail the use of online tools as fun games. From these results, they should develop a plan for reducing their group footprint.
3. **Carry out an energy audit within the classroom:** Taking stock of how and where you are using energy by conducting an assessment on where most of the classroom energy gets wasted. Factors to consider include the number of lights that are on, presence or absence of heat or alternating current, and determining where cuts are necessary. The Kill-a-Watt measurement tool is one of the best ways of physically measuring the energy used.

9.12 Individual Matter: Irving Stowe

Irving Stowe (see Fig. 9.3) co-founded Greenpeace alongside Dorothy Stowe and other pro-environmentalists to promote environmental ethics and deter environmental degradation. Currently, Greenpeace has offices in more than 39 nations and a global coordinating body in Amsterdam, The Netherlands (Weyler 2004). The company states that its objective is to make sure that the planet maintains its ability to nurture life. It places focus on global issues, including climate change, overfishing, commercial whaling, anti-nuclear issues, and deforestation. Irving Stowe uses strategies such as lobbying, research, ecotage as well as a direct action for achieving its goals. Stowe's activities have been depicted as the most visible because of the direct actions that the company engages in (Easterlin 2016). The organization has been successful in raising environmental issues into the public

Fig. 9.3 Irving Stowe, *Source:* Alchetron (2018) via Stowe Archives

consciousness. Due to his criticism of genetically modified organisms, Stowe has been the recipient of an open letter from at least 100 Nobel laureates who have urged him to end the campaign against GMOs (Weyler 2004). His direct action has also led to legal actions, including fines and suspended sentences. Over the years, the organization, through the leadership of the founders, has been able to end nuclear threats, protect all forms of biodiversity, deter pollution, promote peace and non-violence, and end the abuse of environmental resources. Stowe has been among the proponents of environmental ethics who have been quite successful. Arguably, his efforts continue to change the world and spur environmental sustainability.

9.13 Future Trends

9.13.1 Africa

Ecocentrism will become central to solving the current unprecedented environmental crisis through the recognition of the responsibility of humans towards nature. Africa will realize that transforming towards an ecocentric worldview as well as other value systems is necessary for the flourishing of life on Earth. However, increased consumerism and affinity for material things could deter the continent's countries from embracing ethical values that are necessary for environment protection and conservation.

9.13.2 Asia

Future environmental ethicists will consider nature as an end in itself, thus resulting in more challenges. Intragenerational and intergenerational equity will become common concepts as the countries in the continent strive to develop measures towards achieving the same. With countries like China facing challenges related to animal rights, there will be further call from activists about the need to institute measures geared towards protecting them. The government could also be at the forefront of ensuring that industries are not only compliant but also ethical in their undertakings.

9.13.3 Australia

Ethics and policy will go hand-in-hand to solve environmental challenges. Humans will be attracted to a better quality of life, and environmental ethics could persuade many people on the need to participate in conservational efforts. The continent will embrace a comprehensive and more inclusive human welfare concept.

9.13.4 Europe

Culture will be the main determinant of the fate of the Earth compared to nature with the emergence of the Anthropocene. The concept of environmental justice will become increasingly common in the quest for environmental ethics. Ethics will continue being the main bedrock of environmental conservation and protection efforts, with more attention shifting towards development of policies that can be beneficial to both the environment and people. Animal welfare will also continue being prioritized by leaders to ensure that their rights are protected.

9.13.5 North America

There will be an emergence of a new millennium dubbed the super-industrial age or high-technology age. If embraced appropriately, this age may involve the use of technology in solving environmental issues. Ethical approaches to the environment will be further embraced to ensure that policies and programs lead to environmental stewardship at individual, community, and national levels.

9.13.6 South America

Greener environmental policies pushing towards curbing global warming by engaging in ethical activities will become common. There will be more stringent enforcement of regulations governing the Polluter Pays Principle and User Pays Principle. This will be geared towards ensuring that users and polluters are ethically responsible for their actions to reduce adverse harm to the environment.

9.14 Summary

- Environmental risk assessment helps in estimating the likelihood of human activities harming the environment.
- Risk management is crucial for determining the risks associated with a specific event. It has three main stages: preparation, implementing the assessment, and application of its results.
- The six main stages involved in ERA are establishing the context, identifying hazards, determining who will be harmed, assessing the risks, recording findings, and reviewing the assessment.
- Environmental risk management undergoes various stages: identifying threats, assessing the vulnerability of critical assets, determining the risk, identifying ways for reducing those risks, and prioritizing risk reduction measures.
- Risk management techniques are categorized into avoidance, reduction, sharing, and retention.
- The ethics of waste is about being ethical in waste management and generation.
- Intergenerational and intragenerational equity are major considerations in the ethics of waste.
- The seven major principles of environmental management applicable to the ethics of waste are: Polluter Pays Principle, User Pays Principle, Principle of Proportionality, Principle of Participation, Principle of Responsibility, Principle of Effectiveness and Efficiency, and Precautionary Principle
- Many Nigerian environmental policies are founded on the country's 1999 Constitution.
- The major health and safety measures that can be adapted to prevent risks are appointing someone, writing a health and safety policy, risk assessment, training and information, providing workplace facilities, and fire safety.
- The major benefits of EIA are the provision of opportunities for raising environmental issues, creation of alternatives, mitigation of impacts, and monitoring and collecting data.

- Alan Marshall proposes three major approaches that govern environmental ethics: Libertarian Extension, Conservation Ethics, and the Ecologic Extension.
- Apple's programs depict an organization that is committed to ensuring environmental sustainability.
- In Africa, ecocentrism will be critical for solving the environmental challenges while in Europe, culture will be critical for environmental conservation as environmental justice becomes embedded in management efforts.
- Asia will prioritize on intergenerational and intragenerational equity concepts, while a high-technology age will characterize North America.
- For South America, greener environmental policies will increasingly be adopted, and more regulations developed to ensure that the Polluter Pays Principle is strictly enforced. For Australia, humans will become more responsible and get involved in conservational efforts.

9.15 Discussion Questions

1. What are PPEs, and what is their significance at the workplace?
2. State two objectives of the Occupational Health and Safety act or policy.
3. Discuss the methods for addressing environmental risks.
4. What is the significance of screening in EIA?
5. Differentiate between social impact assessment and environmental impact assessment.
6. Explain how Nigeria uses risk-avoidance to control constructions.
7. Identify the environmental policies that govern the following sectors or aspects in Nigeria: water, air, noise/sound, effluents, and emissions.
8. Has the NESREA act been effective in protecting the environment in Nigeria? Illustrate your answers with relevant examples.
9. What is the major challenge facing environmental conservation in Nigeria?
10. As a policy expert, what environmental policies would you recommend for Nigeria?
11. Has CSR been effectively implemented by organizations? Identify and elaborate on three organizations in Nigeria that have effectively embedded CSR into its programs.
12. Despite various agencies and policies, environmental pollution remains a significant concern in Nigeria. How can the government promote sustainability?
13. State the effluent emission standards for Abuja.
14. Environmental impact assessment (EIA) in Nigeria faces numerous challenges. Identify and discuss three significant challenges.
15. How does politics influence EIA in Nigeria?
16. What mechanisms should Nigeria put in place to ensure that EIA is successful?
17. The Curse of Oil in Nigeria, especially in the Niger Delta, is a sign that EIA either failed or was deliberately neglected by the authorities. Using an EIA model, predict the aspects of the EIA process that were neglected during the

implementation of Oil mining in the Niger Delta. Propose mechanisms for rectifying the situation.
18. Community participation is an important aspect of EIA. Identify two successful projects in Nigeria where communities have been involved prior to, during, and after the project. Which communities were involved? What are their experiences? Compare these experiences to those of the Ilaje and Ogoni communities found close to the Niger Delta, where oil mining takes place.
19. Identify three contributions of Marshall to environmental ethics. How is modern-day conservation a reflection of Marshall's ideas?
20. In what ways can humans honor the three approaches proposed by Marshall?
21. Peter Singer is a key proponent of animal rights. Do you agree with his arguments? Explain the arguments that you agree with and discuss the ones that you oppose.
22. How has Irving Stowe contributed towards environmental ethics?
23. Can waste management be ethical? Briefly discuss. Discuss the concept of ethical disposal.
24. How can intergenerational and intragenerational equity be used as guidelines for waste management?
25. Explain how Nigeria could use the seven principles of environmental management to manage its wastes.

References

Ahuja, S. (Ed.). (2013). *Monitoring water quality: Pollution assessment, analysis, and remediation*. Amsterdam: Elsevier.

Alchetron. (2018). *Irving Stowe*. Retrieved January 10, 2019, from https://alchetron.com/Irving-Stowe/.

Attfield, R. (2014). *Environmental ethics: An overview for the twenty-first century*. Hoboken: Wiley.

Ayanlade, A., & Proske, U. (2016). Assessing wetland degradation and loss of ecosystem services in the Niger Delta, Nigeria. *Marine and Freshwater Research, 67*(6), 828–836.

Baker, M. (2018). *How Tim Cook brought corporate social responsibility to Apple*. Retrieved October 11, 2018, from http://mallenbaker.net/article/clear-reflection/how-tim-cook-brought-corporate-social-responsibility-to-apple/.

Biswas, A. K., & Agarwal, S. B. C. (Eds.). (2013). *Environmental impact assessment for developing countries*. Amsterdam: Elsevier.

Bowen, B. W., & Roman, J. O. E. (2005). Gaia's handmaidens: The Orlog model for conservation biology. *Conservation Biology, 19*(4), 1037–1043.

Bowie, N. E. (2017). *Business ethics: A Kantian perspective*. Cambridge: Cambridge University Press.

Brechin1, S. R. (2003). Comparative public opinion and knowledge on global climatic change and the Kyoto Protocol: The US versus the World? *International Journal of Sociology and Social Policy, 23*(10), 106–134.

Cooper, D. F., Grey, S., Raymond, G., & Walker, P. (2005). *Project risk management guidelines: Managing risk in large projects and complex procurements* (p. 400). Chichester: Wiley.

Easterlin, N. (2016). What is literature worth? Narration, cognition, and ethics. *Interdisciplinary Literary Studies, 18*(2), 291–300.

Environmental Responsibility Report. (2018). *Progress report, covering fiscal year 2017. Environmental responsibility at Apple. To ask less of the planet, we ask more of ourselves*. Retrieved October 11, 2018, from https://www.apple.co-m/nz/environment/pdf/Apple_Environmental_Responsibility_Report_2018.pdf/.

Giannakis, M., & Papadopoulos, T. (2016). Supply chain sustainability: A risk management approach. *International Journal of Production Economics, 171*, 455–470.

Hawkins, G. (2006). *The ethics of waste: How we relate to rubbish*. Lanham: Rowman & Littlefield.

Hernando, M. D., Mezcua, M., Fernández-Alba, A. R., & Barceló, D. (2006). Environmental risk assessment of pharmaceutical residues in wastewater effluents, surface waters and sediments. *Talanta, 69*(2), 334–342.

Hussen, A. (2018). *Principles of environmental economics and sustainability: An integrated economic and ecological approach*. Abingdon: Routledge.

Khan, M. R. (2015). Polluter-Pays-principle: The cardinal instrument for addressing climate change. *Laws, 4*(3), 638–653.

Kosma, C. I., Lambropoulou, D. A., & Albanis, T. A. (2014). Investigation of PPCPs in wastewater treatment plants in Greece: Occurrence, removal and environmental risk assessment. *Science of the total environment, 466*, 421–438.

Marshall, A. (1993). Ethics and the extraterrestrial environment. *Journal of Applied Philosophy, 10*(2), 227–236.

Mihelcic, J. R., Zimmerman, J. B., & Auer, M. T. (2014). *Environmental engineering: Fundamentals, sustainability, design* (Vol. 1). Hoboken: Wiley.

Okereke, C. (2006). Global environmental sustainability: Intragenerational equity and conceptions of justice in multilateral environmental regimes. *Geoforum, 37*(5), 725–738.

Pritchard, P. (2014). *Environmental risk management*. Abingdon: Routledge.

Rosner, D., & Markowitz, G. (2016). Educate the individual... to a sane appreciation of the risk: A history of industry's responsibility to warn of job dangers before the occupational safety and health administration. *American Journal of Public Health, 106*(1), 28–35.

Sandler, R. L. (2009). *Character and environment: A virtue-oriented approach to environmental ethics*. Columbia: Columbia University Press.

Singer, P. (2017). All animals are equal. In *Animal rights* (pp. 3–16). Abingdon: Routledge.

Weyler, R. (2004). *Greenpeace: How a group of ecologists, journalists, and visionaries changed the world*. Pennsylvania: Rodale.

Yang, T. (2006). Towards an egalitarian global environmental ethics. *Environmental Ethics and International Policy, 8*, 23–45.

Part III
Principles and Practices of Sustainability

Chapter 10
Sustainability

Abstract Sustainability has become a common term and concept in our society because of the effects of climate change that are being experienced on a wider scale. To understand this concept, this chapter explores the three major pillars of sustainability, primary goals of environmental sustainability, social development, economic development, and environmental protection as well as their applications and major steps required for maintaining a sustainable economy. The Sustainable Development Goals (SDGs) are used as one of the case studies for understanding the three pillars and the concept of sustainability. The other case studies used are the Green School in Bali, spiritual ecology, and Grameen Bank while Mohammed Yunus is explored as one of the major contributors to social and economic development. This is one of the most important chapters of the book since it explores a concept that is central to the going green movement. Because of this significance, future trends are examined to paint an image of the outlooks of sustainability.

Keywords Environmental sustainability · Economic sustainability · Social sustainability · Sustainable development goals · Mohammed Yunus

10.1 What is Sustainability?

What is sustainability? What are its forms? How do policy makers use it and how are businesses and scientists involved in it? What is the role of religious leaders? These are questions we attempt to answer in this chapter. Echoing the title of the report of the World Commission on Environment and Development, "Our common future," sustainability is our goal for the future.

First, we need to define sustainability . The idea of sustainability emerged during the 1980s from the Brundtland Commission that represented 22 countries from five continents. The Brundtland Commission defined sustainable development as "meets our current needs without compromising the ability of future generations to meet their own needs" (International Institute for Sustainable Development n.d). Thus sustainability is about fostering equality among generations. It is also about the

A. A. Akinsemolu, *The Principles of Green and Sustainability Science*,
https://doi.org/10.1007/978-981-15-2493-6_10

"capacity of our society to endure," that is to maintain a certain quality of living for all the inhabitants of our planet (Sustainability.com n.d).

This capacity is questioned not only by the current depletion of natural resources, the loss of biodiversity, and the rising levels of greenhouse gas emissions, but also by social inequalities and injustices, which pertain to our way of life. These problems are important because sustainability, according to contemporary thinkers, is not only about the environment. It also refers to societal objectives, such as inclusion and the alleviation of hunger and poverty.

This definition is the basis for the 17 United Nations' Sustainable Development Goals (SDGs). The SDGs span across several thematic areas, such as poverty, water management, energy, innovation, and climate action. These domains are interrelated, and we cannot tackle one without touching upon the others. Therefore, sustainability is a comprehensive agenda for change at the economic, social, and environmental levels.

The way to make the transition towards sustainability is through strategies known as greening or green growth. Such strategies aim to adjust all sectors of the economy to make sure that they "respect environmental limits while fulfilling social wants and needs" (Sustainability.com n.d). Businesses have begun to understand that creating green and ethically sourced brands is a strategy to attract customers: those who care about the environment, fairness, and human rights. Eco-friendly products, such as organic and fair-trade food, are the outcome of this process.

Besides economic players, scientists are expected to contribute to the sustainability agenda. To start with, sustainability requires hard science; facts and evidence to delineate the problems we need to address. For example, thanks to the scientific findings, we know that an increase in the average temperature of the planet above 2 °C will have dramatic consequences, such as extreme weather phenomena and a significant rise in sea levels. Such knowledge is a critical first step towards action. The next step is to choose alternative solutions and this is where science can contribute the most. From renewable energy and green devices to the fight against diseases, such as malaria, scientists across the world work to resolve some of the toughest problems facing humanity. For example, scientists from the University of Alexandria, Egypt, discovered a way to desalinate and purify seawater to make it drinkable (Nield 2015). This technology has the potential to transform the lives of millions of people.

Lastly, religious leaders have a role to play in advancing the cause of sustainability since most religions refer to nature as something to be preserved (1Million Women.com 2016). Christianity considers nature as the home of God. In Islam, people of faith are foreseen to be stewards of the earth. Buddhism, in turn, calls for the protection of all living beings. From this standpoint, it is up to religious leaders to use these teachings to promote sustainability by influencing both world leaders and the general public to recognize their roles in promoting sustainability. They are also expected to lead by example. The Vatican has responded to this plea by becoming the first carbon neutral country.

Overall, the commitment of every individual on the planet is essential to sustain a livable planet and to ensure the highest possible living standards for all humans, wherever they reside and whatever their socioeconomic status. This is humanity's shared challenge of sustainability.

10.2 The Sustainable Development Goals (SDGs)

In 2012, the United Nations Conference on Sustainable Development met with the intention of discussing this and developing a specific set of goals and targets. At this conference, it was noted that the "Millennium Development Goals" (MDG) was successful in some areas, but that there was more to do (Moldan et al. 2012). Building on the successes and addressing the shortcomings of the previous goals, the SDGs were established. Officially called "Transforming our World: the 2030 Agenda for Sustainable Development," a set of 17 global goals aiming at environmental sustainability were outlined. SDGs are also called the Global Goals (see Fig. 10.1). They are a universal call to action for poverty eradication, protection of the planet, and ensuring that all people experience peace and prosperity. The SDGs build on the achievements of the Millennium Development Goals, while simultaneously including new areas of focus. The expansion puts economic inequality, climate change, innovation, peace and justice, and sustainable consumption on the agenda. The 17 Global Goals are inherently interconnected and interdependent and are based on the spirit of pragmatism and partnership for appropriate decision-making necessary to improve life and insure sustainability for future generations. As global objectives, they offer guidelines and targets for all nations. The SDGs came into effect in January 2016 and will continue to guide the United Nations Development Policy and funding focus until the proposed end of the program in 2030. As the lead development agency, the UNDP is poised to assist in implementation of the goals and is instrumental for supporting the three pillars of sustainability.

Fig. 10.1 Sustainable development goals

10.2.1 Goal 1: No Poverty

The main objective of this goal is to "end poverty in all its forms everywhere." Since 1990, it is estimated that extreme poverty has been reduced by more than 50%. Despite this progress, it is estimated that there is still more than one in five individuals living on less than $1.25 per day (Le Blanc 2015). Poverty goes beyond the lack of income resources, and includes living without basic services, including security, education, and healthcare. In addition, people experience social discrimination and marginalization as the result of key decision-making processes. Children make up the larger proportion of people who live in extreme poverty, and because of their vulnerability, this goal plays a crucial role in liberating young people from the constraints of malnutrition, poor healthcare, and lack of education. United Nations Development Policy (UNDP) stated in 2013 that about 385 million children lived on no more than $1.90 a day (Le Blanc 2015) (see Fig. 10.2).However, these figures remain largely unreliable because of the lack of data regarding the status of children globally. This is particularly evident in developing states where data is inaccurate, outdated, and uncollected. Some even rely on figures provided by institutions such as the UNDP and World Bank. It is also known that 97% of nations lack data to ascertain the state of impoverished children and projections to meet SDG Goal 1, while 63% lack data about child poverty (Le Blanc 2015).

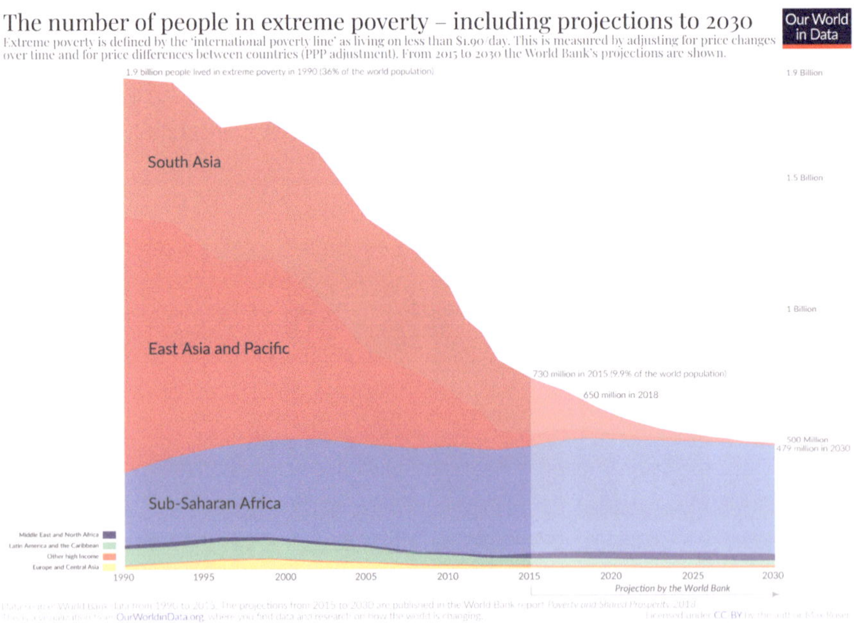

Fig. 10.2 The number of people living in extreme poverty, *Source:* World Bank (2018a); Roser and Ortiz-Ospina (2019a)

Gender inequality plays a critical role in the perpetuation of poverty and its associated risks. Women are at an increased risk of poverty compared to men and the negative effects of poverty are most damaging during and prior to pregnancy. This has a net effect on the educational outcomes, participation, and chances of a better income for women. The main challenges undermining achievement of poverty eradication are growing inequality, fragile statehood, and climate change effects. Achieving this goal would be a step toward achieving sustainability.

How has your community benefitted from the Goal 1? What steps can you take to support the efforts being made towards achieving this goal?

10.2.2 Goal 2: Zero Hunger

The major objective of this goal is to "end hunger, achieve improved nutrition and food security, and enhance sustainable agriculture" (United Nations nd) The goal seeks to enable all nations to end all forms of malnutrition and hunger by 2030. To end hunger, the best way has been identified as doubling agricultural productivity and the income of food producers, particularly indigenous peoples and women, by ensuring sustainable food production systems, and enhancing the quality of land and soil. Globally, agriculture remains a major employer, and provides livelihoods for at least 40% of people globally (Le Blanc 2015). In developing states, it is estimated that 43% of the agricultural labor force is composed of women, and this number increases to 50% in Africa and Asia. Unfortunately, inequity undermines development partly because women only possess 20% of the land (Le Blanc 2015).

The other targets in this goal tackle the maintenance of genetic seed diversity, increased access to land, the prevention of trade restriction as well as the elimination of waste. One in nine undernourished children lives within developing states (see Fig. 10.3). Undernutrition is among the major causes of child mortality (45%) in children less than 5 years. On the other hand, chronic malnutrition affects about 155 million children globally and stunts their development. Regarding the progress of this goal, it is estimated that only twenty-two UN member states were on track to meet the goal of eliminating malnourishment and undernourishment by the end of 2017 (Le Blanc 2015). Of these countries, 20% have not made any substantial progress and about 70% do not have sufficient data to ascertain their progress (Le Blanc 2015). A 2013 report made by the International Food Policy Research Institute (IFPRI) suggested that there is need to re-orientate the focus of SDGs from ending poverty by 2030 to the elimination of undernutrition and hunger by 2025 (IFPRI 2013). This recommendation is founded on experiences drawn from Vietnam, Thailand, Brazil, and China. The main objectives for achieving this are identified as: improving nutrition and social protection, expanding agriculture, and

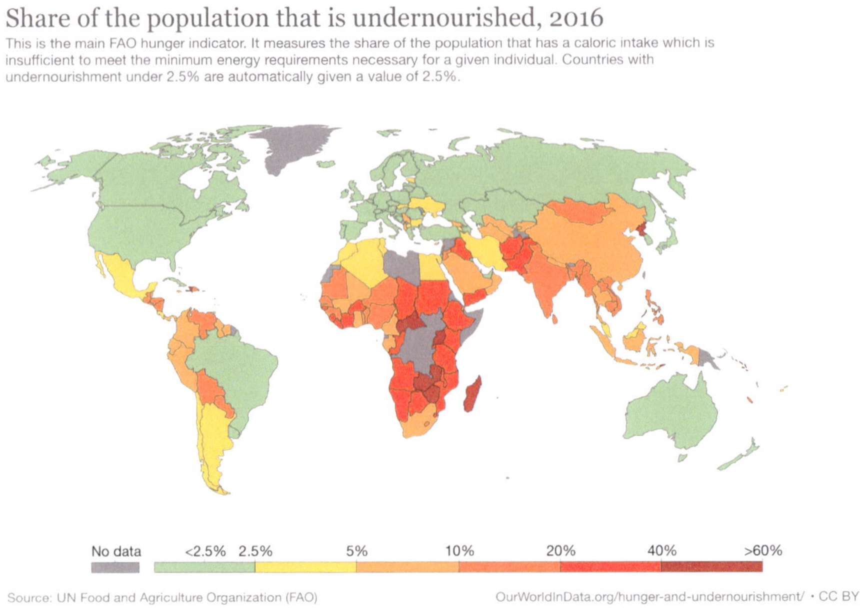

Fig. 10.3 Share of the population that is undernourished, 2016, *Source:* UN, FAO (2016); Roser and Ritchie (2019a)

an amalgamation of these approaches. Africa is among the most affected regions in terms of Goal 2's progress.

Identify two African nations that have made progress on this target and briefly explain how this has been made possible. What is the progress of Africa in achieving zero hunger? Discuss three major challenges that act as impediments towards achieving this goal in Africa.

10.2.3 Goal 3: Good Health and Well-Being for People

The major objective of this goal is to "ensure and promote healthy lives and foster the well-being of all individuals at all ages." Globally, there has been significant progress in raising the life expectancy and eliminating some of the contributors to maternal and child mortality. Between 2000 and 2016, there was a global reduction in cases of under-five mortality rate by 47%, a reduction from 78 to 41 deaths in every 1000 live births (Griggs et al. 2013). Nevertheless, the number of children who die before their fifth birthday is still quite high. For instance, in 2016 alone, 5.6

million children died. Newborns make up an increasing number of these deaths. This goal seeks to reduce the under-five mortality to the lowest possible threshold (25 per 1000 live births) (Griggs et al. 2013). However, in case current trends continue as they are, at least 60 nations will not meet the target by 2030. About 50% of these states will not meet the target even if the period is extended up to 2050. This goal also seeks to minimize maternal mortality to no more than 70 deaths for every 100,000 live births (Griggs et al. 2013). Despite declining maternal mortality rate by 37% between 2000 and 2015, about 303,000 maternal deaths occurred in 2015 by preventable causes (see Fig. 10.4). During 2015, maternal health conditions were also the leading causes of deaths among girls aged between 15 and 19 years. The main strategies for achieving Goal 3 will be reducing adolescent pregnancy, obtaining better data for all girls and women, and attaining universal coverage of skilled birth attendants. There has also been tremendous progress in the access to clean sanitation and water as well as the reduction of malaria, tuberculosis, polio, and HIV/AIDs transmissions. Statistics show that between 2000 and 2016, new HIV infections were reduced by 66% in children under 15 years, and 45% among adolescents between 15 and 19 years (Griggs et al. 2013). Despite this, current trends imply that one in four nations will still not meet the sustainable development target to eliminate AIDS among children under 5 years (Griggs et al. 2013). In addition, only 50% of women in developing states have the healthcare they require. In spite of numerous challenges, the goal's aim to achieve universal health coverage

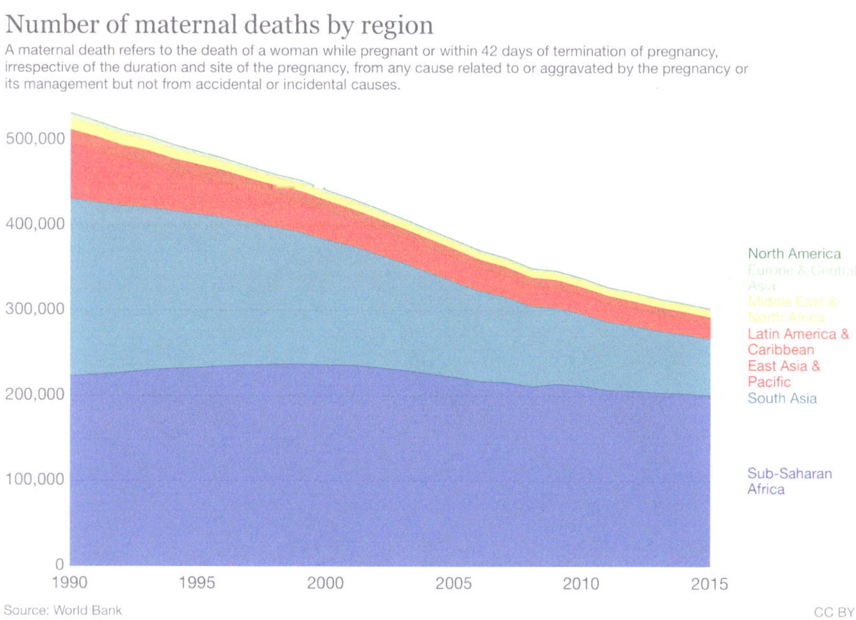

Fig. 10.4 Number of maternal deaths by region, *Source:* World Bank (2015); Roser and Ritchie (2019b)

is a step in the right direction and will be critical for enhancing sustainability (Griggs et al. 2013).Good health is a key determinant for the progress of a society.

Discuss how health has been an impediment in sustaining production in Africa. Provide four states in Africa that have set up crucial sanitation programs.

10.2.4 Goal 4: Quality Education

The main objective of this goal is to "ensure that there is equitable and inclusive education as well as promote lifelong learning opportunities for all the individuals." There has been tremendous progress in providing access to education, especially at the primary school level (see Fig. 10.5). There are still major challenges because of the complex nature of education. In addition, access does not always imply a good quality of education or primary school completion. It is estimated that 103 million young individuals lack basic literacy skills and at least 60% of them are women (Richter et al. 2017). There is also difficulty in tracking progress and ensuring that

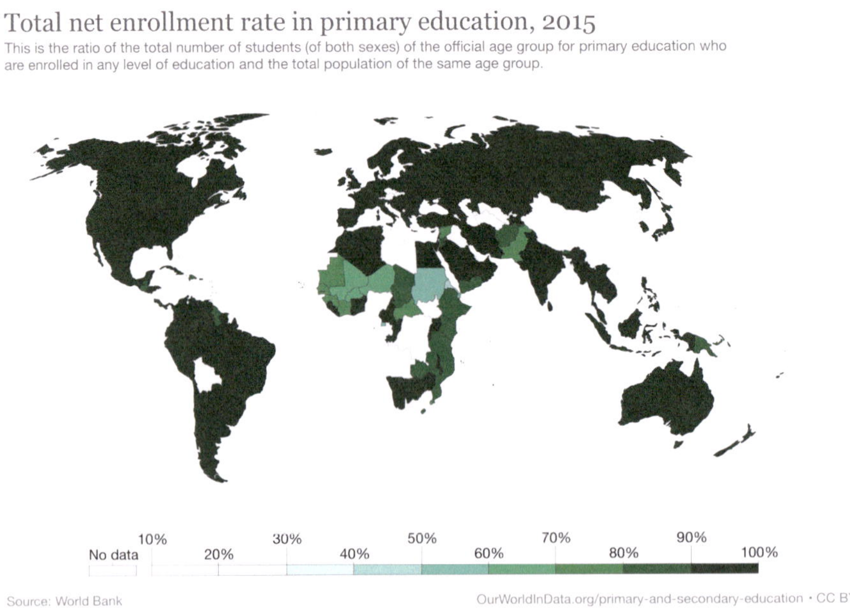

Fig. 10.5 Total net enrollment rate in primary education, 2005, *Source:* World Bank EdStats (2015); Roser and Ortiz-Ospina (2019b)

this goal is achieved. These challenges should only act as an impetus regarding the need to improve reporting on the progress and achieve the quality education targets. Africa has placed its attention on the quantity rather than quality of education.

Discuss how the continent can use the Education for All insights to achieve quality education.

10.2.5 Goal 5: Gender Equality

The main objective of this goal is to "achieve gender equality and empower all girls and women." It recognizes gender equality as a basic human right necessary for a sustainable, peaceful, and prosperous world. The provision of girls and women with equal access to political and economic representation, education, decent work, and healthcare is beneficial to humanity and society. This is one of the targets that has made tremendous progress with a record 143 nations guaranteeing equality between women and men through their constitutions. The worrying aspect is that 52 of them have not committed themselves to this target (Sachs 2012). Many nations still embed gender discrimination in their social norms and legal systems. One of the achievements of this goal is the reduction of child marriages, which has declined over the last decade. Specifically, child marriage in some African and Asian countries. These regions have been reported to have the highest rate of child marriage (see Fig. 10.6). Despite gender equality remaining elusive, significant steps and commitments have been made. There is a need for more enforceable legislation that can empower women. Achieving gender equality is one of the key steps towards delivering transformative and sustainable change.

Devise strategies for achieving gender equality by 2030. Why do you think your strategies will be successful?

10.2.6 Goal 6: Clean Water and Sanitation

The main objective of this goal is to "ensure sustainable sanitation and water management for all people." This goal has eight targets along with 11 indicators for monitoring progress. The first target concerns sanitation and drinking water supply. As of 2012, it is estimated that 60% of people do not have safely managed sanitation services, while 30% lack access to safely managed water services globally (Sachs

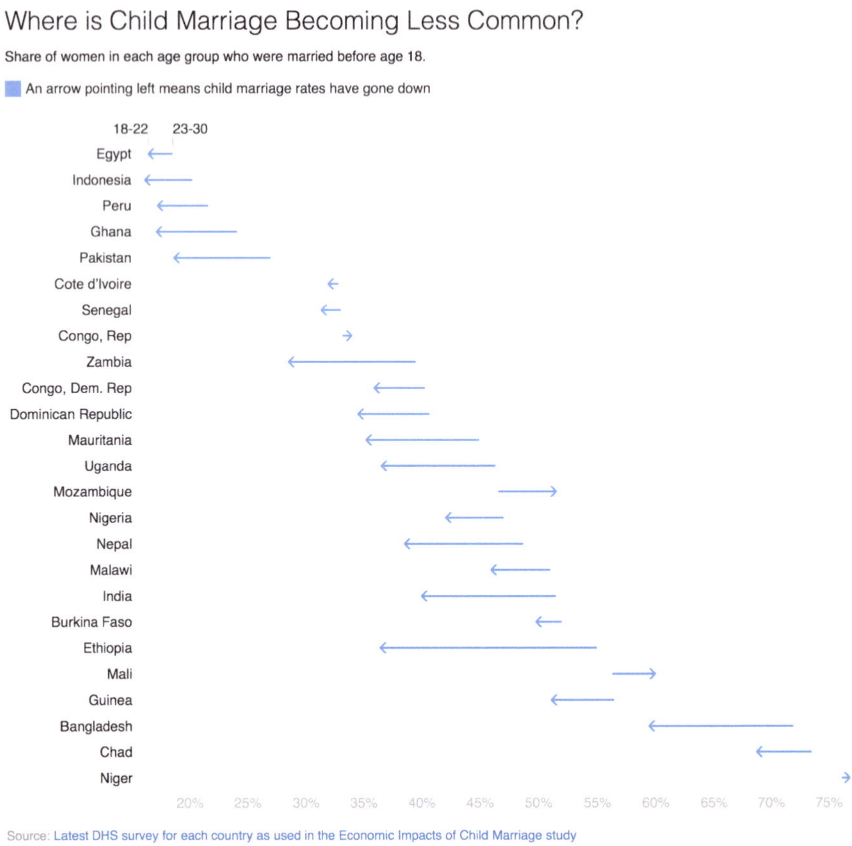

Fig. 10.6 Child marriage becoming less common, *Source:* World Bank (2017a,b); Ortiz-Ospina and Roser (2019)

2012). However, access to improved water sources has increased to 91% in 2015, although access remains lowest in sub-Saharan Africa (see Fig. 10.7). This goal recognizes that safe drinking water as well as hygienic toilets provides protection from disease and allows societies to be economically productive. There is a need for more toilets and sanitation, behavior change, and cooperation between the various sectors. To meet the sanitation targets by 2030, there is a need to hasten progress towards elimination of open defecation in nations such as Nigeria, Ethiopia, India, and Brazil. The Sustainable Sanitation Alliance should continue playing a crucial role in Goal 6. In tandem with the alliance's position, sustainable development goals are highly interdependent. As such, the provision of sanitation and clean water for all populations is fundamental for achieving other SDGs.

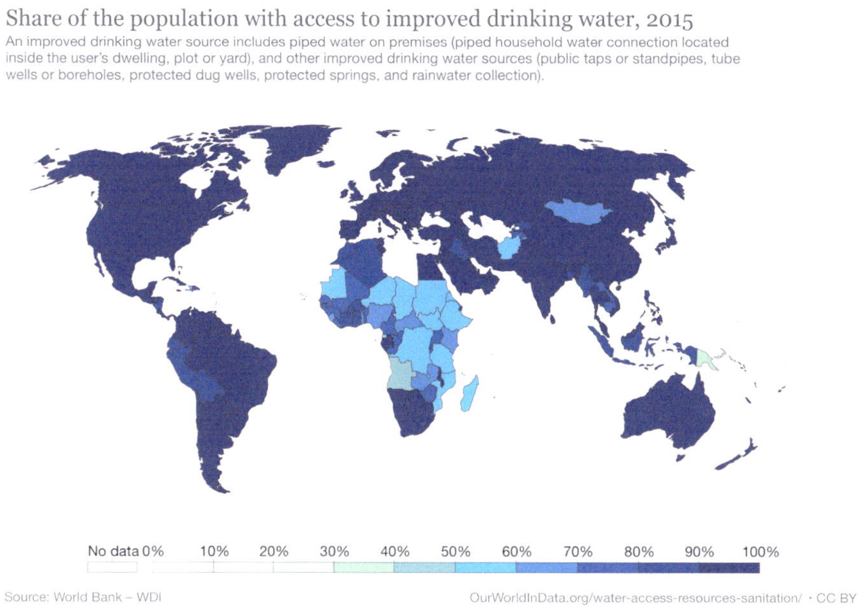

Fig. 10.7 Share of the population with access to improved drinking water 2015, *Source:* World Bank and World Development Report (2008); Ritchie and Roser (2019a)

Explore the best ways of ensuring that we meet the Clean Water and Sanitation objectives.

10.2.7 Goal 7: Affordable and Clean Energy

The main objective of SDG 7 is to "ensure access to affordable, reliable, sustainable and modern energy for all" (Sachs 2012). The targets for this goal are access to reliable and affordable energy by 2030. This entails the improvement of energy efficiency as well as international cooperation for better access to technology for clean energy. There is also a need for more investments that can increase the number of clean energy infrastructure projects. Figure 10.8 shows the global investments in clean energy technologies from 2004 to 2015. There is a remarkable growth in investment across all regions, but at significant different rate. The disadvantaged or least developed nations as well as developing countries should be given core priority.

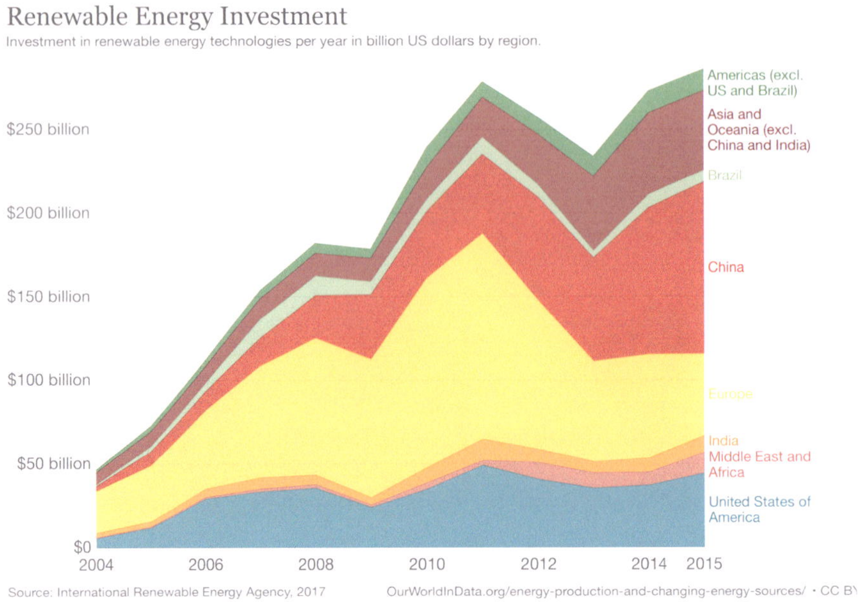

Fig. 10.8 Global investment on clean energy technologies, *Source:* IRENA (2017); Ritchie and Roser (2019b)

Critically discuss the major targets of this goal. What are the major challenges that deter us from adopting clean energy? How can we overcome these challenges?

10.2.8 Goal 8: Decent Work and Economic Growth

The main objective of SDG 8 is to promote decent work as well as inclusive and sustainable economic growth for all. There is a need to prioritize infrastructure and long-term economic growth. Diversification and better technology are also required for reducing youth unemployment and rationalizing a global strategy for solving unemployment (Sachs 2012). Figure 10.9 shows the annual change in real gross domestic product per employed person from 2000 to 2016. By 2030, Goal 8 seeks to develop policies that emphasize the need for sustainable tourism that can result in the creation of jobs.

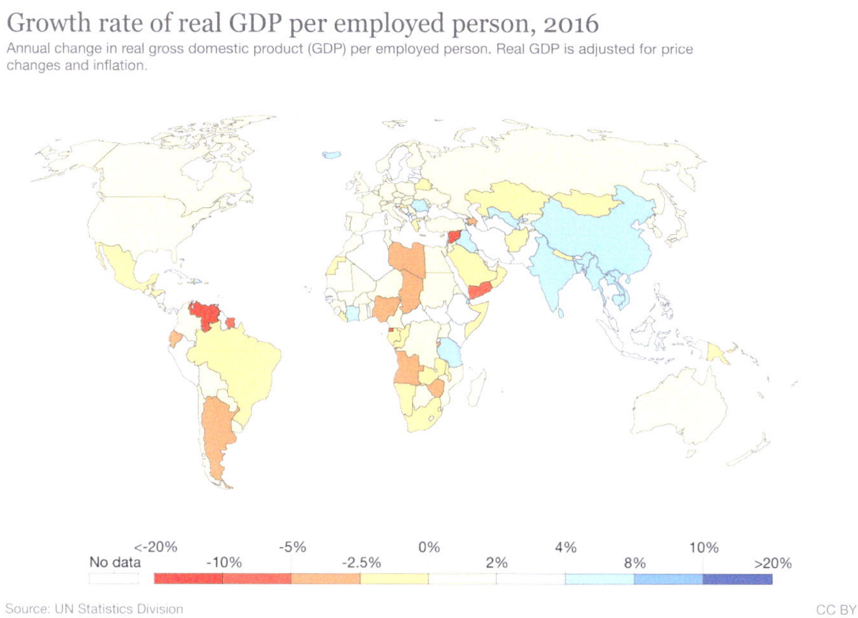

Fig. 10.9 Growth rate real GDP per employed person, 2016, *Source:* Roser (2019); United Nations Statistics Division (2019)

What is meant by decent work? How does economic growth lead to the creation of jobs in Africa?

10.2.9 Goal 9: Industry, Innovation, and Infrastructure

The core objective of SDG 9 is to "build resilient infrastructure, promote sustainable as well as inclusive industrialization and foster innovation." Through manufacturing, countries can provide viable environments for employment (Sachs 2012). Figure 10.10 shows the contribution of the manufacturing sector to the total gross domestic product (GDP) of various countries from 1960 to 2017. Today, there has been significant improvement of mobile-cellular signal coverage and there are signs that manufacturing will improve in a substantial way. Many countries are investing heavily in the manufacturing sector as a way of creating jobs. Also, many areas that had poor network coverage are currently covered because of the increased investment in mobile networks as a result of many people owning mobile phones.

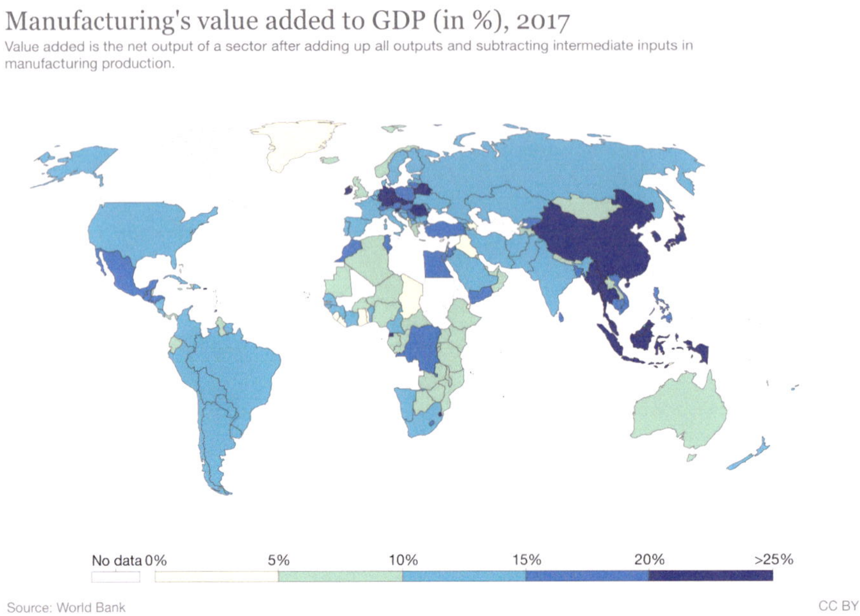

Fig. 10.10 Manufacturing value added to GDP (in%), 2017, *Source:* World Bank (2017a,b); Ritchie et al. (2018a,b)

What is resilient infrastructure? Discuss three Green innovations that can drive this goal.

10.2.10 Goal 10: Reducing Inequalities

The major objective of SDG 10 is to "reduce income inequality in nations." One of the targets is to minimize the cost of commodity exportation from the Least Developing Countries, LDCs. Least Developing Countries (LDCs) are developing nations that have the lowest indicators of socioeconomic development. Examples of LDCs today are Democratic Republic of Congo, Afghanistan, Cambodia, Comoros, Burkina Faso, and Angola. Due to this target, there has been substantial expansion of duty-free treatment as 65% of products emanating from LDCs were duty-free by 2015 (Sachs 2012).

Figure 10.11 shows the global income distribution in 1800, 1975, and 2015. In 1800, over 80% of the world population lived in extreme poverty. However, about 175 years later, the world income distribution became bimodal which represents the developing and developed world. The developed world was ten times richer

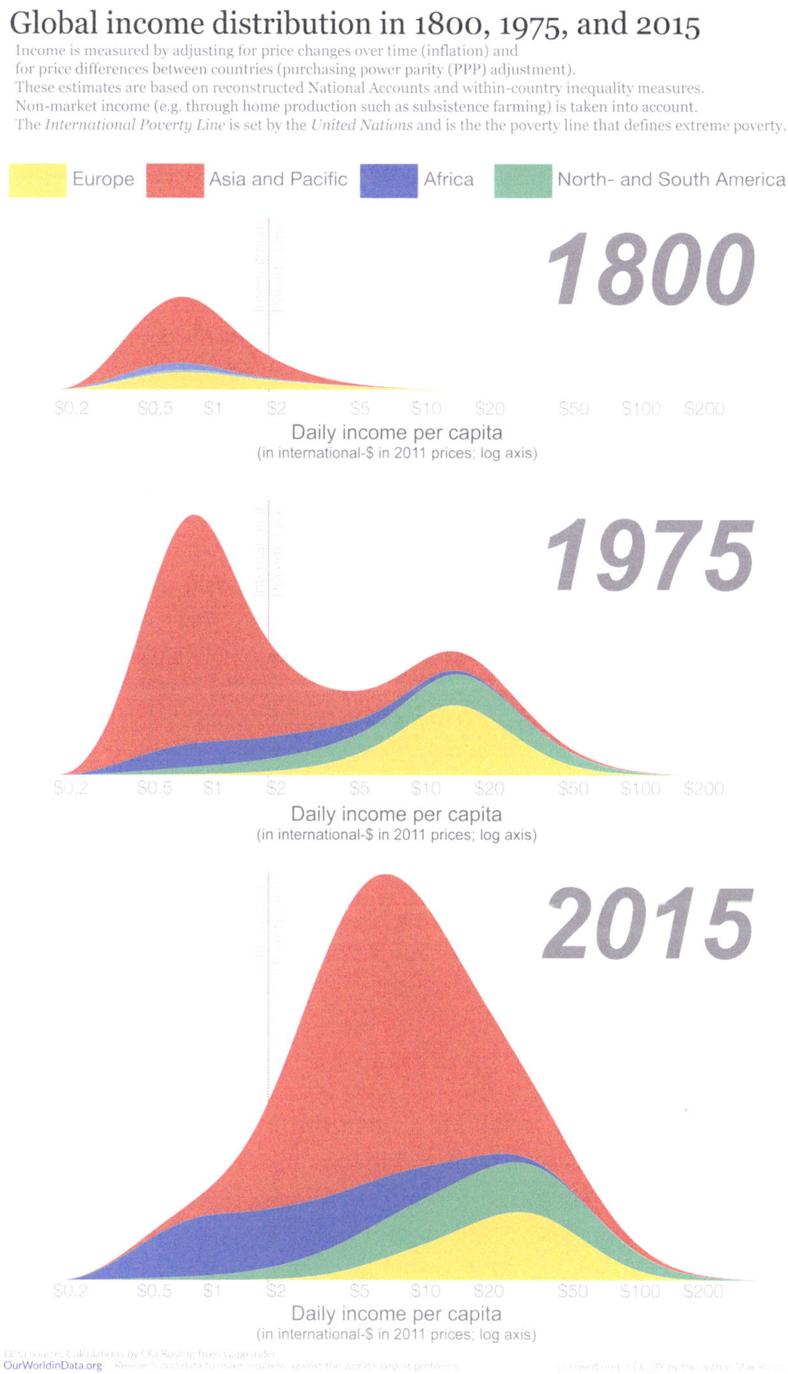

Fig. 10.11 Global income distribution in 1800, 1975, and 2015, *Source:* Rosling (2018); Roser and Ortiz-Ospina (2019c)

than the developing world. Four decades later, local income distribution changed dramatically. Incomes have grown faster in power countries in South-East Asia than they have in rich countries. Although there is a considerable income difference between developed and developing countries, the world income distribution is no longer bimodal. Extreme poverty has fallen faster than ever before in human history (Roser and Ortiz-Ospina 2019b). The reduction of inequalities is a key resource for sustainability because it enhances productivity and minimizes poverty.

Which country has the highest and lowest income inequality (a) in the world and (b) in Africa? What are the sustainable strategies for reducing inequalities globally and in Africa?

10.2.11 Goal 11: Sustainable Cities and Communities

The core objective of SDG 11 is to make human settlements and cities safe, sustainable, and resilient. The target is to make sure that people can access affordable and safe housing. The determinant or indicator of measure for this goal is the number of people who live in slum, informal settlements, or inadequate housing. There has been significant progress made towards this target as this population has been reduced from 39 to 30% between 2000 and 2014 (Sachs 2012). See Fig. 10.12. With population growth, the movement of people from rural to urban regions has exponentially grown. Consequently, there is need to ensure that we have resilient communities and sustainable cities.

Identify and discuss 6 major ways of making cities sustainable and communities resilient.

10.2.12 Goal 12: Responsible Consumption and Production

The core objective of this goal is to ensure that there are sustainable production and consumption patterns. Some of the targets of SDG 12 are the use of eco-friendly production methods as well as reduction of waste (Sachs 2012). Interestingly, there are a number of international multilateral agreements such as the Montreal protocol, Rotterdam convention, Basel convention, and Stockholm convention (see Fig. 10.13). By 2020, an increase in global recycling rates is anticipated in accordance with the agreed international frameworks. Towards this end, organizations

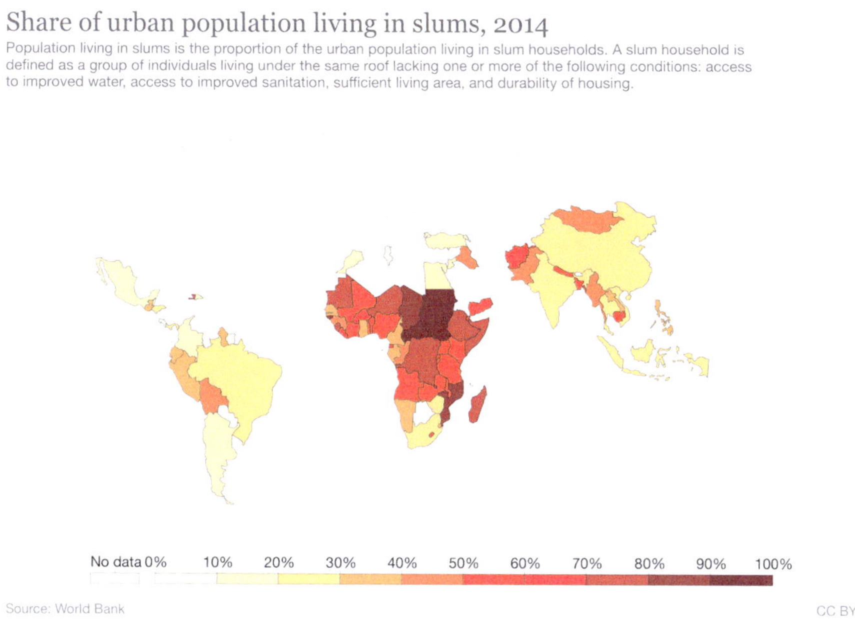

Fig. 10.12 Share of urban population living in slums, 2014, *Source:* World Bank. (2014)

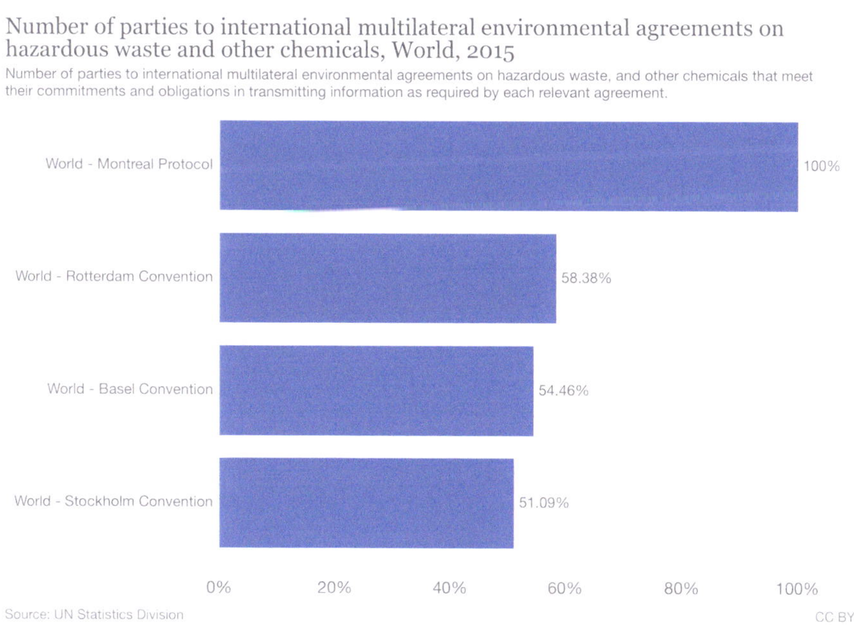

Fig. 10.13 Multilateral environmental agreement on hazardous waste and other chemicals, World, 2015, *Source:* United Nations Statistics Division (2015); Ritchie et al. (2018a,b)

need to adapt sustainable products as well as publish sustainability reports to inform the public and circulate best practices.

What are consumption patterns? What is meant by sustainable consumption patterns? How can we embed sustainable consumption patterns today?

10.2.13 Goal 13: Climate Action

The core objective of SDG 13 is to tackle climate change and its effects through emission regulation and promotion of renewable energy developments. This target acknowledges that climate change and economic developments are inter-connected, especially in considerations of energy production, poverty reduction, and gender equity. There have been multilateral agreements such as the Kyoto protocol and the United Framework Convention on climate change (UNFCCC) signed by global parties to address trans-boundary environmental issues (see Fig. 10.14). However, there is a need for the public sector to step up its efforts in reducing negative effects on the environment (Sachs 2012). Tackling climate change is critical because of

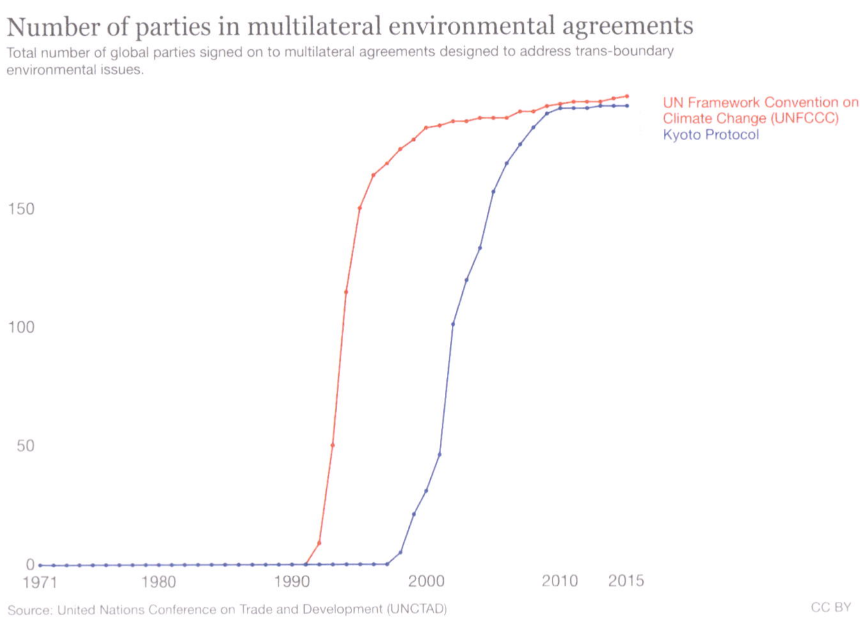

Fig. 10.14 Number of parties in multilateral environment agreements, *Source:* UNCTAD (2016); Ritchie and Roser (2019c)

its effect on food production, women's livelihoods, children's development, poverty reduction, and socioeconomic sustainability.

Why is climate action critical for achieving Vision 2030 and sustainable development goals? Provide four reasons.

10.2.14 Goal 14: Life Below Water

The core objective of SDG 14 is to "conserve and use seas, oceans and marine resources in a sustainable way to promote sustainable development." As oceans cover 70% of the Earth's surface, they are critical for ensuring that the planet is liveable (Sachs 2012). Currents and ocean temperatures regulate rain water and drinking water. More than three billion individuals rely on marine life to sustain their livelihoods, thus making these ecosystems important. Consequently, there is a need to protect and conserve these ecosystems to ensure that resources are used efficiently and sustainably. There is also a need to reduce acidification caused by carbon emissions that began to accelerate during the industrial revolution . Additionally, it is crucial to reduce the destruction caused by the overexploitation of fish stocks (see Fig. 10.15). Efforts should also be put in place to reverse marine pollution. People can assist oceans by minimizing their consumption of energy and plastic use. Nations should also take action. For instance, Norway earns money using a web page, finn.no, for collecting plastics from the beach. Various nations, including Kenya, have made progress by banning the use of plastic bags. The improvement of oceans leads to poverty reduction as it provides low-income families with a source of income and nutritious food.

Identify and discuss seven sustainable ways for conserving and using marine resources.

10.2.15 Goal 15: Life on Land

The core objective of this goal is to protect, restore, and promote sustainable utilization of terrestrial ecosystems as well as promote sustainable management of forests, combat desertification, halt biodiversity loss, and reverse land degradation. Figure 10.16 shows important biodiversity sites protected by natural authorities mainly for sustainable use. However, more effort needs to be made to reduce

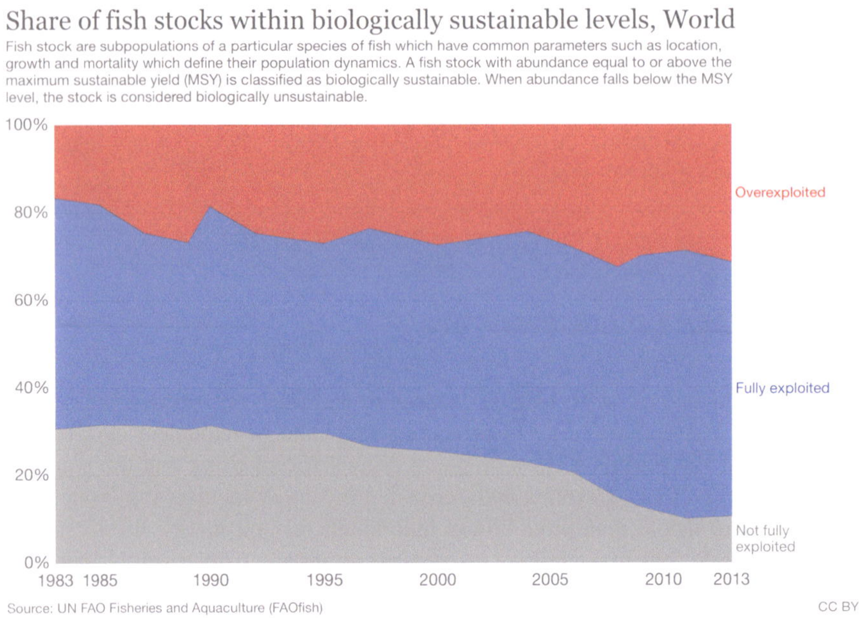

Fig. 10.15 Share of fishstock within biologically sustainable levels, *Source:* United Nations Food & Agriculture Division (FAO) (2013); Ritchie and Roser (2019d)

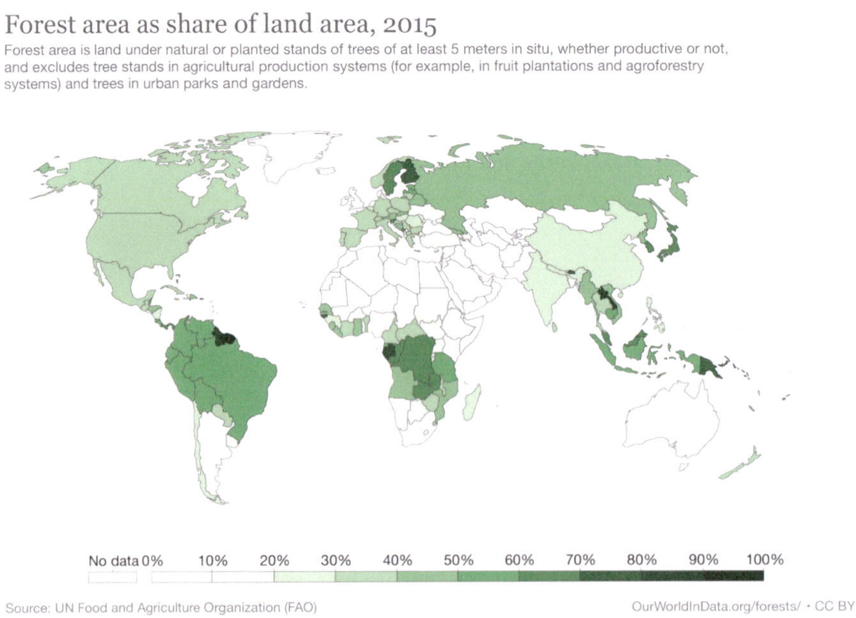

Fig. 10.16 Forest area as share of land area, 2015, *Source:* FAO (2012); Rosner (n.d)

degradation. To achieve a world devoid of degradation, there is a need to restore degraded forests and land lost to floods and drought. SDG 15 advocates for more attention in preventing the invasion of alien species and protection of endangered wildlife.

You have been employed as a policy expert in Africa for devising ways ensuring sustainable use of terrestrial ecosystems. What model countries would rely on your report? Discuss 10 points you would include in your report.

10.2.16 Goal 16: Peace, Justice, and Strong Institutions

The main objective of Goal 16 is to "develop peaceful and inclusive societies for sustainability, enhance access to justice and build strong institutions at all the levels" (Sachs 2012). The reduction of sex trafficking, violent crime, compulsory labor, and child abuse are important global goals. There has been a decline in the number of individuals exposed to sexual exploitation. Ending sex trafficking, forced labor as well as all forms of violence is critical for enhancing sustainable development. The lack of sufficient data poses a major challenge and there is a need for countries to improve their reporting (see Figs. 10.17 and 10.18). This goal also seeks to achieve universal legal identity and registration of births by ensuring that all rights are upheld and protected.

Africa faces a major challenge in building strong institutions. How does the lack of strong institutions in the continent impede progress and sustainability?

10.2.17 Goal 17: Partnerships for the Goals

The core objective of this goal is to "strengthen and revitalize global partnerships for sustainable development" (Sachs 2012). International cooperation is critical for achieving all previous goals that have been discussed. This goal is critical for ensuring that organizations and states cooperate rather than compete. The development of multi-stakeholder partnerships for sharing expertise, knowledge, and financial support is crucial for the overall success of the 17 SDGs. There is a need for public–private partnerships (PPPs), which engage civil societies to

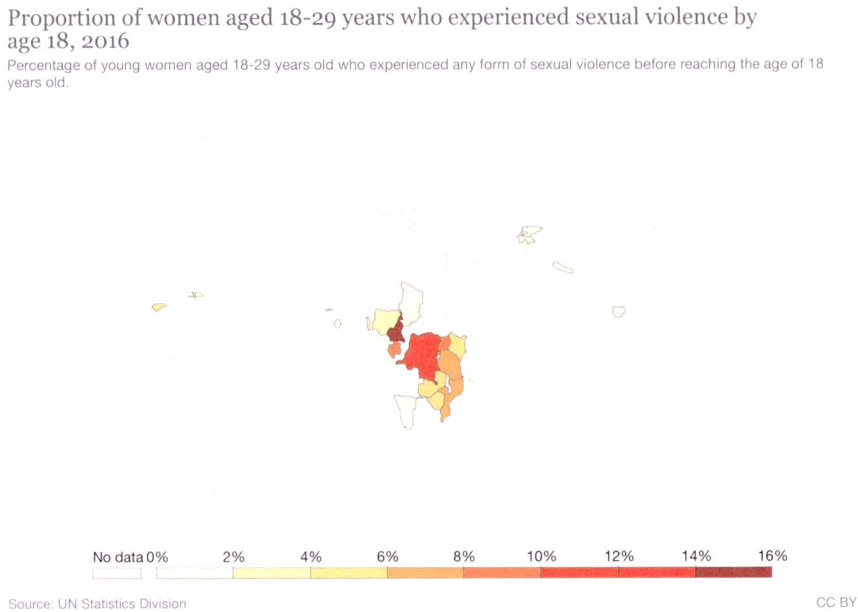

Fig. 10.17 Proportion of women aged 18–29 years who experienced sexual violence by age 18, 2016, *Source:* United Nations Statistics Division (2016); Ortiz-Ospina and Roser (2019)

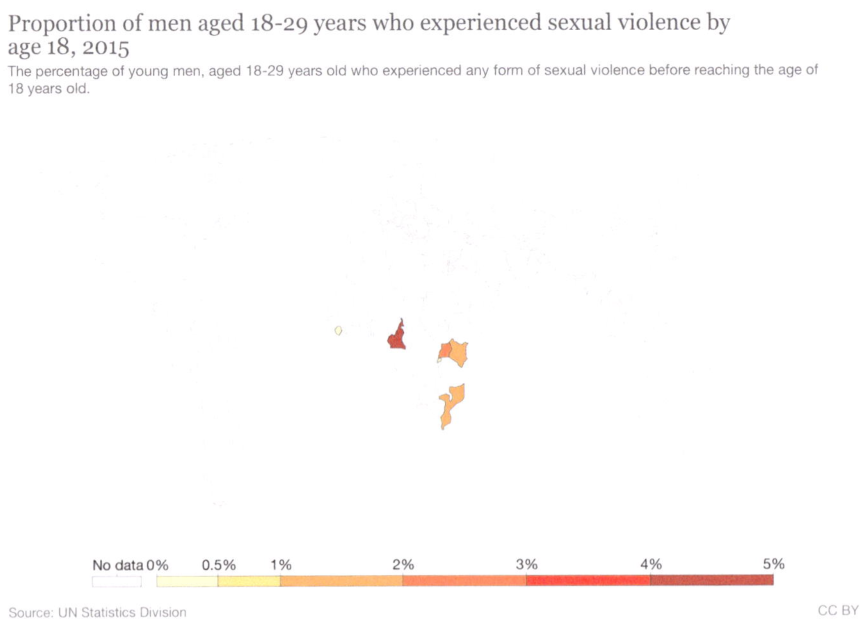

Fig. 10.18 Proportion of men aged 18–29 years who experienced sexual violence by age 18, 2015, *Source:* United Nations Statistics Division (2016); Ortiz-Ospina and Roser (2019)

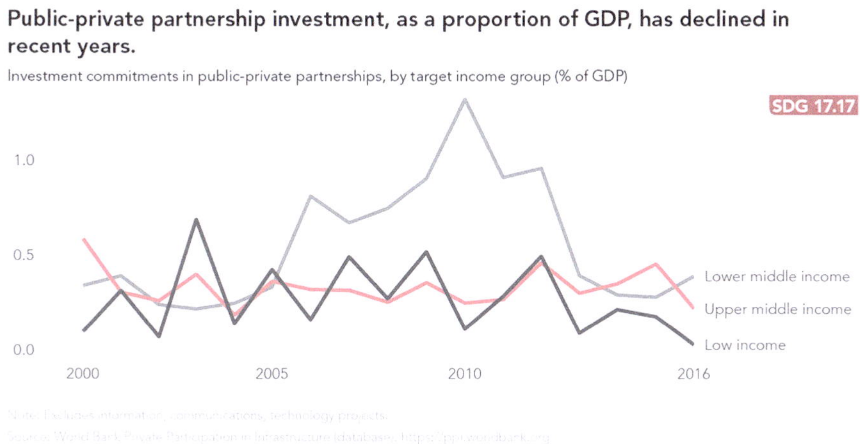

Fig. 10.19 Investment commitments in public–private partnerships, by target income group (% of GDP), *Source:* World Bank (2018b)

enhance sustainable development. Public–private partnerships (PPPs) are cooperative arrangements between public and private sectors, and are often long-term. A public–private partnership is a cooperative arrangement between two or more public and private sectors, typically of a long-term nature. Governments have used such a mix of public and private endeavors throughout history. However, public–private partnership investments have declined in recent years (see Fig. 10.19).

What are the major barriers that impede the development of partnerships, and how can they be tackled?

10.3 The Pillars of Sustainability

The term sustainability was first coined in the 1980s, when the United Nations introduced the concept in the Brundtland Commission, which raised the issue of human sustainability on Earth. This was further developed in 2005 at the World Summit on Social Development. At that time, this was the largest ever gathering of world leaders, who discussed the issues (of which the environment plays a central role) that face our planet. Following the summit, three core areas of sustainability were identified that are critical in the underlining philosophy of sustainable development. These three "pillars" of sustainability form the backbone of many national and international schemes that tackle the core problems of sustainability the world faces (Khoshnava et al. 2016) (Fig. 10.20).

Fig. 10.20 Pillars of Sustainability

Which year was the concept of Sustainability coined? What is the difference between the definition provided then (1980s) and today?

10.3.1 Pillar 1: Economic Development

This issue of sustainability is widely considered the most problematic. People generally disagree on the political and economic ideologies, as any sustainability issues within this area will affect businesses, and by extension, employment. Sustainability choices within this pillar involve incentivizing businesses and other organizations to aspire to sustainable operations, and adhere to guidelines that are beyond their normal legal requirements. It also aims to encourage individuals to contribute, wherever and whenever they are able (Khoshnava et al. 2016). One person can very rarely achieve anything momentous on their own, however, when contributing as part of a larger group, effects are cumulative, and much more can be achieved.

As we already know, in our consumerist lifestyles, our modern-day way of life demands the consumption of substantial natural resources on a day-to-day basis. Therefore, it is critical that we understand the quantities that we consume, so that we are able to manage consumption in order to protect the environment (Khoshnava et al. 2016). The economic development pillar is therefore essentially meeting the needs and wants of the people, without compromising the environment. This is especially important when considering the developing world by allowing them to develop in a way that reduces the financial burden on future generations.

How does consumerism deter us from achieving and/or promoting sustainability?

10.3.2 Pillar 2: Social Development

This pillar takes a number of different components into consideration. The most important is the encouragement of awareness, legislation, and protection in order to promote people's health. Additionally, it is a safeguard against the damaging effects of pollution and other harmful or potentially future harmful activities that result from the operations of businesses and other organizations.

In the developed world, such as Europe and North America, there are strict legislations to ensure that society's health and well-being is protected. The social development pillar is also concerned with the maintenance of access to basic human needs and resources without the overexploitation of nature. The final element is education; encouraging people to participate in environmental sustainability by teaching people about the effects of their actions on the environment, the effects that protecting it can bring, and the consequences if we do not (Khoshnava et al. 2016).

Discuss how education is a prerequisite for the success of social development.

10.3.3 Pillar 3: Environmental Protection

We all know what is required from our individual contributions to protect the environment. Be it recycling, reduction of power consumption by switching off our appliances, or walking short journeys instead of using the car. However, these are personal choices and governments are reducing taxes on renewable energy source imports to encourage the adoption of renewable power installation in homes.

Protection of the environment is the third and final pillar of sustainability, and one that many people consider the most important for the future of the human race. Environmental protection dictates how we should study our ecosystems, the air quality around us, and sustainable extraction of our natural resources. There is a focus on individual elements and actions that we inflict on the environment and how it puts it under stress. There is also a focus on the need for technology to help drive a greener future, with the recognition that development in technology and biotechnology will be the key to achieving this sustainable future (Khoshnava et al. 2016). This strategy must also plan on protecting the environment from the potential damage that our technological advances may bring (Fig. 10.21).

Identify the major ecosystems on the planet. Using these ecosystems, explain measures that can be used for protecting them.

Case Study 10.1: Green school in Bali

Fig. 10.21 Green School, Bali, *Source:* Ibuku (2019)

The Green School in Bali has taken teaching to a whole new level. Located in Indonesia, Green School provides holistic and green education in an amazing environment (Macrory 2013). Its objective is to develop a generation of citizens with knowledge and inspiration to be responsible for the planet's sustainability.

(continued)

The school achieves this through:

- Emphasis on learning by doing.
- Development of social responsibility.
- Appreciation of integrity, ethics, compassion, and honesty as major underlying values.
- Recognition of the significance of holistic education.

They also achieve this through a curriculum that:

- Trains relevant skills and science to a high level.
- Evolves from ecology, environment, nature, and sustainability through a hands-on approach.
- Acknowledges and appreciates the local cultural and geographical context.
- Emphasizes progression and continuity (Macrory 2013).

The major aims to the curriculum are to:

- Instill an understanding and respect for the environment.
- Heighten environmental awareness and promote stewardship.
- Develop ecological values from practical learning.
- Gain insight and practice permaculture (development of self-sufficient and sustainable agricultural systems), including the use of appropriate aquatic systems, rice agriculture, and animal care.
- Integrate learning curiosity and multiple levels of intelligence.
- Understand sustainability as an intricate web of community relationships.
- Enhance collaboration between schools and community members.
- Integrate subject matter in meaningful experiential projects.
- Instill pride and cooperation among the students as future world leaders (Macrory 2013).

Arguably, Green School is a great project that pioneers sustainability in education. It pushes boundaries in schools when the world is faced with daunting environmental challenges. The curriculum combines the academic strength expected in schools and higher learning institutions with practical experiential learning. It implies that by developing important core subjects of mathematics, science, and English, Green School learners will open doors for whichever form of learning and careers they choose (Macrory 2013). The Green School is attended by students from across the world, and each brings with them knowledge and experiences that help them change perceptions about environmental management. Arguably, the school has played a critical role in imparting skills, knowledge, and values that are critical for environmental sustainability.

10.4 Environmental Sustainability

10.4.1 What Is Environmental Sustainability?

Human civilization is now a fully integrated and interwoven part of the global society. With growing human populations and vast economic growth, we have become an integral and crucial component of the Earth's ecosystems. The concept of sustainability lays foundation for understanding the concept of environmental sustainability. Therefore, environmental sustainability can be defined as responsibly interacting with the environment to deter degradation or depletion of natural resources, and sustain long-term environmental quality. These ecosystems are the foundation for all life on Earth (Moldan et al. 2012). Through our constant search for the acquisition of natural resources from the planet that sustain our modern-day way of life, we are manipulating, changing, and often harming our planet's natural order and the way these ecosystems function.

Some argue that humanity has already created an imbalance or negatively affected the natural balance of the world's systems. In order to leverage such imbalance, it is crucial that we ensure all of human civilization's endeavors are environmentally sustainable. However, environmental sustainability is a broad and complex term that encompasses a host of disciplines ranging from business and technology to the social sciences and the environment itself.

Sustainability can be thought of as the ability to continue a given behavior, or task indefinitely. In relation to environmental sustainability, it is therefore how human civilization can take resources from the environment in order to sustain its own expansion, while allowing natural systems to produce everything they need to function normally and remain ecologically diverse (Moldan et al. 2012).

Humans find it difficult to develop environmentally responsible behaviors. What are the contributing factors?

Our modern-day consumerist and predominantly urban existence throughout most of the developed world result in the consumption of vast amounts of natural resources. Electricity is produced through the burning of coal, while oil and gas generate heat. This creates energy that keeps our streets and buildings lit, our modern-day appliances working, and our homes warm. With such a vast range of processes that require natural resources, an imbalance has arisen between the amount we take out and the amount we put back into the environment. It is estimated that we take 40% more resources from the environment than we are capable of putting back every year (Moldan et al. 2012); therefore, illustrating the problem facing humanity in our pursuit of sustainability. Environmental sustainability therefore puts a focus on restoring this balance, where there is an incredibly thin line between the competing needs of our civilization moving forward in technological and economical advancements, as well as the need to protect the environment within which our society lives.

Explain how the triple-line approach could be beneficial for environmental sustainability?

However, environmental sustainability is not just exclusively about protecting the environment. It also recognizes the importance of incorporating human health, and economic as well as social well-being to ensure that people, or other forms of life, do not suffer as a result of laws enacted to protect the environment. This is all within a time frame that addresses both the immediate and longer-term effects of human actions, and how we can seek to improve them now and for future generations.

10.4.2 The Primary Goals of Environmental Sustainability

Environmental sustainability is a multi-faceted and complex combination of issues that must be tackled simultaneously. Given the complexity of the problem, it is essential that we set a number of goals to strive towards finding equilibrium between sustaining human civilization and the proper functioning of the environment. In 2012, the United Nations Conference on Sustainable Development met with the intention of discussing this and developing a specific set of goals and targets.

Which SDG do you believe to be the most important?

One crucial component of the SDGs was the novel concept of nature having rights. We as a society have stewardship of our planet and it is important that we place people at the forefront of solving these global issues, through environmental management and managing resource consumption. Now that we fully understand the full scope of what must be done and the angles that must be addressed to ensure sustainability, we will discuss specific areas of human behavior, how they affect the world we live in, and how environmental sustainability can be used to solve the problems they produce. This must all be considered with the overall mantra of environmental sustainability as development that can meet our current needs but at the same time ensure that future generations' needs are not compromised.

10.4.2.1 Constructions and Human Settlements

Throughout the last decades, we have experienced unprecedented and exponential urban growth. Close to four billion people (54% of the world's population) lived in cities in 2015, and this number is only projected to increase to around five billion people by 2030. This rapid urbanization has resulted in tremendous pressure on the environment, and with over 4% of the Earth's entire land area now occupied by cities, it has created a number of challenges (Moldan et al. 2012). Increased air pollution and unplanned urban sprawl into the environment are the most damaging consequences. The safe management and removal of waste from our cities also present a further set of environmental challenges. However, these will be discussed in further detail later.

While it may sound counterintuitive, urbanization can actually benefit the environment. People that live in dense cities generally drive less and their homes require less energy. This results in decreased resource consumption, and as such, leads to less pollution. However, while as a general rule this may be true, there are a number of aspects of urban life that can cause negative consequences. With a densely populated city, there is greater traffic congestion that leaves cars stationary, often for a long period of time, which leads to the burning of gasoline while idling, a fossil fuel and energy source that is responsible for a large amount of pollution across the world.

Many cities also often construct their buildings using materials from non-renewable sources, much of which is imported from a great distance, therefore increasing the amount of energy consumed via transportation. Indeed, the energy that powers a city may also come from a distant location, where a power station may not be within a city's limits, further increasing the energy cost for our cities to operate.

Identify three world cities that construct their buildings from sustainable resources. How can this be emulated globally?

In order to address this unsustainable nature, a new concept is the creation of "eco-city." These are essentially cities designed with careful urban planning that consider their environmental impact and place an emphasis on the minimization of water, energy, and food inputs as well as the minimization of waste outputs such as air pollution and heat. For a city to achieve environmental sustainability, it needs to be able to feed itself with a sustainable reliance on the nearby countryside and should be able to power itself on sustainable renewable energy sources. These eco-cities can be achieved through a variety of methods, many of which have already begun to be incorporated into many greener cities around the world. An example is Copenhagen which is touted to be Europe's greenest city.

Road occupancy levels are set to increase six-fold in some countries around the world in the next 50 years. To tackle this, incorporating better and more efficient mass transit systems can help achieve sustainability, while not sacrificing quality of life for urban settlers. An improvement in public transport and an increase in pedestrianization would allow for the reduction in car pollution. While there has been a big push for this around the world, it has not been at the pace that experts say is required. By implementing more strategies to minimize traffic on city roads, the effects of emission gasses, such as carbon dioxide (CO_2) and methane (CH_4), will be lowered.

What is the major challenge in the adoption of green transportation modes? Orient these challenges to the African context.

Cities also need to address the issue of energy efficiency within their infrastructure. Throughout the developed world, over 40% of carbon emissions come from cooling, heating, and powering buildings (Bettina 2011). Air conditioning units in particular are a huge energy burden. Through better design and planning, city buildings can be made more efficient to lower these energy demands (Bettina 2011). Trees can be planted, and buildings painted in lighter colors to reflect light and heat better as well as the installation of natural ventilation systems that do not require energy. Additionally, cities could be planned with more green open spaces and water features that reduce what is known as the "heat island effect," which results from the abundance of tarmac and asphalt making urban areas significantly warmer than rural areas. Experts say these measures could save us as much as 30% of energy usage (Bettina 2011).

Enumerate six constructions in major cities that have embraced energy efficiency in their designs and constructions.

To meet the stipulation of an eco-city, it must be able to feed itself sustainably from nearby surroundings, such as new agricultural systems designed within cities. These urban farms would build plots within the city either in the suburbs or in the city center and would reduce the distance that food would need to travel from the field to the people. This could be done either in small scale or private farm plots, or on a larger scale, such as on building rooms (farm scrapers). For instance, Tianjin Eco-City has designed to depict many latest green amenities and technologies, including solar-scrapes, wind power, wastewater treatment, rolling landscapes, and light powered rail. The second stipulation of an eco-city is of course its ability to sustain itself through renewable and clean energy sources. Again, renewable energy sources will only be touched on briefly as it will be discussed in a later section, but the installation of renewable energy sources would make cities greener and put less strain on the environment. Wind turbines could be installed near cities to provide a regular energy supply in close proximity to minimize energy consumption. In hot countries, solar panels could be installed on every building, which would massively boost environmental sustainability by exploiting an energy source that would otherwise be completely wasted. Additionally, cities could also utilize biogas by using the cities' own sewage. This would tackle the issue of removing waste while also providing a much greener energy source.

How can Africa embrace solar energy and ensure that it is sustainable?

While this is but a mere snapshot of the possibilities for creating more environmentally sustainable human settlements, it clearly highlights the broad spectrum of solutions that can be implemented to make our cities greener. Clearly, given the busy and complex nature of cities, it may appear that achieving environmental sustainability in such an environment is difficult. However, even with the examples provided, it is possible to be done in tandem with meeting the needs of the inhabitants without the sacrifice of modern-day lifestyles. Indeed, not only will they make our cities more sustainable, but many strategies will also contribute in meeting the targets in other areas of environmental sustainability.

10.4.2.2 Emissions and Wastes

One aspect of environmental sustainability that we must address which is intrinsically linked with our need for "eco-cities" (among other areas) is the control and maintenance of our emissions and waste management. Fossil fuels are formed through natural processes, such as the decomposition of organisms. These fuels typically contain high percentages of carbon—such as coal, petroleum, and natural gas. Every year, over 21.3 billion tons of CO_2 is emitted into the atmosphere through the burning of fossil fuels. This is rising every year and is over 60% above the level of emissions recorded in 1990 (Hosseini and Kaneko 2012). These emissions are coming from a huge variety of sources, although primarily from industry and the use of fossil fuels, which account for around 91% of all human-induced CO_2 emissions, with the other 9% arising from land use changes (Hosseini and Kaneko 2012). Of the emitted CO_2, the biggest contributor is coal with 41%, while 34% came from oil, 19% from gas, and 5.6% arising through cement production (Hosseini and Kaneko 2012).

While there have been a number of new protocols, targets, and implemented legislations set to curb our emissions and bring them under control, our emissions are still increasing. In 2014 our global fossil fuel emissions grew by 0.7%, and held steady through 2015, before only seeing a small rise of 0.2% in 2016 (Le Quéré et al.

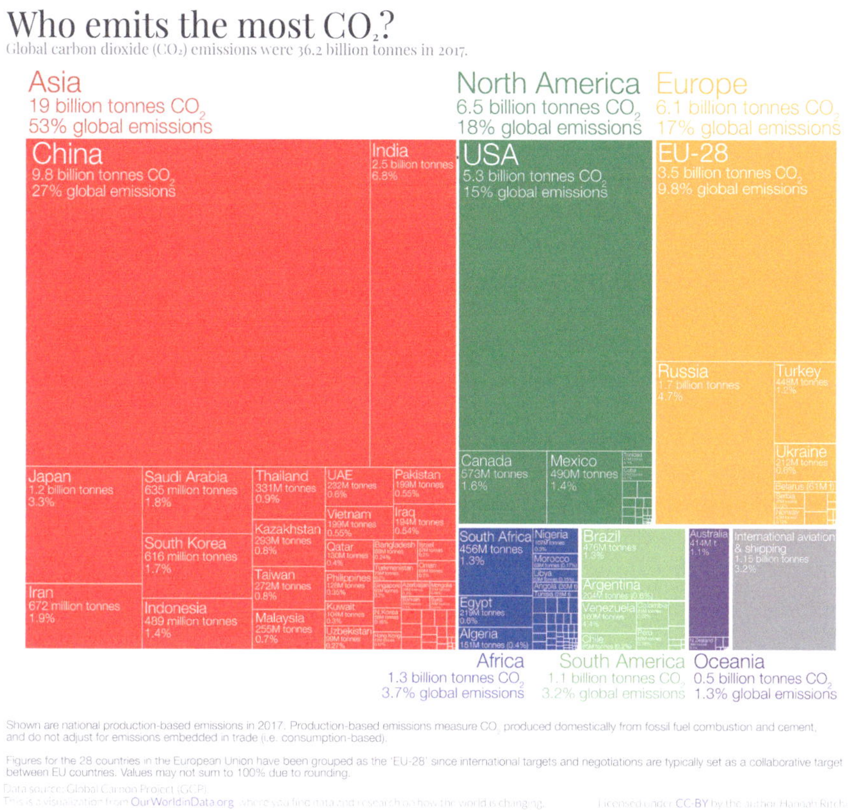

Fig. 10.22 CO_2 emission rate based on geographical region, *Source:* Global Carbon Project (2018); Ritchie and Roser (2019a)

2016). Figure 10.22 shows the annual CO_2 emission by geographical region, with Asia being the largest emitter accounting for 53% of global emissions and Oceania being the lowest emitter accounting for 1.3%. Therefore, if we are to tackle global warming and climate change, we must introduce new measures that bring down and reverse these figures drastically in the coming few decades.

Additionally, waste that is created from the multitude of human enterprises which we employ is also of increasing concern for the prosperity of environmental sustainability. The pollution of soil, water, and air poses a serious threat to sustainable development, our health and the environment. Set in a backdrop of raw material scarcity, more expensive energy, and pollution risks, our waste management and issues of disposal are exacerbated by urbanization, industrial development, and changing patterns of consumption. Our traditional systems that accommodate the disposal of solid waste as well as current recycling methods may no longer be appropriate.

Waste is being used increasingly to produce many materials and as an energy source. While recycling has reached a point where it saves more GHG than it originally generates, countries that are developing and emerging face the challenge of needing to improve their unsustainable and ultimately inadequate waste management systems. We must look for other alternatives so that waste is no longer simply deposited within residential areas, dumped into landfills that are uncontrolled, that can ultimately end up in waterways.

How can governments help in controlling waste generation and prevention?

These methods of waste disposal cause waste to accumulate as marine litter within the world's oceans. Plastic waste causes major damage to the world's wildlife and eventually finds its way through the food chain in the form of micro particles. A large proportion of the waste in the ocean and waterways originates from terrestrial sources and waste deposits caused by non-existent and highly inefficient waste management regimes.

Devise three waste disposal policies that can help alleviate bad waste management practices.

The development of cooperation projects is simply not enough to analyze technical, financial, and organizational strategies for environmentally sustainable resources as well as waste management. These factors, therefore, should feature prominently in discussions at both international and national levels. Management of waste and emissions is linked with many areas and must be tackled with a multidisciplinary approach. The following subsections address some of the key areas of emissions and waste with a discussion into the various strategies that either are or can be implemented to reduce their environmental impact and increase their sustainability.

10.4.2.3 Agriculture

Farms were responsible for six billion tons of GHG emissions in 2011, equating to around 13% of total global emissions, making the agricultural industry the second largest emitter after the energy sector (Oliver et al. 2013). The use of natural and/or synthetic crop fertilizers for our soils and cattle belching are the largest sources of farm originated emissions, making up 65% of all agricultural emissions globally (Oliver et al. 2013). Other smaller sources include those related to manure management, field burning of crop residue, fuel use for machinery, and rice cultivation.

The biggest opportunities to reduce these agricultural emissions come from changing our modern-day dietary preferences and habits as well as evaluating our agricultural practices. According to the World Resources Institute (WRI), a global research organization, we waste 24% of all calories that are produced by the agricultural industry for human consumption (Oliver et al. 2013). This is either through loss in the food supply chain or simply wasted by excessive spending and overindulgent habits. To address this, we can alter our food habits by reducing our meat intake particularly beef. With a reduced demand for beef, a substantially reduced amount of methane emissions will follow. Combined with a more efficient and stricter spending as well as eating regime, this will result in a decreased amount of energy wastage and as a result, significantly less GHG emissions.

What is your individual contribution to reducing agricultural emissions?

This can also be augmented through an improvement in our agricultural practices. On the supply side, utilizing crop management practices to include improved management of fertilizer application and a more conservative tillage approach would offer the largest reduction in emissions at a relatively low cost. In California, for example, an extensive water transfer and storage system has been established that allows crops to be produced in arid regions. During drought periods, a limited surface water supply has led to an overdraft of groundwater, which consequently allows the intrusion of saltwater, or worse, permanent collapse of aquifers. To improve on such practices, we could develop drought-resistant farming systems even in "normal" years such as:

- An improvement to water conservation and storage methods.
- Transition into a reduced-volume irrigation system.
- Providing incentives for the planting of drought-tolerant species.
- Not planting in drought seasons.
- Managing crops for a reduction in water loss.

Of course, it is not just farming agriculture that needs to be evaluated and addressed for us to achieve environmental sustainability. Other forms of food production and harvesting are also leading to unsustainable environmental consequences. Fish account for 17% of the global population's protein intake, however as a result, the world's fisheries are vastly overfished (Oliver et al. 2013). Over 70% of all fish species are currently either fully exploited, depleted or overexploited (Oliver et al. 2013). It is predicted that all fisheries will collapse before 2048 (Oliver et al. 2013). By modifying our fishing practices, such as tackling destructive fishing practices (blast and cyanide fishing), illegal fishing, and implementing individual fishing quotas, we could reduce the pressure on these fisheries and allow them to rebound and recover for future prosperity.

Changing our land management strategies is also crucial for tackling the environmental impact within this industry. Palm oil is widespread throughout the international food industry, having become very popular on account of its lower production costs. On average, an individual consumes 17 pounds of palm oil per year. This staggeringly high use, along with the high yield of oil palms, has encouraged its wider cultivation, leading to the clearing of large tracts of forests, particularly in Malaysia and Indonesia. These oil palm monocultures lead to significant losses in natural habitat and deforestation that has resulted in severe repercussions on local wildlife. These include the Sumatran orangutan and the Sumatran tiger, both of which are now listed as critically endangered as a result of habitat loss.

Given the high yields of these plantations, we must develop more sustainable practices for their cultivations. Many have suggested simply avoiding purchasing products that contain palm oil. However, research suggests that switching to other types of oil, such as vegetable oil, could lead to even worse environmental consequences, as they provide lower yields and would require even greater cultivation sizes for the same quantity. Instead, we should seek to implement policies for greater sustainability in our land usage, such as those set out by the Roundtable on Sustainable Palm Oil, an organization that operates to make sustainable palm oil the norm. They introduce policies that ensure no deforestation occurs, no peat development, and no overexploitation. These can be achieved by incentivizing farmers to stop using slash and burn techniques to clear their land, thereby reducing overall GHG emissions and land access for its conservation value and carbon stock. By incorporating these strategies into the wider framework of agricultural sustainability, we can strive for a reduced environmental impact and a greater efficiency of our land usage.

How can vertical farming be a solution to the decline in the size of arable lands?

10.4.2.4 Water

Water is the basis for all life on Earth and there is an ever-increasing pressure on its resources. Our exponential population growth and ongoing economic and technological expansion have resulted in a continued and increasing need for water resources. Many important aquifers are over-pumped, which has led to a significant widespread decline in groundwater levels. Several major rivers around the globe, such as the Yellow River in China and the western US Colorado River, now no longer reach the sea most years (Wang et al. 2013a,b). Over half the world's wetlands have been completely lost, and the quality of the world's water is consistently being degraded, thus threatening the health of local ecosystems and people with ever-increasing costs associated with its treatment. These problems are deeply rooted in our wasteful use of water, which is characterized by inefficient management systems, lack of economic incentives, underinvestment, and our failure to implement new technologies for superior management. Our water wastage is further exacerbated from a political standing as groundwater aquifers and watersheds do not respect political borders, meaning conflict between countries and municipalities for its ownership often leads to over-extraction. With a water demand that grows day-by-day, we are further draining our existing supplies, creating a greater debt on groundwater use. The rate at which groundwater stores are depleted has doubled since 1960. With such a high level of drainage from aquifers, we are causing major problems to the world's aquatic ecosystems, such as rivers, floodplains, and watersheds. These ecosystems bear the brunt of the repercussions following water reductions, by entering a state of degradation, or completely disappearing all together. This further increases the environmental costs, as we often rely on these ecosystems for a number of other ecosystem services besides water provision. There are many opportunities to conserve water, such as implementing more efficient management of our resources. A large quantity of global water usage is for agriculture, so this should be the first area we look for better management solutions. Through the implementation of better agricultural techniques, such as drip irrigation and growing crops that are more suitable to their local climate, vast sums of water can be saved. This can be further augmented by offering financial incentives to farmers to encourage them to incorporate more efficient water practices into their farming methods.

What activities can put in place to maintain the efficient management of water?

It is also imperative that we safeguard against water contamination to preserve the quality of water that we still have. Agricultural runoff, which leads to leeching of fertilizers, pesticides, and herbicides into the soil and groundwater supplies has resulted in contaminated local water sources. By switching to more sustainable alternatives, such as some organic sources, we can decrease contamination levels and preserve water quality. Of course, it is not just farming practices that can be altered to save water. Other industries must also incorporate better water management strategies so that their enterprises can be more sustainable. Businesses should begin to conduct audits that will reveal areas where major water costs are occurring (particularly those that use large amounts of water such as the hospitality industries, such as hotels). This will highlight areas that can be improved and government incentives should be offered to allow investment in new technology and water reduction schemes. All these methods of improved water management, can also be incorporated within a wider framework of our future "eco-cities." Many residential areas have employed a number of strategies to contain and save water, such as water butts in the garden that are connected with a guttering system that traps and stores water for later use. Additionally, education campaigns will help raise awareness on the scarcity of water and ways that citizens can reduce their individual water consumption and wastage.

What impedes Africa from developing eco-cities? How can the challenges you have identified be solved?

10.4.2.5 Energy

As we know, the burning of fossil fuels as a source of energy is causing large amounts of CO_2 emission into the atmosphere. It is estimated that only half of the emitted CO_2 is absorbed back by natural processes, leaving an excess in the atmosphere which, given its GHG status, contributes to global warming. The Energy Information Administration has estimated that 36% of global energy comes from petroleum, 23% from natural gas, and 27.4% from coal (Dincer and Rosen 2012). With world energy consumption growing at a rate of 2.3% per year, the use of these fossil fuels is unsustainable on two fronts. Firstly, due to the level of pollution and contribution to global warming that they incur, and secondly, they are non-renewable and are set to run out in the not too distant future. With pressure to reduce our fossil fuel usage, we must strive to find more sustainable and renewable options that can still meet our growing energy demands. Sustainable energy is the provision of energy that meets the needs of the present without any compromise in the ability for future generations to be able to meet their own needs. We have developed a number of different technologies that promote and produce sustainable energy through renewable energy sources such as the wind, solar radiation, wave power, hydroelectricity, geothermal energy, bioenergy, and tidal power.

Identify critical measures for developing a robust infrastructure for the development of sustainable energy.

Solar energy is the best form of sustainable energy. It comes in two forms, light and heat, both of which can be harnessed for energy and has the widest application of any renewable energy source. Through installation of solar panels, we can trap the sun's rays and use it to generate electricity and heating. The energy produced from the panels can be stored for later use, meaning that if a network of solar panels were installed on every home and building, a completely green energy source could be provided from a power grid across our towns and cities. Of course, the efficiency of this energy source will vary depending on geographical location as well as the amount of sunlight received year-round at different latitudes and seasons.

Wind energy is another renewable source that has the capacity to provide vast quantities of power. Alongside solar power, this energy has become heavily commercialized in today's modern world. Many companies are investing in power grids and wind turbines that allow us to tap into this energy source. There is some disagreement in the use of wind turbines, as many people see them as unsightly, and do not want them "on the doorstep", blocking their views and becoming an eye sore. However, the development of offshore wind farms, where wind turbines are built in large numbers in the ocean and out of sight, provides us with an answer to this problem of aesthetics. In the past, the cost of power produced from offshore wind farms has been much higher than from onshore farms. However, since 2016 this has decreased substantially and there are now many more offshore farms in development around Europe. If these offshore wind farms can become a mainstay in the future, they could provide vast quantities of clean, green energy without any negative effects on the beauty of our natural environment.

What are the major benefits of wind as a sustainable source of energy over other sources?

Geothermal energy allows us to fetch energy from below the earth. By installing geothermal power stations, we can use the heat rising from the Earth's core to generate electricity. The temperature below earth at around 10,000 meters deep can be high enough to boil water in some locations. This heat can be used to produce steam and move turbines that produce electricity. However, while there are vast sums of this heat rising up through the Earth, its use as a renewable energy source is more limited in its application (Kaygusuz 2012). Geothermal temperatures vary at different locations around the globe and may not be high enough to move the turbines. However, for regions that have high seismic and volcanic activity, it can be harnessed in an environmentally friendly way to produce electricity on a 24/7 basis.

What deters countries from fully harnessing geothermal energy?

Biomass energy is produced through the burning of timber, wood, and municipal, agricultural, and landfill waste. It is completely renewable and does not produce harmful gases. Unlike other sources of renewable energy, it also has the added benefit of aiding in our waste management systems. It is considered a carbon neutral energy source, as while CO_2 is released through the burning of this fuel, it is taken back up in equal measure by the plants that are grown to produce the biomass. While this energy source may be renewable, it is unlikely that we can provide energy consistently enough for it to be used as a sole energy source.

This gives a small snapshot of the types of renewable and sustainable energy sources that we have at our disposal. While they all have various advantages and disadvantages, it is most likely the incorporation of several types of energy that will provide us with the energy that we require, while still being green and sustainable. There is, however, another type of clean and ultimately infinite source of energy that is highly efficient in its production, but that has produced a significant amount of controversy.

Nuclear energy can be obtained from an atom. These are very small particles, which make up everything throughout the universe and large amounts of energy are found within the bonds holding these atoms together. Through the processes of nuclear fusion (where energy is released when two atoms are fused to make a larger particle atom) or nuclear fission (energy release when atoms are split), a nuclear power plant can produce electricity. However, large amount of nuclear radiation is released during these processes, which has led to significant environmental concerns from the production of this energy. Waste disposal is one primary concern. Burying the irradiated waste is one option, however, this can cause serious ecological consequences as the radiation may seep into local water supplies and intoxicate local wildlife as well as ecosystems (Kaygusuz 2012). Other methods of waste disposal are being explored, such as ejecting the waste into space. Additionally, there are numerous safety concerns using nuclear energy, as is evidence from history, through incidences such as the Chernobyl nuclear disaster in Ukraine. However, nuclear power is becoming safer and once these environmental concerns are addressed, it could be the future of both sustainable and renewable energy. Nuclear energy could also be completely renewable if the source of uranium changes from ore mining to seawater. For example, China, USA, and Japan are competing to be the first countries to ensure that nuclear energy is completely renewable. The rationale behind making nuclear energy renewable is that uranium in seawater cannot be exhausted.

What are the disadvantages of nuclear energy sources? How can we tackle the cost challenge associated with many sustainable sources of energy?

10.4.2.6 Manufacturing

Manufacturing has historically been an extremely unsustainable area. Following the industrial revolution and the commencement of fossil fuels as a means of energy, the manufacturing industry has boomed. With its boom has come a number of environmental issues, many of which still remain unaddressed. Cleaning agents used in many manufacturing industries are pollutants that can seep into the environment during waste disposal and cause serious degradation of the natural environment around it.

Pharmaceuticals and personal care products can also cause serious side effects to the environment. Microbeads, for example, are common in many healthcare and beauty products. They are essentially manufactured plastic particles, less than one millimeter in size. However, these have now been proven to have little or no beneficial health effect, but can cause a huge buildup in plastic concentrations within aquatic environments, particularly the ocean. Microbeads are particularly damaging to the environment, as unlike some other plastic sources, they are so small that they cannot be filtered by our sewage systems.

The manufacturing industry as a whole must take into account of such examples and make changes to their waste management and resource use to make their enterprises more sustainable. A large number of manufacturers are now just beginning to realize that there are significant environmental benefits to incorporating more environmentally friendly and sustainable business practices that, by association, also come with financial benefits too.

Sustainable manufacturing is the creation of manufactured products through processes that minimize negative environmental impacts and conserve natural resources, while still remaining economically sound (Despeisse et al. 2012). Manufacturing in this way also enhances employee, product, and community safety. A growing number of companies are now treating "environmental sustainability" as an extremely important objective in their strategies and operations as a means to increase their growth and competitiveness.

Discuss companies that have embedded sustainable manufacturing into their practices.

There are a number of measures that can be used to make manufacturing more sustainable, predominantly in three areas:

Area 1: Reduction of Production Inputs
Many companies can opt for pre-used materials, which reduces waste production. Additionally, alternative materials can be used that are considered less environmentally damaging. A reduction in the use of solvents in building insulation, for example, can cut down chemical waste that ends up in the environment and decreases employees' exposure to hazardous materials.

Area 2: Improving Facility Operations Efficiency
There is a new concept in business known as "the factory of the 21st century in harmony with nature." This essentially goes by the notion that a manufacturing plant is built in a location using local supplies, such as timber, within the space that it is replacing. Additionally, it is designed so that it sustains itself and neutralizes

its emissions from the local area, such as by designs that incorporate large forest and green tracts to offset its emissions. One electronics corporation in Japan that incorporated this concept benefits economically by more than USD $6.5 million, as it reduced its use of concrete and the cost of waste management (Despeisse et al. 2012).

Area 3: Reducing Product Environmental Impact Throughout Its Lifespan Many manufacturers can seek to reduce the environmental impact of their products, both during the use of the product by the consumer and disposal. Using microbeads as an example, many manufacturers have now ceased their use. This means at the end of the product's life, they have a reduced environmental impact as the plastic will no longer seep through waste systems and into the ocean. Some strategies can also result in an economical and financial benefit (Despeisse et al. 2012). In this case, the microbeads no longer need to be made and the manufacturer will attract a new "eco-friendly" consumer base who specifically look for products that do not contain microbeads.

While these are just a few small areas in which the manufacturing industry can seek to improve upon its environmental sustainability, it illustrates that simple measures can have a huge effect. There are already strict laws and legislation in certain states that demand companies and manufacturers meet certain environmental targets for their emissions and waste. However, implementing stricter laws for environmental protection, there could be a substantial decrease in the impact of manufacturing on the environment. What measures can you recommend for your community or county to ensure that it develops sustainable manufacturing approaches? The Cancer Villages in China is one of the case studies that highlight the way poor manufacturing practices can lead to nation-wide epidemics. Cancer villages are the areas close to factories that cause pollution with unusually high levels of deaths.

Discuss this concept and its effect on the population.

10.4.2.7 Mining

Similar to the manufacturing industry, mining has also historically been an industry that has substantial detrimental environmental effects. With the continued discovery of new oil, mineral reserves, and coal deposits, an improvement in material recycling and our technological advances in recent years, our concerns over running out of non-renewable resources have eased. This has meant the mining industry continues to boom with no signs of slowing down (Despeisse et al. 2012). While mining itself occurs only on a relatively small area of land, the associated pollution and infrastructure from its activities have a vastly disproportionate effect on ecosystem health.

Water usage at mining sites is one area that must be addressed. Water is used for a number of purposes during mining operations. Through a diversion of both surface water and the pumping of groundwater supplies, mining can reduce both water quality and quantity at downstream sites for aquatic ecosystems, especially in arid climate areas. Mining operations can implement a number of water conservation practices to reduce their overall water use. Canada reduced mining water intake by 33% simply by implementing a new innovative system (Asif and Chen 2016).

The process of mining and metal processing can also be very energy demanding. Fossil fuels, such as diesel, are used by excavators and trucks during the mining process and electricity is used to grind ore deposits. This intensive energy usage is further exacerbated by the indirect consequences of fossil fuel extraction and infrastructure construction that is required for these processes. Again, innovative processes can be implemented to reduce these energy demands, such as the Alcoa's RopeCon transport system in Jamaica. This generates electricity, while transporting mined ore downhill from the mine site to the rail transport station. Through investigation of renewable energy sources that reduce mining reliance on traditional fossil fuel sources, such as solar power and wind turbines, much of the energy that a mining operation needs can be satisfied by on-site energy production.

Land disruption is also a substantial problem incurred through mining activities. The local area is disturbed at every state of the mining cycle, through exploration, construction, and the operation itself, all the way to closure and post-closure of the mine. Vegetation is cleared to allow space for the construction of roads, powerlines, and buildings, as well as tunnels and pits dug for ore access. This can lead to severe deforestation and habitat loss.

To address this issue, future mining operations could reduce their land use impacts, by designing more efficient mining sites that minimize their overall size and minimize waste production. Additionally, this can be supplemented by better storage management and an effort to maintain local biodiversity. This can be achieved by better planning of mines around pre-existing infrastructure, or through culturing and/or transplanting any endangered life found on site. Technological advancements may also help in this endeavor, where research into areas such as biomining, which utilizes microorganisms in the recovery of metals, also offers the possibility of mining with minimal disruption to land.

Waste management, as mentioned, is also key to increasing environmental sustainability. Mining waste includes mine water, air particles, and solid waste, which can vary drastically in their potential for environmental contamination. Waste management plans must be selected and designed that are tailored for each individual mining location, depending on the mined material. This means appropriate storage facilities that contain the large volume of waste produced at each individual mining site. Generally, by preventing pollution, we provide a more effective and economic way of reducing mining environmental impacts, rather than cleaning it up later and leaving an environmental degradation legacy for our future generations.

Following the cessation of a mining operation, environmental impact can be further minimized by facilitating environmental function restoration. Given that mining is a temporary activity that ceases once the desired material has been fully extracted, by restoring the disturbed land to an acceptable state for re-use by the environment, this can also increase their long-term sustainability and limit their environmental impact.

Explain how mining is among the major contributors to environmental degradation in developing countries. Land dereliction is among the negatives associated with mining. Explain this concept and describe how it leads to environmental degradation. How can mining be made sustainable in Africa?

10.4.2.8 Transportation

We do not need an expert to tell us the world's current transport systems are highly unsustainable. With examples of recent 60-mile long traffic jams in parts of China and traffic congestion costing the Texas Transportation Institute an estimated USD $87.2 billion every year, these "traffic snarls" illustrate that we must develop more sustainable alternatives (Schuman and Glancy 2015). The environmental costs of our inefficient transportation methods are very sobering, with the transport sector estimated to account for over one third of the US's CO_2 emissions every year. It is the fastest-growing source of GHGs and accounts for half of the net increase in total US emissions since 1990.

Given the extent of these figures, it is imperative that we find new and innovative ways to make auto travel more efficient and more sustainable.

One of the most promising solutions to this problem is through the development of better rail transport. While this mode of transport has been around for over 150 years, today's trains are faster, cleaner, and more efficient than ever. Traveling by train produces between three to ten times less CO_2 emissions, when compared to road or air transport. By developing and installing better rail links, our overall GHG emissions will decrease substantially and ease the pressure on road usage, all without sacrificing our ability to travel easily and efficiently.

There has already been some action taken to make a transition to greener transportation services. The US congress has recently authorized USD $13 billion to kick start 13 new high-speed rail links across the USA, and the new trains are expected to reach speeds of up to 170 mph (Perl and Goetz 2015). If rail links like this were incorporated on a wider scale across the world, it would also take the emphasis off other means of long-distance transport, such as aviation.

With aviation as a mode of transport now cheaper than ever, it is one of the most common ways for us to get around, with passenger numbers expected to double by 2050. Given such a vast number of people opting to fly, GHG emissions from this mode of transport has increased by a staggering 87% since 1990 (Pojani and Stead 2015). Considering there will be such a substantial increase in use, it is unlikely that we will be able to increase its environmental sustainability in the short term. As such, we must seek ways to limit and reduce its environmental impacts as much as we can.

Currently, over 130 airlines use "frequent flyer programmes" that offer incentives for flying regularly (De Boer and Gudmundsson 2012). By introducing legislation to reduce these incentives to fly, we can cut down on the number of people flying and reduce pollution. This could be complemented with further incentives for other modes of transport, such as carpooling, lowered public transport costs, and improved bicycle paths on our roads.

How can the world transit to a greener and environmentally friendly transport system?

Management of Invasive Species and Conservation of Biodiversity Our planet is now more traversed than ever before. With cheap, long-haul flights, and powerful cruise ships capable of transporting us and our possessions great distances, we can now visit exotic locations. With this unprecedented capability for long-distance travel, it has allowed our human society to establish an intricate web of trade across the entire globe. This has allowed humanity to flourish, and has impacted beneficially on our health, economic, and technological development.

However, while it has brought many advantages to our modern-day living and way of life, it has also brought a number of profound consequences to the environment. Through this intricate web of trade and travel, our transportation enterprises have allowed, either intentionally or unintentionally, the transmission of foreign wildlife from one region to another, usually where it would not naturally exist. This transportation of foreign wildlife can have profound impacts on local ecosystems.

If a given species establishes itself in a new environment, it is known as an invasive species. An invasive is defined as a species of plant, fungi, or animal that is not native to a specific region or location in which it exists, but has the capacity to establish in such a way that causes damage to the local environment, human health, or human economy (Guisan et al. 2013). The capability of such a species to establish itself in an environment is associated with possessing a specific trait or combination of traits that allow them to outcompete native species. This competition can be related to growth rates and reproduction, or a more direct version of competition through predation, where the invasive species either has no natural predator or predates easily on the local wildlife.

The best way to visualize and understand how an invasive species establishes itself, and the damage it can cause, is through an example. One common invasive species that is often discussed is the rabbit, a well-known and serious mammalian pest in Australia. It has caused millions of dollars of damage to crops and local wildlife since its introduction to the country in the eighteenth century. It was deliberately introduced to the country when the first colonial fleet arrived on Australia's shores as a source of meat for the early settlers. They reproduced exponentially to the point that they are now causing serious repercussions on Australia's natural ecology.

Rabbits have caused overgrazing of native and sown pastures, which has led to a significant loss in plant biodiversity and a reduced yield from agricultural crops. They also outcompete native animals and domestic livestock for food resources and shelter. Given their abundance, they have become a plentiful food source for predators, which has led to a dramatic increase in predator populations (Guisan et al. 2013). As a consequence, other small native mammal populations have reduced and an increase in domestic animal predation and disease prevalence. Additionally, they have also caused the spreading of other invasive species of weeds. The culmination of these impacts has resulted in what people now recognize as the rabbit being the most significant known factor of species and biodiversity loss in Australia's history.

Such an example of an invasive species causing a host of problems to both the environment and the economy is not uncommon, and there are many other examples throughout the world where humans have acted as the catalyst for their geographical expansion. Several other examples include the Asian tiger mosquito, which naturally exists only in South-East Asia, but is now known to have spread to at least 28 countries outside its natural range in the last 20 years. This is thought to have come from our international car tire trade, which is often a breeding ground for mosquito larvae if left outside before transportation. This type of mosquito is known to carry many diseases, including yellow fever, chikungunya fever, dengue fever, and the Zika virus (Campos et al. 2015). As such, it is responsible for the transmission of many human diseases into regions where the diseases would not normally be prevalent.

By evaluating case studies such as Australia's rabbits and the Asian tiger mosquito, it is clear that introducing foreign wildlife to a new area may potentially be severe. In the USA, the estimated damage and costs to control invasive species has risen to over USD $138 billion per year in the form of recreational, tourism, and agricultural damage as well as the costs for management. Through the recognition of such extravagant levels of damage, we must learn from past mistakes and to understand what precautions must be taken to avoid them in the future. This will also help us to understand how to rectify these past mistakes and alleviate the environmental consequences of already established invasive species. Applying the concept of environmental sustainability as a means to control and manage invasive species is extremely challenging. However, it is absolutely crucial, given the increasing worldwide rates of invasion and the high cost of their impact on the environment.

Management of invasive species is of course inherently unique to each species and the specific environment that it invades. As such, each case must be evaluated and tackled independently for the most appropriate and effective strategy. An example of where there has been a big push in implementing invasive species management in recent years is in the shipping industry. The industry is worth well over USD $200 billion and its value is increasing substantially every year (Shin and Thai 2015). With ships transporting our goods and products all over the world, there is an inherent risk of accidental transportation of wildlife in ship hulls. As a means of balancing themselves in the ocean, ships take on water at different stages of their journey to help counteract the weight from the cargo it is transporting. This "ballast water" is released later on in the journey at different times to help maintain balance. However, water taken up by a ship in the Atlantic Ocean will have very different marine wildlife than water in the Indian Ocean, and thus presents a serious threat of new species introductions worldwide.

In order to counteract this, management strategies have involved the implementation of a number of strict regulations and incorporation of new technology to inhibit the potential establishment of new species. Shipping companies are strictly regulated on where and when they are allowed to dump their excess ballast water. This has been in conjunction with a huge push in the development of new technologies that "treat" the ballast water before it is expelled. These water treatments essentially use environmentally safe chemicals to kill any micro-wildlife in the water, so that they cannot be transmitted to a new location. This combination of strategies illustrates how a combination of techniques can be utilized in order to develop an environmentally sustainable solution to conserve a natural biodiversity equilibrium without damaging the economic potential of a human enterprise.

While the establishment of invasive species has become a severe problem in damaging our planet's biodiversity, it is not the only way that we are damaging our world's beautiful wildlife. We are upsetting the Earth's natural order in our quest for natural resource acquisition. We are causing significant environmental damage through deforestation, pollution, and hunting, all of which directly affect local wildlife. Increasing human populations and economic growth in many tropical countries has resulted in a substantial increase in human activity, such as agricultural expansion and commercial timber extraction. This results in significant changes to tropical forest ecosystems, through large-scale habitat losses. It is estimated that an astonishing 150 acres of forest are lost every minute around the world. A large proportion of this is in the Amazon rainforest, where it is thought that 20% of its total coverage has been lost (Ellison et al. 2012).

Additionally, the substantial pollution that our society emits to the environment is also having profound effects on our planet's biodiversity. By looking at any given ecosystem, there could be a host of different contaminations. Streams can contain a spectrum of toxic chemicals leaked into water supplies from our industrial processes. Rivers can become overloaded with agricultural nutrients, trash can blow away from our landfills, and the skies above our cities are rife with smog from our gas emissions. These types of pollution can muddy our beautiful landscapes, poison our soils and waterways, and ultimately, kill our plants and animals even in environments thousands of miles away from the initial pollution source.

To acerbate and place further pressure on the world's biodiversity even further, the overhunting of Earth's creatures is placing many species at risk of extinction. Poaching in many areas throughout the world has led to a severe decline in many species' populations. The black and white African rhinoceros are both at risk, due to a surge in illegal poaching to meet the demand for illegal rhinoceros' horns used in several traditional Asian medicines. Since 2007, there has been an unbelievable 9000% increase in rhinoceros poaching, from 13 to a record 1215 in 2014 (Gonzalez 2016). It is now thought to be as little as 5000 black rhinos left in the world and is listed as critically endangered on the International Union for Conservation of Nature (IUCN) list of endangered species (Gonzalez 2016). This story is the case for many other beautiful creatures. There are only around 3200 tigers left in the world, as a result of hunting. Only 4000 of the critically endangered eastern lowland gorillas are left, primarily in the Democratic Republic of Congo (Gonzalez 2016). And both the Sumatran and Bornean Orangutan are listed as critically endangered, due to animal trafficking and exploitation.

These three human enterprises are culminating into the primary causes for species loss and extinctions worldwide. The very shocking and sobering fact is that it is estimated that we lose up to 5% of worldwide species per decade as a direct consequence from human activity (Barnosky et al. 2012). With so much of the deep Amazon rainforest left to discover, many of these species that we are losing are yet to be discovered. Some scientific studies have even predicted a doomsday scenario where we could lose as many as 50% of all worldwide species within the next 50 years. With such a wide spectrum of animals now endangered and at risk of extinction, alongside such dramatic figures of species loss for the near future, it would seem clear that we have entered the dawn of a new extinction event that scientists are now calling "the 6th mass extinction." It is therefore imperative that we act now in an attempt to reverse these trends and save our wildlife.

Similar to the management of invasive species, there is not a "one size fits all" approach to the conservation of endangered species and biodiversity. We must adopt a tailored strategy to each animal at risk. In the case of the rhinoceros, there have been a number of suggestions on how to combat the poaching of rhinoceros horns. One suggestion involves the creation of a controlled market, where rhinoceros' horns are cut off in a humane and safe way, so that there is no incentive for illegal poachers to kill the animals (Barnosky et al. 2012). This would create a sustainable solution, where the market could be satisfied without any risk to the animal. Another is through the implementation of trackers into rhinoceros' horns, so that they can be tracked back to the main dealers in the trade, in order to tackle the problem at its source.

How can politics play a role in ending the hunting of endangered species? What should be done to end the ivory trade that threatens the life of some species?

In order for conservation work to achieve the long-lasting and large-scale success that is needed to safeguard the world's biodiversity, it is imperative that we tackle the underlying causing of these types of environmental degradation. We must strive for collaborative, science-based, and innovative techniques and approaches, to tackle the threats to biodiversity at their roots. New techniques must be developed for more sustainable approaches to managing forest and wetlands, crop growing, and dealing with waste. Additionally, we must look at opportunities to transform markets, and develop strong partnerships between governments, businesses, and local communities, that all strive to drive change.

This is particularly the case for overexploitation of both animal products and deforestation, where the problem must be tackled on a wider scale. Collaboration is crucial for the implementation of stricter regulations, stronger deterrents for illegal activity, and a worldwide consumer attitude change, to shift towards more sustainable and renewable materials. By understanding such key factors behind the impacts of these detrimental enterprises on biodiversity, and engaging key stakeholders such as businesses, organizations, communities, and individuals, we will only then ensure a long-term biodiversity conservation plan, which will leave us with a sustainable ecological footprint.

How can we manage invasive species? The water hyacinth is among the invasive species that bring about increased harm not only on ecosystems, but also on people's livelihoods. Using the Lake Victorian water hyacinth in Kenya as an example, briefly discuss the trueness of this statement. How can we increase the number of carbon sinks in Africa? Describe the role of community culture in conservation of beneficial species.

10.5 Social Sustainability

During the last couple of years, the term sustainability has reached an important peak in terms of recognition within the academia, the government, and civil society. Social sustainability, alongside economic and environmental sustainability, is one of the three pillars of sustainability. However, the social side has often remained left without much consideration because it is the most difficult of the three items to quantify or to study its impact. This theory of social sustainability and organization tries to identify the link between poverty levels, the environment, colonization, and sustained natural resources exploitation. So, it is easily argued that environmental sustainability acts as a prerequisite for poverty alleviation through sustainable economic growth. This is relevant because nowadays there is an individualistic lifestyle obsessed with personal needs, status, work, destruction, and materialistic wealth, without regarding the environment and its disintegration. When an investigation tries to demonstrate themes such as employment or the levels of poverty, they are often referred to as "hard" social problems because they are tangible, one can easily see how impoverished a community is. However, social sustainability often studies "soft" elements that are more difficult to measure, for example, whether or not a community is happy. Moreover, the ways in which the indicators are portrayed could often be done by specific companies or political groups, thus making it more difficult to obtain independent results.

Social sustainability is similar to sustainable development because they work towards the same goal: to provide a sound sustainable environment for the individuals that form part of the communities of a specific area (Vallance et al. 2011). This type of development tries to improve the capacity to create healthy and livable communities while also maintaining human and labor rights and keeping an eye on how corporate governance works towards a better working environment. The aim of this is to ensure that future generations can have access to the same amount or greater social resources. When a community improves the quality of living , it is also helping future generations understand the core of what is meant by a good community. So, in broader terms, social sustainability's aim is to have a supported infrastructure for social, cultural, working, and health systems in order to allow individuals to evolve by providing them with the necessary tools to do so.

Also, within social sustainability there is a great amount of energy spent on identifying the positive and negative aspects that businesses have on people. This is usually done when reviewing how companies influence employees or how they treat the environment and its customers. It is fair to say that by improving a business' sustainability , the greater sustainability within that community will also be achieved because a conscious customer will consume from a conscious business. Furthermore, by improving the community's well-being, it gives an opportunity for individuals to become resilient , so a constant flow of energy is consciously portrayed on health, housing , transport, relationships, work, education, spirituality, governance, and environment because the community has a deep commitment to improving the quality of life for all its citizens.

Thus the main objective is to constantly improve these things, because a strong and inclusive society can be built, but only if the responsibility is shared between the government, the community, and of course, businesses. For example, if a government sets a good and viable option for businesses to develop, and this business takes advantage of the scheme and performs on a sustainable matter, it will create conscious customers who are acting in a sustainable manner too. In other words, it is a win-win situation for everybody involved. When one studies social development, several key definitions and concepts need to be understood. They are discussed below.

10.5.1 Equity and Diversity

It is fair to say that the success of a sustainable development program is based on a community's ability to provide equitable opportunities and outcomes for all its members, particularly the poorest and the most vulnerable, in order to achieve a community development that will benefit the greater amount of people. Equity and diversity work hand in hand, because they support each other. If every single member of a community is given an opportunity to have access without discrimination, it means the basic human and civil rights of individuals are being respected. Thus, equity involves all people, while diversity will ensure that different people are represented.

One of the aims of a sustainable social development environment is to increase the living standards of individuals by providing affordable (and sometimes even free) health, education, transport, housing, and recreation with the least amount of environmental degradation . This also means that there should be transparency in government and policies are in accordance with the sustainability principles that the community is trying to achieve. Moreover, instead of disintegrating and fractioning groups, sustainable communities across the world are trying to promote and encourage diversity. This is done by generating an interdisciplinary model in which social, economic, and environmental elements are positioned as a whole to promote an equitable and just society. At the same time, there should also be a cultural respect in regard to the diversity of sexuality and gender within the community. So, diversity plays an important role on this specific point because discrimination towards individuals will not be tolerated and the differences that one could possibly find will always be accentuated in a positive manner.

In 1943, a psychologist named Maslow proposed a new psychology theory based on a hierarchy of needs, which he called "A Theory of Human Motivation," which basically stated that individuals are motivated to achieve specific needs at a specific pace (Maslow 2013). For instance, a person will first achieve a physiological need and will then move onto safety needs, which could include an economic need for stability. After completing the safety needs, the person will search for social belonging and this is the area most relevant in terms of social sustainability because it is where people become more aware of their surroundings in terms of sharing with the community they reside. As they move on, the need for the next level to be fulfilled will be created and used as a motivation.

Does Maslow's Hierarchy of Needs provide an accurate account of human needs? What are the major criticisms of this theory? In what ways can the theory inform policy formulation and development?

10.5.2 Quality of Life

Social sustainability must take into consideration some relevant factors that could possibly fall under the economic and environmental sustainability outlook. By doing this, the community ensures that basic needs are met while fostering a good quality of life for all members at the individual, group, and community level. It is important to understand that the term "quality of life" englobes a multi-attribute value concept, meaning that it has many layers of how it can truly be portrayed. For example, in terms of employment, it could question the number of people with jobs, the level of unemployment, whether there are equal opportunities in an area, if there are more of fewer well-paying jobs, or whether there are real opportunities for education and training. If the government makes education a priority and there are more jobs available, one can clearly see the correlation. Also, if there are better paid jobs, are people more optimistic about the housing situation? Are they going to obtain a good health service in the area? The basic principle is to question everything associated with employment to obtain the pros and the cons of a specific situation.

One must truly understand the concept of "sustainability" to achieve balance among the economic, environmental, and social parts of a community as they are portrayed as a whole. In other words, the economy exists because there are humans who interact among each other and who are willing to exchange money for knowledge of the services or goods, thus these humans create a society where music, art, religion, the environment, friends, and family. At the same time, society is created within an environment because humans, according to Maslow, have basic needs that must be fulfilled, which come from the environment humans depend on (Maslow 2013). Moreover, social sustainability exposes how positive management can be arranged to ensure that the economy, the society, and the environment are all working alongside each other to preserve a good quality of life. Therefore, it is an issue that involves all communities regardless of their size.

> What are the major challenges experienced in addressing social sustainability? Can a socialism paradigm help in enhancing social sustainability? Why or why not?

10.5.3 Spirituality

When sustainability is discussed, spirituality must also be taken into consideration. This field, called spiritual ecology , is a response to the present ecological crisis worldwide (Gardner 2002). It is a mixture of ecology, environmentalism, sustainability, and the conscious awareness of the sacred within nature and creation. It does not involve a particular religion per se, but usually involves individuals who are trying to resolve environmental issues, such as global warming, overconsumption, or the eradication of species.

Spiritual ecology understands that the present environmental crisis comes from a reflection of the inner spiritual crisis most societies are living in. Now more than ever before one shall look at the interconnection between life and death, and while both are a necessity in the environment, death should be natural and not created by an unbalanced society (Sponsel 2001, 2007, 2010). Firstly, there should be a collective pursuit of wellness, happiness, and respect. Not many people are looking for what is sacred within all life and without devastating effects on others or on the environment. This planet was only lent to humanity, the biggest community there has ever been, and little by little new destruction is occurring worldwide. Secondly, one must understand the laws of action and reaction. Every single action performed has a reaction, whether one sees it or not. Sustainability is necessary in order to achieve a spiritual awareness and when achieving this, a longer cycle of sustainability is created as the interdependence of ecosystems becomes apparent.

When individuals feel at peace and secure in their own beliefs, culture, identity, and spirituality, they are less likely to feel threatened or uneased by differences. Where there is a place of trust, love, and understanding, there is less room for racism or other forms of prejudice. This is why spirituality is relevant within sustainability as it removes those elements that are not beneficial within the community. If one works towards the benefit of oneself, the benefit of the community will come along as well.

Spiritual ecology provides a specific opportunity for a balance between the extremes of mentalism and materialism. There is a need for hard evidence to determine the effectiveness of spiritual ecology. Various religions contribute to an array of relationships between nature and humans with diverse effects on biodiversity, ecological processes as well as ecosystems. Spiritual ecology entails spiritual and sociopolitical movements that hold substantial promise to help minimize or resolve eco-crises globally. Many years of diverse secular approaches for resolving environmental issues have largely been insufficient. As such, we could argue that spiritual ecology could be the last possibility for turning things around and addressing environmental and social sustainability. Spiritual value is among the various forms of biodiversity values that drive people to become conservationists. Many biologists believe that if the importance of biodiversity is felt in the soul, pocket, and brain, there might be permanent conservation.

What is the major role on spiritual ecology in enhancing conservation and social sustainability? In what ways can Africa embrace spirituality as a way for social sustainability?

The environmental conservation movement has been led and influenced by people who have significant experiences of the sacredness of nature and have struggled to protect it. Appreciating the intimacy between nature and the human soul can play a crucial role in changing the way people think about society and the environment. Today, numerous facets of the movement geared towards environmental conservation are often empowered by cooperation and spiritual principles.

Robon Wall Kimmerer is a professor of Environmental and Forest Biology who recently played a pivotal role in the foundation of the Centre for Native Peoples (Rich 2012). This incorporates the ideas of ecology, environment, and spirituality. She believes that traditional ecological knowledge has increasingly become important as a potential source of inspiration for ecosystem management and ecological restoration. Admittedly, traditional ecological knowledge is not only present to Native American culture, but globally irrespective of ethnicity. It arises when people have material and spiritual connection with their landscape. Recently, the World Wildlife Fund (WWF) has come up with a programme called Sacred Earth: Faiths for Conservation, a collaboration between faith communities and spiritual leaders from diverse spiritual traditions globally. The programme is aimed at facing various environmental issues, including pollution, deforestation as well as unsustainable extraction. Khoryug is among the conservation projects that the programme has developed. It is based in the Eastern Himalayas and is an association of various Tibetan Buddhist monasteries. The project has been instrumental for the publication of various environmental guidelines that Buddhists use for addressing water loss, waste, climate change, and forest degradation. In addition, Krishna Kant Shukla is a musician and physicist who has provided beneficial lectures about Indian villages as sustainable development models. He has played a critical role in the establishment of Saha Astitva; model eco-village as well as organic farm found in Maharashtra, India. There is also a need to appreciate the role played by the nature and instinct of women and commitment to provide protection to our resources. This is perfectly depicted in the life of Wangari Maathai who founded the Green Belt Movement. This movement initially consisted of women who planted trees. Similarly, Vandana Shiva is an Indian feminist who has actively worked on various issues, including protection of small Indian farms and seed saving (Shiva 2014). Other environmentalists who fall under the category of spiritual ecologists are Wendell Berry and Satish Kumar. Admittedly, spiritual ecology has and will continue to play a crucial role in social and environmental sustainability.

10.5.4 Democracy and Governance

There is a strong relationship between democracy, governance, and sustainability, thanks to the community which provides democratic processes and offers an open and accountable governance structures by participating in governance and rule making. An issue has emerged over the last two decades whether a sustainable life could be achieved through democratic forms of governance by disempowered communities and disenchanted citizens. There must be a strong sense of engagement and identity between communities in order to organize themselves and form relevant governance where institutions are democratic and remain true to the sustainable economic growth instead of simply following an old ideal of the global market economy. When individuals come together and form this type of "front," one can argue that a widespread political organization, participation, and representation of citizens exists on levels that not only involve electoral procedures but also participation in a sustainable matter. If individuals are represented, a sense of integration within the community is also presented and transmitted to the next generation through some sorts of dialogues, activities, and actions, which in turn will help next generations to see and understand the strengths and weaknesses of current communities. Ultimately, the way governance works in a sustainable environment is through values such as transparency. Transparency can lead to trust because people view decisions and leaders as accountable. This prevents conflicts that often emanate from societies that lack transparency and accountability. The more at peace the community is, the better chance the leaders they choose will follow these values too.

In 1995, the Commission on Global Governance (CGG) proposed seven core values for sustainable governance: mutual respect, equity, respect for life, caring, liberty, justice, and integrity (Bosselmann et al. 2008). It proposed a global civic ethic described as a new philosophy to be introduced. However, there were many complaints that these values were too often looked at under the eyes of the western world, thus making it impossible for other cultures to comprehend or even participate on this project. Even though "soft" and "hard" themes were previously described, the relevant aspects of social sustainability and how one can achieve them as a society must be considered. Increased analysis has created an instructional sphere on social sustainability, but this debate will continue for the foreseeable future as larger communities and societies become more affluent and others become less protected by current governments. It is also important to know that in order to obtain a just society, the basic needs of individuals must first be met. Furthermore, even though sometimes it is difficult to obtain, there must be a selection of indicators that monitor the framework of different policies across communities worldwide, thus deeper analysis and studies must be made.

Lastly, all of these definitions and measures must be taken as a whole to obtain a positive result from a socially sustainable environment. In terms of policies, an open discussion and planning approach should be necessary in order to reestablish what the community really wants to accomplish. Thus, an integral commitment from individuals within a society is also a must. A societal consensus could be formed to establish an urban development that forms part of the natural environment. Within sustainability, a balance between the three pillars is necessary to live in a genuine healthy environment, where future generations understand the basics of economics and how it can destroy communities and nature if they are addressed in an aggressive matter, instead of linking the cities to sustainable natural systems.

Governance is a critical pillar for achieving social sustainability. Explain. Identify governance models that have been successful in promoting social sustainability.

10.6 Economic Sustainability

10.6.1 What Is Economic Sustainability?

The term sustainability involves three basic pillars that encompass a healthy ecosystem, these are economic, social, and environmental activities or ways that englobe a truly sustainable approach by remaining interdependent and holistic in order to evolve towards a better and efficient economy. On a level that deals purely with the economy, this has led businesses to reflect on the output of their industries.

The basic principle of economic sustainability refers to the capacity industries hold in order to continue their activities or processes in an indefinitely way for an indefinite period of time (Jänicke 2012). Communities play an important role within the sustainability sphere because communities are formed by individuals and those that consume the goods made by the industries. For example, people from the western hemisphere tend to consume more than people from the eastern hemisphere and sometimes the levels of consumption are so elevated that they overpass the levels they can afford. Moreover, another significant phenomenon occurs in developing countries, because even though the levels of consumption are low (within the industries sector), the levels of growth in terms of population are increasing high and fast, in sharp contrast to the western hemisphere, thus also creating an unfair distribution of resources within their populations.

Therefore, economic sustainability has gained a relevant position in today's business talks because it is highly required to encourage responsible use of resources. while keeping environmental concerns relatively low or non-existent. By doing this, it creates a sense of belonging within the community, thus creating locally investments to help industries maintain their businesses.

Why is economic sustainability of utmost significance?

10.6.2 Characteristics of a Sustainable Economy

Under a capitalist economy, it is cheaper to buy a product that has a high cost and impact on the environment when it is produced compared to a product that perhaps does less harm to the environment. An example is the current consumerism trend in various societies that places more value on goods rather than the environment. What is worse is the fact that this does not necessarily mean the customer will pay more if the product is made on a higher-cost scale on the environment; quite the opposite occurs because businesses and customers alike do not take the toll on externalizing these expenditures (Jänicke 2012). In other words, usually those tariffs are not represented when deciding to create and produce a product. In other words, a capitalist economy is not sustainable because it prioritizes on consumption of materials that satisfy their needs without considering their carbon footprint and the harms they have on other people.

Capitalism is blamed as one of the causes of degradation of the environment. Why do you think this is the case?

Therefore, a sustainable economy comprehends that the economy must move forward, which is only made possible when an economy fully includes jobs and wages, and technology that is appropriate and relevant to provide healthy stability of businesses. Only when this occurs will a Sustainable Economy be truly vibrant. Nevertheless, under a sustainable economy, these costs are indeed quantified due to the fact that when producing a product at a low price the planet and communities nearby will be thoroughly thought of in order to maintain low levels of harm within the region. This would automatically create a win-win situation, wherein customers can buy something affordable, businesses would sell their stocks and the environment would be respected.

One of the consequences of a capitalist economy is that individuals are entitled to buy whatever they want without understanding where it might have originated. Thus, there is need for consistent awareness to inform people about the impact of their consumption patterns or lifestyle choices on the environment. Under a sustainable economy approach, there is a higher responsibility to encourage the three R's policies: reduce, recycle, and re-use (Jänicke 2012).

What impedes us from embracing this approach?

Therefore, there is a strong correlation between a sustainable economy and a sustainable society because one cannot exist without the other. A healthy economy and production can be sustained by individuals who express their social and ethical values , and this will be represented by the collective choices and actions they make. Thus, businesses will not be able to ignore these requests as it would mean losing money in the long run. An example of the success between a sustainable economy is the work done by Muhammad Yunus in establishing the Grameen Bank (Yunus 1998, 1999; Yunus et al. 2010). As a social entrepreneur, he depicts the value of sustainable economies in creating sustainable societies. This is extrapolated under the Individual Matter section. Grameen Bank is a sustainable economy since it created a market niche and reached the poor women. The bank has developed a decentralized management structure and impacted rural wages positively.

One of the major debates about economic sustainability is about the limits of growth. Is there a limit to economic growth for a particular country? Does Rostow's theory of development accelerate or impede economic sustainability? Provide reasons and examples from Africa.

10.6.3 Green Growth and Economy

In 2005, during the fifth "Ministerial Conference on Environment and Development" (MCED) held in the Republic of Korea, the name "Green Growth and Economy" (GGE) started appearing, where it was adopted as a strategy for achieving sustainable development (John 2014). According to the Organization for Economic Co-operation and Development (OECD), "Green growth means fostering economic growth and development while ensuring that natural assets continue to provide the resources and environmental services on which our well-being relies." Though, it is fair to say that both green growth and economy act together within the sustainable economy in order to advance, as a whole, the economy, the environment, and social well-being. In addition, considering that GGE represents a non-specific set of strategies, countries are encouraged to design their own paths to create jobs that are beneficial for the twenty-first century society, and to move away from old economic methods, such as a non-beneficial industrial development that continuously increases the greenhouse gas emissions and destroys unpreserved ecosystems.

As a result, GGE is required to maintain low levels of greenhouse gas emissions and pollution, while having high levels of resource productivity, and ideally operating with renewable energy. It also must have (or intend to create) a circular economy, in order to recreate the 3Rs approach. Therefore, even though one talks about green growth and economy, worldwide economics cannot lead the way to implement a system simply based on actual economic models. In fact, it is quite the contrary, because in order to achieve a sustainable green transformation, a political strategy is very relevant indeed.

This also means that firstly, states must create regulations on local businesses, whether it is for environmental taxation or recycling strategies, while keeping the public informed throughout possible changes. This in turn will increase the chances of creating a healthy environment where there is a reduced usage of food, energy, and/or water resources, thus keeping a balance in nature. Secondly, once public and private regulations are placed, states should be required to cooperate with other nearby states to achieve a level of awareness that the world needs today; the capacity of its individuals, communities, enterprises, and states to maintain a healthy world where all societies are able to respect one another and live in a sustainable way.

Identify five key areas Africa should focus its attention on to promote green growth and economy. Economic development and environmental sustainability are intangible. Evaluate the trueness of this statement.

10.7 Applications

10.7.1 Production and Resourcing

Under a sustainable approach both resourcing and production work together for the creation of a beneficial balance on the environment. With this view in mind, resourcing must be efficient to use less of the available resources to achieve a product. In other words, the inputs (resources) must create the same (or even better) outputs (products). So, the keyword is "less" because industries must lessen the amount of input they once thought was necessary in order to truly be sustainable. It is also fundamental to consider the product's life cycle, which goes from the moment the product was designed to its finalization, including an added value if the product can be reused. This is where economic sustainability also plays an important role because both businesses and consumers should act together in order to maintain an efficient sustainable development.

Explain how this can be achieved in Nigeria. Identify and explain at least four ways.

10.7.2 Agriculture and Food Systems

Within the economic sustainability sphere, the food system and agriculture practices are key indicators of how well production is going, how all systems are developing, and whether the new changes implemented are being used in a respectful manner towards the environment and local economies. Thus, these agricultural and food systems try to represent farming in a sustainable way, solely based on the needs presented by the community, and how they find respectful ways to act in parcels of land to avoid overexploitation. Alongside this, they should also present a clear and efficient use of water by ensuring that water is used sparingly and from when necessary, and the understanding of the ecosystem that will give place to agriculture.

A capitalist economy requires a profitable agriculture industry even if it means a total detrimental way of acting in terms of social and environmental integrities. What it fails to understand is that under an economically sustainable approach, one could generate a steady income through the design and implementation of local food systems (Lipper et al. 2014). This could be achieved by using agriculture as an extension of life, where farmers can produce and distribute their crops within the community because the food that is sold locally does not require the extra time consumption involved in transportation.

How can Africa improve the livelihoods of farmers using sustainable economic production systems?

10.7.3 Exchange and Transfer

With regard to a sustainable economy , exchange and transfer of resources mainly deals with assets being held by current societies and how to maintain a level that is either equal or greater for future generations (Hamari et al. 2016). The importance of resource allocation within a sustainable economy is putting a great emphasis on infrastructure, human, technology, and financial resource as they must work together in order to achieve development towards a sustainable society.

The results of a positive integration will be highlighted in the long-term development by incorporating the population while also achieving equality in terms of resource allocation. This can all be made by reorganizing the current system of quotas, hierarchy, and centralization of resources based merely on the financial aspect, and redirecting priorities towards correct planning and allocation according to local needs. This would create a holistic vision of how a sustainable economy should be developed.

How can we improve resource allocation in Africa?

10.7.4 Accounting and Regulation

In terms of economic sustainability , it is necessary to develop a resource efficient economy that will form the basis of Green growth and economy. There are three levels (environment, economic, and social) that form the basis of what is accountable within a sustainable framework. What this means is that policies need to be created in order to improve productivity, manage natural resources and materials, while also protecting resources. Some regulations could appear in terms of avoiding waste (or minimizing waste) and the life cycle of the materials used must also be considered.

These types of policies could be drawn out on a local, national, or international level. However, if they are presented on an international level, independent and local policies must be considered in order to create a peaceful development, wherein supply chains are diversified and completed. What's more is the fact that even though accounting and regulating could fall under an economic perspective, these objectives also work hand in hand with other areas of sustainability. Innovation and technology must be regulated as well as different types of trade, wealth distribution, and environmental costs in which a sustainable economy develops. Even though it is sometimes difficult to achieve a level of commitment between all spheres of sustainability, it can be argued that industries must try to find a balance between the social, economic, and environmental aspects.

Describe three policies that directly relate to economic sustainability in Nigeria.

10.7.5 Consumption and Use

Nowadays, consumption of natural resources is increasing rapidly which in turn makes it unsustainable. In terms of regulation, governments are required to rethink the incentives that are being offered to diverse companies. Businesses should think about the environment in the long-term, instead of thinking of short-term profits.

If governments and the public want to curb this trend and create a different consumption pattern for future generations, actions are required immediately. This could range from promoting education on sustainable farming and production to promoting alternative ways of consumption within areas where large amounts of people gather or simply introducing the 3Rs scheme. People should always be informed about the impact their consumption has on the environment. One relevant example is how the Japanese education system has taught children to be responsible for their own schools by cleaning at the end of the day. They are not obliged to do this, in fact, it is quite the opposite, they do this because they feel a sense of belonging and tradition to that specific place. If one sees how the local environment is being destroyed by local consumption, they may be influenced to begin creating new ideas on how to slow down consumption.

How should reduce, reuse, and recycle guide our consumption patterns? What is our role as consumers in economic sustainability?

10.7.6 Labor and Welfare

Welfare, in terms of socioeconomic terms, can at times be referred to as equal distribution of wealth within populations. Thus, economic sustainability can offer a sound response to the main issues involved in income inequality of both capital and labor resources in a society. It can be argued that if a state is playing an active role in ensuring the growth of their own economies in an economically and socially sustainable way, a sustainable and advanced welfare state could be created alongside people who are willing to work hard towards necessary change. A welfare state is a system whereby the state is committed in protecting the well-being and health of citizens, particularly the ones who require social and financial needs through grants, pensions as well as other benefits. An example is the United Kingdom that is committed towards meeting the social and financial needs of its citizens.

On the other hand, sometimes the most profound examples are focused on individuals and not the state. Some industries have implemented programs that improve labor and productivity, which in turn improve waste reduction and lower energy consumption . In this case, businesses are contributing directly to the development of a healthier and educated society, and also presenting new opportunities for the population to help themselves through support given to socially disadvantaged groups. Also, the promotion and encouragement of industries' leaders to create Labor Unions within their enterprises can result in a more horizontal way of communication. This can be the basis of a circular enterprise where those who produce can freely bring ideas or ways of improvement to others in the industry. Circular enterprises are founded on closed-loop production-consumption models that can convert dirt or trash into wealth or treasure.

Is it possible to achieve a Utopian society ? What is an ideal Utopian society according to you? In what ways can we reduce societal inequalities today?

10.7.7 Technology and Infrastructure

One of the biggest concerns in terms of environmental protection is how to use technologies accordingly in order to preserve natural resources. It is safe to say that sustainability depends on the evolution of technology . In particular, alternative technologies that will ultimately help with the creation of a sustainable economy. Globalization has also played an important role within technology as some companies opt for a reduction of workers on the ground, while others try to reduce production costs by outsourcing (Jain and Lee 2012). In other words, they will sign contracts with individuals that are in other areas of work in order to fulfill a job. More often than not, this type of outsourcing involves paying cheaper wages to candidates abroad. However, it can also be argued that outsourcing from abroad can lead to a bigger understanding of international problems that other societies face and without this information, it would be impossible or at least very difficult to even find out what is truly happening in other areas of the world. Therefore, proper management of technology and information becomes valuable as it can help find solutions to sustainability problems in areas such as health, education, and management of natural resources on an international community level.

Furthermore, a sustainable infrastructure tries to fulfill social needs while reducing the impact of production on future generations. This might be done through appropriate design, construction, and operation of enterprises because in order to achieve sustainable economic growth, capitalist industrial processes that are so harmful to the earth must be abandoned. Examples of appropriate design that led to sustainable infrastructure are the "zero-carbon" desert city of Masdar and the BRT-lite system that seeks to reduce congestion and offer alternative mode of transport to the poor in Lagos.

How can we create a society that has sustainable infrastructure and technology?

10.7.8 *Wealth and Distribution*

It can be said that these two concepts are the basis of economic sustainability. One could argue that the disparity in wealth and its distribution is majorly seen in developing countries, where usually there is a higher percentage of poverty. Wealth income and thus its distribution relates to social justice, which is a mixture of political and economic policies as well as sustainability placed by the state towards different social groups. However, it can also mean the lack of correct policies for both the political and economic stability.

One clear example is the growing numbers of "Green hotels," which adhere to Green and sustainable regulations found internationally. There have been, however, negative aspects when it comes to these types of hotels. From direct environmental impacts to people-orientated impacts, the reality is that some areas should be discussed to avoid major and complicated damages in the long run. Usually eco-hotels or Green hotels are intended for small groups of people, whether this is because of the elevated price or that they only serve a few patrons at a time (Wang et al. 2013a,b). However temporary the duration of the visit is, it puts an extra pressure on both the local community and its environment. On the other hand, most hotels are owned or invested by foreign individuals or corporations. This means reinvestment in both the local economy or people decreases significantly due to poverty and unemployment, thus, leading to degradation of morality and lack of trust within the community.

Wealth and distribution have been cited as one of the triggers of inequalities and unsustainable production. Discuss.

10.7.9 *Urban and Rural (Sustainable Cities)*

Sustainable cities have become of great importance during the last decade due to the fact that cities continue to grow and available space has been reduced. A sustainable city will usually be combined alongside sustainable agriculture, which in turn will produce balanced economical sustainability, for example, when creating food production systems using a vertical farm (Nevens et al. 2013). A vertical farm uses stacked layers, vertically inclined surfaces or integration in other structures to produce food and medicine. (Refer to Chap. 3). In this type of context, it is easy to see how one could invest in both production and profit in equal terms as one produces their own food but is also the main consumer, even if they are living in the city center. In this way, not only is the land space being reduced, the profit margin of the person producing grows and one will start living in a sustainable coherent manner.

Another economically sustainable example within urban and rural sustainable cities involves the hospitality field. Within this field, creating a Green hotel means that everything involved has been carefully thought of, partly because it introduces programs that are cautiously sustainably designed and have a safer impact on the neighboring surroundings, environment, and communities in high contrast to a regular hotel. Green lodgings are environmentally friendly because they introduce measures to lower water and solid waste, food production, and consumption, and promote direct relationships with local communities and educate travelers on recycling, conservation of biological and cultural diversity. Overall, they help build up environmental awareness.

Sustainable cities remain elusive globally. How can architecture be reoriented to achieve sustainable cities?

10.8 Twelve Steps to a Sustainable Economy

There are, however, twelve steps that can be used as a guideline to form a sustainable economy. These are:

1. Developing macro-economic capability
2. Investing in jobs, assets, and infrastructures
3. Increasing financial and fiscal prudence
4. Improving macro-economic accounting
5. Sharing the work and improving work–life balance
6. Tackling systemic inequality
7. Measuring prosperity
8. Strengthening human and social capital
9. Reversing the culture of consumerism (James 2014).
10. Imposing clearly defined resource or emissions caps
11. Fiscal reform for sustainability
12. Promoting technology transfer and ecosystem protection.

10.9 Practical Application: Sustainability Project Ideas

Sustainability is a crucial concept to all people on the planet. People or groups can gain deeper insight when they engage in projects or activities that demonstrate ways they can involve in sustainability in their day-to-day lives. The following practical ideas are critical in learning about sustainability.

10.9.1 Recycled Journal

Journals can be created using recycled paper. Neatly tear pages from old notebooks with extra pages. Notebooks in the classroom or at home could be used. All the paper should be stacked to ensure that it is appropriately lined up and turned the appropriate way. Card stock should then be positioned on the top and bottom of each paper stack. Re-check to ensure that all pages are lined up and staple them on the left side to bind them. After creating the books, draw a design on the back or front covers. Draw things that are related to sustainability, including trees, animals, plants as well as other naturally occurring organisms. As you work on the activity, use the recycling knowledge to explore more types of materials that could be recycled and gain insight on the importance of recycling.

10.9.2 Help the Community

Community members or groups should participate in cleanup activities within their community. Select a specific area in the community for the exercise or go to a local park. Pick up trash and tend to plants or a garden around the area. This helps groups be proud of their environment and creates awareness about an area they see every day.

10.9.3 Upcycling

Upcycling entails taking something old and converting it into something that is usable again. It can be used as a home project with some of the examples being turning milk jugs into bird houses or turning old clothes into backpacks. Students can also decorate old glass jars to become small storage containers. The knowledge on how to recycle can enable you to understand the significance of minimizing consumption and reusing commodities.

10.10 Individual Matter: Muhammad Yunus

Yunus was born on June 28, 1940 and is an economist, banker, leader of civil society, and social entrepreneur (see Fig. 10.23). His creation of Grameen Bank and ideas of microfinance/microcredit (type of banking service offered to low-income or unemployed persons or groups that lack access to other financial services) awarded him the Nobel Peace Prize in 2006. The bank gives loans to entrepreneurs whose income status does not enable them to qualify for contemporary bank loans.

Fig. 10.23 Muhammad Yunus during the Building Social Business Summit, *Source:* Salford University (2013)

In 2011, he co-founded Yunus Social Business-Global Initiatives. The initiative empowers social businesses to tackle social issues globally. Professor Muhammad Yunus is acknowledged because of the work he has carried out in alleviating poverty and empowering poor women. He has been successful in combining capitalism with social responsibility with the Grameen Bank (Yunus et al. 2010). Since its foundation in 1976, the bank has expanded to offer collateral-free loans to more than 7.5 million customers in not less than 82,000 villages, with 97% of them being women (Yunus et al. 2010). Furthermore, it has maintained a repayment rate of more than 98%. The microcredit movement inspired by Yunus has triggered an international movement that reaches millions of poor women across the world. The role played by microfinance institutions in economic development is immense, and can indirectly linked to Professor Yunus.

10.11 Future Trends

10.11.1 Africa

Implementation of the SDGs has stagnated in many countries, and failure to achieve them will have implications in all places since some of them have a transboundary nature including the ones touching on effective partnerships, life below water, and climate change. The continent will increasingly focus on the Common Africa Position, which is well-enshrined in Agenda 2063 (UN 2015). This will continue being based on six major pillars, partnerships and finance, science, innovation and technology, peace and security, people-centered development, environmental sustainability, disaster risk management, and natural resource management.

10.11.2 Asia

By 2030, it is predicted that Asia may account for more than 50% of the global middle class, thus influencing sustainability efforts (Fernandez 2019). Conscious consumerism will increasingly emerge, particularly for millennials who have lower disposable incomes, with more willingness to purchase ethically produced goods. The next generation of conscious leaders in the Middle East could support sustainability. With an increasingly international outlook as well as a legacy to uphold, the chance for this generation to drive sustainability is ripe. A significant proportion of the population will access the Internet with South-East Asia leading with at least 370 million users (Fernandez 2019). Hyperconnectivity will allow people to develop better transparency opportunities for mobilizing around problems that are part of them. More countries in the Middle East will continue embracing sustainability, with Saudi Arabia leading through the construction of probably the largest wind farm within the Gulf. More countries in the Middle East will invest in renewable energy and engage in sustainable mining. With proceeds obtained from oil and other minerals, countries could prioritize investing green mining technology.

10.11.3 Australia

More millennials will be willing to spend more on goods and services produced sustainably. The conscious consumer choices will drive brands to rethink their social and environmental impact. This trend will intensify as organizations transform from sustainability as a necessary marketing expense to become a core part of businesses.

10.11.4 Europe

Europe will be characterized by the adoption of more common strategic goals that will be set by companies, including energy efficiency, sustainable supply chains, and climate change. Higher investments in energy will also continue in this region. Artificial intelligence will further be used for enhancing sustainability efforts. Technological innovation could incubate the new emerging business models that may support future waste management systems of the world.

10.11.5 North America

Adoption of United Nation sustainable development goals has been progressing a slow pace pertaining to organizations founded in North America. However, 41% of businesses will integrate SDGs into strategy and the manner they conduct business within 5 years (Centre for Sustainability and Excellence 2017). Carbon footprint minimization will become priority of organizations that have highest Sustainability Ranking.

10.11.6 South America

Artificial Intelligence (AI) will increasingly gain more followers as it provides more beneficial tools in almost any industry. AI will play a great role in facilitating and scaling sustainability solutions. Also, corporate sustainability reporting will be significantly transformed.

10.12 Summary

- This chapter has defined sustainability as the development/activities/ processes that cater for our present needs without compromising the ability of future generations to meet their own needs.
- Individual commitment is crucial for ensuring the development of a livable planet.
- The SDGs discussed in order are: ending poverty, ending hunger, promoting good health and well-being for people, quality education, gender equality, clean water and sanitation, affordable and clean energy, decent work and economic growth, industry innovation and infrastructure and reducing inequalities.
- SDGs 11–17 are responsible for consumption and production, tackling climate change, life below water, life on land, peace, justice and strong institutions and partnerships for the goals.
- The three major pillars of sustainability are economic development, social development, and environmental protection.
- The main objectives of environmental sustainability are clean energy supply, sustainable human settlements, sustainable consumption patterns and production of resources.
- There is need for the world to move towards a more sustainable and strong foundation for construction and human settlements.
- Waste and emissions is a critical aspect for achieving environmental sustainability.
- Agricultural practices and habits need to be oriented in a way that enhances sustainability.
- Various forms of sustainable energy should be adopted including solar energy, wind energy, and geothermal energy.
- Sustainable manufacturing is critical for minimizing negative environmental effects and conserving natural resources.

- Measures for achieving sustainable manufacturing are reduction of production inputs, improvement of facility operational efficiency, and reducing environmental impact in product lifespan.
- Steps should be put in place to safeguard the planet's biodiversity.
- Social sustainability should enhance the living standards of people by providing affordable education, housing, and health.
- The major concepts in social sustainability are equity and diversity, quality of life, spirituality, democracy, and governance.
- Economic sustainability is critical for the prudent use of resources and economic models that conserve the environment.
- Economic sustainability is instrumental for promoting green growth. The applications of economic sustainability discussed are production and resourcing, agriculture and food systems, exchange and transfer, accounting and regulation, labor and welfare, technology and infrastructure, wealth and distribution, and sustainable cities.
- The chapter has also identified 12 crucial steps necessary for achieving a sustainable economy.
- SDG implementation has stunted in Africa and this could impact sustainability efforts. For Europe, countries will embrace common strategic goals and technological innovation will be a major driver of sustainability. Asia's sustainability efforts will be supported by conscious leaders while hyperconnectivity will drive the sustainability agenda.
- Organizations in North America will prioritize the integration of SDGs into strategy and minimization of carbon footprint while AI will offer sustainable solutions for South America.
- In Australia, more millennials will use consumerism to transform sustainability.

10.13 Discussion Questions

1. After reading this chapter, what is your definition of sustainable development?
2. How can our society meet today's economic needs without diminishing economic opportunities for our future?
3. Identify and discuss the major ecological principles of sustainability.
4. Why is sustainability so elusive for the world and Africa?
5. Which countries have made significant progress in the social pillar of sustainability?
6. What is a sustainable model?
7. Briefly discuss the Keynesian economic model.
8. How can we strike a balance between the environmental and economic pillar?
9. What roles can sustainability play in the overall strategic plan of an organization?
10. What is the role of the Kyoto protocol in sustainability?
11. Which SDG is directly linked to the Kyoto Protocol?

12. Which is the best definition of social sustainability?
13. How many people globally lack access to clean drinking water?
14. Electric cars are perceived as environmentally friendly, but unsustainable. Discuss.
15. How can Africa tackle climate change?
16. What is the role of gender in sustainability?
17. Which is the most important pillar of sustainability? Provide reasons for your answer.
18. Muhammad Yunus' contribution has been engulfed in controversies. Explain any criticism labeled against him.
19. Discuss the 12 steps necessary for a sustainable economy.
20. Describe four key roles of spiritual ecology in environmental sustainability.
21. What is the role of CSR in sustainability?
22. What are our roles in promoting social sustainability?
23. Democracy is heralded as a key determinant for social sustainability. Briefly explain.
24. Discuss how Africa can rely on wind energy as a source of economic sustainability.
25. Which SDGs should Africa focus on to achieve the environmental pillar?

References

1 Million Women.com (2016). *What do the world's religions have to say about sustainability?* Retrieved November 4, 2018, from https://www.1millionwomen.co-m.au/blog/what-do-worlds-religions-have-say-about-sustainability/.

Asif, Z., & Chen, Z. (2016). Environmental management in North American mining sector. *Environmental Science and Pollution Research, 23*(1), 167–179.

Barnosky, A. D., Hadly, E. A., Bascompte, J., Berlow, E. L., Brown, J. H., Fortelius, M., et al. (2012). Approaching a state shift in Earth's biosphere. *Nature, 486*(7401), 52.

Bettina, W. (2011). Bali school makes sustainability a way of life. *New York times*. Retrieved November 4, 2018, from https://www.nytimes.com/2011/01-/03/business/energy-environment/03iht-green.html/.

Bosselmann, K., Engel, R., & Taylor, P. (2008). *Governance for sustainability: Issues, challenges, successes* (No. 70). Gland: International Union of Conservation of Nature, IUCN.

Campos, G. S., Bandeira, A. C., & Sardi, S. I. (2015). Zika virus outbreak, Bahia, Brazil. *Emerging Infectious Diseases, 21*(10), 1885.

Centre for Sustainability and Excellence (2017). *2017 Centre for sustainability & excellence (CSE) sustainability reporting trends in north America*. Retrieved November 4, 2018, from https://www.cse-net.org/wpcontent/uploads/documents/Sustainability-Reporting-Trends-in-North%20America%20_RS.pdf/.

De Boer, E. R., & Gudmundsson, S. V. (2012). 30 years of frequent flyer programs. *Journal of Air Transport Management, 24*, 18–24.

Despeisse, M., Mbaye, F., Ball, P. D., & Levers, A. (2012). The emergence of sustainable manufacturing practices. *Production Planning & Control, 23*(5), 354–376.

Dincer, I., & Rosen, M. A. (2012). *Exergy: Energy, environment and sustainable development*. Amsterdam: Elsevier.

Ellison, D., Futter, M. N., & Bishop, K. (2012). On the forest cover–water yield debate: from demand-to supply-side thinking. *Global Change Biology*, *18*(3), 806–820.
Fernandez, H. (2019). *4 trends that will shape sustainability in Asia by 2030*. Retrieved November 6, 2019, from https://www.eco-business.com/news/4-trends-that-will-shape-sustainability-in-asia-by-2030/.
Food & Agriculture Division, FAO. (2012). *The state of the world's forests*. Retrieved September 20, 2018, from http://www.fao.org/forestry/fra/80298/en/.
Gardner, G. (2002). *Invoking the spirit: Religion and spirituality in the quest for a sustainable world*. Washington: Worldwatch Institute Paper 164.
Global Carbon Project. (2018). *Carbon dioxide*. Retrieved November 09, 2019, from https://www.globalcarbonproject.org/carbonbudget/18/data.htm/.
Gonzalez, J. (2016). Putting the illegal wildlife Trade in the crosshairs: How the global conservation crisis demonstrates the need for Lacey Act Enforcement of Foreign laws. *William & Mary Environmental Law and Policy Review*, *41*, 321.
Griggs, D., Stafford-Smith, M., Gaffney, O., Rockström, J., Öhman, M. C., Shyamsundar, P., et al. (2013). Policy: Sustainable development goals for people and planet. *Nature*, *495*(7441), 305.
Guisan, A., Tingley, R., Baumgartner, J. B., Naujokaitis-Lewis, I., Sutcliffe, P. R., Tulloch, A. I., et al. (2013). Predicting species distributions for conservation decisions. *Ecology Letters*, *16*(12), 1424–1435.
Hamari, J., Sjöklint, M., & Ukkonen, A. (2016). The sharing economy: Why people participate in collaborative consumption. *Journal of the Association for Information Science and Technology*, *67*(9), 2047–2059.
Hosseini, H. M., & Kaneko, S. (2012). Causality between pillars of sustainable development: Global stylized facts or regional phenomena? *Ecological Indicators*, *14*(1), 197–201.
Ibuku. (2019). *Mepantigan auditorium at green school*. Retrieved November 10, 2019, from https://ibuku.com/mepantigan-auditorium/.
International Food Policy Research Institute, IFPRI (2013). *2013 Global food policy report*. Retrieved November 6, 2018, from http://www.ifpri.org/publication/2013-global-food-policy-report/.
International Institute for Sustainable Development, IISD (n.d). *Sustainable development*. Retrieved November 6, 2018, from https://www.iisd.org/topic/sustainable-development/.
International Renewable Energy Agency, IRENA. (2017). *Rethinking energy: Accelerating the global energy transformation*. Retrieved January 24, 2018, from http://www.irena.org/DocumentDownloads/-Publications-/IRENA_REthinking_-Energy_2017.pdf/.
Jain, R., & Lee, L. (Eds.). (2012). *Fiber reinforced polymer (FRP) composites for infrastructure applications: focusing on innovation, technology implementation and sustainability*. Berlin: Springer Science & Business Media.
James, P. (2014). *Urban sustainability in theory and practice: circles of sustainability*. Abingdon: Routledge.
Jänicke, M. (2012). "Green growth": From a growing eco-industry to economic sustainability. *Energy Policy*, *48*, 13–21.
John, J. V. (2014). Becoming and being a middle power: Exploring a new dimension of South Korea's foreign policy. *China Report*, *50*(4), 325–341.
Kaygusuz, K. (2012). Energy for sustainable development: A case of developing countries. *Renewable and Sustainable Energy Reviews*, *16*(2), 1116–1126.
Khoshnava, S. M., Rostami, R., Valipour, A., Ismail, M., & Rahmat, A. R. (2016). Rank of green building material criteria based on the three pillars of sustainability using the hybrid multi criteria decision making method. *Journal of Cleaner Production*, *173*, 82–99.
Le Blanc, D. (2015). Towards integration at last? The sustainable development goals as a network of targets. *Sustainable Development*, *23*(3), 176–187.
Le Quéré, C., Andrew, R. M., Canadell, J. G., Sitch, S., Korsbakken, J. I., Peters, G. P., et al. (2016). Global carbon budget 2016. *Earth System Science Data*, *8*(2), 605–649.
Lipper, L., Thornton, P., Campbell, B. M., Baedeker, T., Braimoh, A., Bwalya, M., et al. (2014). Climate-smart agriculture for food security. *Nature Climate Change*, *4*(12), 1068.

Macrory, B. (2013). Green school in Bali. *Green Teacher, 99*, 25–27.
Maslow, A. H. (2013). *A theory of human motivation*. New York: Simon and Schuster.
Moldan, B., Janoušková, S., & Hák, T. (2012). How to understand and measure environmental sustainability: Indicators and targets. *Ecological Indicators, 17*, 4–13.
Nevens, F., Frantzeskaki, N., Gorissen, L., & Loorbach, D. (2013). Urban Transition Labs: co-creating transformative action for sustainable cities. *Journal of Cleaner Production, 50*, 111–122.
Nield, D. (2015). This new technology converts sea water into drinking water in minutes. *Science alert*. Retrieved October 20, 2018, from https://www.science-alert.com/this-new-technology-converts-sea-water-into-drinking-water-in-minutes/.
Oliver, M., Bishop, T., & Marchant, B. (Eds.). (2013). *Precision agriculture for sustainability and environmental protection*. Abingdon, Routledge.
Ortiz-Ospina, E., & Roser, M. (2019). *Violence against children and children's rights*. Retrieved November 09, 2019, from https://ourworldindata.org/violence-against-rights-for-children/.
Perl, A. D., & Goetz, A. R. (2015). Corridors, hybrids and networks: Three global development strategies for high speed rail. *Journal of Transport Geography, 42*, 134–144.
Pojani, D., & Stead, D. (2015). Sustainable urban transport in the developing world: Beyond megacities. *Sustainability, 7*(6), 7784–7805.
Rich, N. (2012). Introduction: why link Indigenous ways of knowing with the teaching of environmental studies and sciences? *Journal of Environmental Studies and Sciences, 2*(4), 308–316.
Richter, L. M., Daelmans, B., Lombardi, J., Heymann, J., Boo, F. L., Behrman, J. R., et al. (2017). Investing in the foundation of sustainable development: Pathways to scale up for early childhood development. *The Lancet, 389*(10064), 103–118.
Ritchie, H., & Roser, M. (2019a). *CO_2 and greenhouse gas emissions*. Retrieved November 05, 2019, from https://ourworldindata.org/co2-and-other-greenhouse-gas-emissions/.
Ritchie, H., & Roser, M. (2019b). *Energy production & changing energy sources*. Published online at OurWorldInData.org. Retrieved October 14, 2019, from https://ourworldindata.org/energy-production-and-changing-energy-sources/.
Ritchie, H., & Roser, M. (2019c). *Meat and dairy production*. Retrieved November 04, 2019, from https://ourworldindata.org/meat-production/.
Ritchie, H., & Roser, M. (2019d). *Water use and stress*. Retrieved August 30, 2019, from https://ourworldindata.org/water-use-stress/.
Ritchie, H., Roser, M., Mispy, J., & Ortiz-Ospina, E. (2018a). *Measuring progress towards the Sustainable Development Goals*. Retrieved July 06, 2018, from http://www.SDG-Tracker.org/.
Ritchie, H., Roser, M., Mispy, J., & Ortiz-Ospina, E. (2018b). *Measuring progress towards the sustainable development goals*. Retrieved July 06, 2018, from http://www.SDG-Tracker.org/.
Roser, M. (2019). *Economic growth*. Retrieved November 11, 2019, from https://our-worldindata.org/economic-growth/.
Roser, M., & Ortiz-Ospina, E. (2019a). *Global extreme poverty*. Retrieved November 09, 2019, from https://ourworldindata.org/extreme-poverty/.
Roser, M., & Ortiz-Ospina, E. (2019b). *Primary and secondary education*. Retrieved November 10, 2019, from https://ourworldindata.org/primary-and-secondary-education/.
Roser, M., & Ortiz-Ospina, E. (2019c) *Income inequality*. Retrieved November 07, 2019, from https://ourworldindata.org/income-inequality/.
Roser, M., & Ritchie, H. (2019a). *Hunger and undernourishment*. Retrieved July 31, 2019, from https://ourworldindata.org/hunger-and-undernourishment/.
Roser, M., & Ritchie, H. (2019b). *Maternal mortality*. Retrieved November 12, 2019, from https://ourworldindata.org/maternal-mortality/.
Rosling, O. (2018). *Income*. Retrieved November 02, 2018, from https://www.gap-minder.org/tools/#$chart-type=mountain/.
Rosner, M. (n.d). *Forests*. Retrieved August, 02, 2019, from https://ourworld-indata.org/forests/.
Sachs, J. D. (2012). From millennium development goals to sustainable development goals. *The Lancet, 379*(9832), 2206–2211.

Salford University. (2013). *Muhammed Yunus*. Retrieved October 18, 2018, from https://upload.wikimedia.org/wikipedia/commons/thumb/7/77/Muhammad_Yunus_%28cropped%29.jpg/1280px-Muhammad_Yunus_%28cropped%29.jpg/.

Schuman, R., & Glancy, R. (2015). *How freight probe data is revolutionizing the industry*. Retrieved October 4, 2018, from http://docs.lib.purdue.edu/cgi/viewcontent.cgi?article=1502&context=roadschool/.

Shin, Y., & Thai, V. V. (2015). The impact of corporate social responsibility on customer satisfaction, relationship maintenance and loyalty in the shipping industry. *Corporate Social Responsibility and Environmental Management*, *22*(6), 381–92.

Shiva, V. (2014). *Close to home: Women reconnect ecology, health and development*. Abingdon: Routledge.

Sponsel, L. E. (2001). Do anthropologists need religion, and vice versa? Adventures and dangers in spiritual ecology. In C. L. Crumley (ed.) *New directions in anthropology and environment: Intersections* (pp. 177–200). Walnut Creek: AltaMira Press.

Sponsel, L. E. (2007). Spiritual Ecology: One Anthropologist's reflections. *Journal for the Study of Religion, Nature & Culture*, *1*(3), 340–350.

Sponsel, L. E. (2010). Religion and environment: Exploring spiritual ecology. *Religion and Society*, *1*(1), 131–145.

Sustainability.com. (n.d). *Can our society endure*? Retrieved October 18, 2018, from http://sustainability.com/sustainability/.

UN Food & Agriculture Organization, FAO. (2016). *Share of the population that is undernourished, 2016*. Retrieved November 11, 2019, from http://data.worldbank.-org/datacatalog/world-development-indicators/.

UNCTAD. (2016). *Number of parties in multilateral environmental agreements*. Retrieved March 28, 2018, from http://stats.unctad.org/Dgff2016/index.html/.

United Nations, UN. (2015). *Agenda 2063. The Africa we want*. https://www.un.org/en-/africa/osaa/pdf/au/agenda2063.pdf/.

United Nations Food & Agriculture Division (FAO). (2013). *Share of fish stocks within biologically sustainable levels*. Retrieved January 28, 2018, from http://-www.fao.org/sustainable-development-goals/indicators/en/.

United Nations Statistics Division. (2015). *Number of parties to international multilateral environmental agreements on hazardous waste and other chemicals*. Retrieved January 16, 2019, from https://unstats.un.org/sdgs/indicators/database/.

United Nations Statistics Division. (2016). *Proportion of women aged 18–29 years who experienced sexual violence by age 18*. Retrieved January 16, 2019, from https://unstats.un.org/sdgs/indicators/database/.

United Nations Statistics Division. (2019). *Growth rate of real GDP per employed person*. Retrieved February 12, 2018, from https://unstats.un.org/sdgs/indicators/database/.

United Nations. (nd). *Goal 2: Zero Hunger*. Retrieved October 18, 2019, from https://www.un.org/sustainabledevelopment/hunger/.

Vallance, S., Perkins, H. C., & Dixon, J. E. (2011). What is social sustainability? A clarification of concepts. *Geoforum*, *42*(3), 342–348.

Wang, W., Shao, Q., Yang, T., Peng, S., Yu, Z., Taylor, J., et al. (2013a). Changes in daily temperature and precipitation extremes in the Yellow River Basin, China. *Stochastic Environmental Research and Risk Assessment*, *27*(2), 401–421.

Wang, Y. F., Chen, S. P., Lee, Y. C., & Tsai, C. T. S. (2013b). Developing green management standards for restaurants: An application of green supply chain management. *International Journal of Hospitality Management*, *34*, 263–273.

World Bank. (2014). *Share of urban population living in slums*. Retrieved November 11, 2018, from http://data.worldbank.org/data-catalog/world-development-indicators/.

World Bank. (2015). *Number of maternal deaths by region*. Retrieved November 11, 2019, from http://data.worldbank.org/data-catalog/world-development-indicators/.

World Bank. (2017a). *Manufacturing's value added to GDP*. Retrieved April 14, 2018, from http://data.worldbank.org/data-catalog/world-development-indicators/.

World Bank. (2017b). *Where is child marriage becoming less common?* Retrieved November 10, 2019, from http://documents.worldbank.org/curated-/en/530891498511398503/Economic-impacts-of-child-marriage-global-synthesis-report/.

World Bank. (2018a). *Extreme poverty: poverty and shared prosperity*. Retrieved November 11, 2019, from https://www.worldbank.org/en/publication/poverty-and-shared-prosperity/.

Word Bank. (2018b). *Partnership for global development*. Retrieved December 19, 2018, from http://datatopics.worldbank.org/sdgatlas/SDG-17-partnership-for-global-development.html/.

World Bank EdStats. (2015). *Total net enrollment rate in primary education, 2015*. Retrieved November 11, 2019, from https://data.worldbank.org/data-catalog/ed-stats/.

World Bank & World Development Report. (2008). *Agriculture for development.* Retrieved November 18, 2018, from http://siteresources.worldbank.org/-INTWDRS/Resources/477365-1327599046334/WDR_00_book.pdf/.

Yunus, M. (1998). Poverty alleviation: Is economics any help? Lessons from the Grameen Bank experience. *Journal of International Affairs, 52*, 47–65.

Yunus, M. (1999). The Grameen bank. *Scientific American*, *281*(5), 114–119.

Yunus, M., Moingeon, B., & Lehmann-Ortega, L. (2010). Building social business models: Lessons from the Grameen experience. *Long Range Planning*, *43*(2–3), 308–325.

Chapter 11
Green Entrepreneurship

Abstract An appropriate waste management will be beneficial for improving environmental quality and provide business opportunities. Greenpreneurship is one of the concepts that can be beneficial for waste management and the quest for environmental sustainability. This chapter seeks to explore the concept of greenpreneurship basing on waste management, examine its fundamentals, look at the current and future trends, explore the role played by green entrepreneurship in attaining sustainable development goals, challenges encountered in developing a green business, and propose solutions for addressing them. Various case studies in architecture, agriculture, environment, and business are used for demonstrating the growth and status of green entrepreneurship today. The practical application is about green business ideas for the aspiring green entrepreneurs, while the individual matter of Achenyo Idachaba is used for more in-depth exploration of the topic. This chapter acts as a timely contribution to the Going Green initiatives by demonstrating the way businesses can orient themselves to promote environmental sustainability.

Keywords Green business · Entrepreneurship · Achenyo Idachaba · Green economy · Green Institute

11.1 Introduction

Many international organizations share the opinion that the world should transition towards a Greener economy. The three major pillars of life sustenance are social, environmental, and economic sustainability, as discussed in Chap. 10. Striking a balance between them needs a high level of entrepreneurial spirit. The first attempt to define a Green entrepreneur is evident in literature of the 1990s. Berle (1993) features the term "Green entrepreneurship" for the first time in The Green Entrepreneur: Business Opportunities That Can Save the Earth and Make You Money (Hall and Rosenberg 2010). Berle provides a short and precise definition, "Green entrepreneurship is taking responsibility to create the world we dream of." Farinelli et al. (2011) argue that a proper understanding and definition of this concept is significant in a situation where large organizations prefer invest-

A. A. Akinsemolu, *The Principles of Green and Sustainability Science*,
https://doi.org/10.1007/978-981-15-2493-6_11

ing in Green marketing instead of Green innovations. Green marketing is the creation of a Green image without significant change in the business processes. Many scholars use the terms eco-entrepreneurship, sustainable entrepreneurship, and Green entrepreneurship interchangeably. There are slight differences in the terms, but they have numerous similarities. It is difficult imagining that Green businesses can only be sustainable from an environmental perspective without being economically and socially sustainable since it relies on the relationship between all the dimensions. This chapter seeks to encourage a new generation of entrepreneurs in sub-Saharan Africa who can identify and tap into Green business opportunities. The viewpoints of developing and developed states on Green entrepreneurship often diverge because of conflicting interests and pending concerns. First, there is concern that developed nations can use the promotion of a Greener economy as a new opportunity or excuse for imposing conditions and protectionism. There is a need to, therefore, reduce the risk that it is not used as another vehicle for developed states to repackage their interests in the pretense of sustainable economic development. Second, there is limited development of Green commodities that can meet the needs of poor consumers. Therefore, it appears that Green entrepreneurship involves the production of expensive Green products that only benefit consumers in developed nations. Third, the majority of funding for Green entrepreneurship is obtained from international donors. There may be a problem when these donors withdraw the funding. Fourth, despite niche markets contributing to sustainable production and consumption, the global transformation towards a Green economy may stall unless green products reach consumers. This chapter addresses this gap by examining the way sub-Saharan Africa can ensure that models deployed in welfare economics consider positive externalities that Green entrepreneurship generates.

In Africa, policymakers have embraced economic transformation as well as industrialization as ways for accelerating inclusive growth. They also acknowledge the importance of economic transformation for the delivery of Green growth. The perception is that Green growth will protect the endangered African natural environment in ways that could minimize the welfare of generations. Entrepreneurs' activities determine Green growth and economic transformation through innovations and risk-taking. Consequently, fostering entrepreneurship is a major policy objective for many African states. Currently, Africa is already facing the negative impacts of environmental deterioration on numerous fronts, including climate change. The rapid growth of the population and land use contributes to degradation and deforestation, aggravates water scarcity, and damages fragile ecosystems. The increased environmental problems and high unemployment rates in sub-Saharan Africa are a concern for many policymakers across the region. However, they can also be perceived as an opportunity for the development of Green entrepreneurship across the region. Intense adoption of Green entrepreneurship would be instrumental in enhancing economic transformation and increasing the availability of jobs. However, there is a need to define and understand the fundamental concepts associated with Green entrepreneurship.

11.2 Definition

Green entrepreneurship can be defined in terms of the output and the process of economic activity. It is the activity of rationally tackling a social or environmental issue or need by realizing the entrepreneurial ideas with a high risk-level that is environmentally friendly and financially sustainable. A Green entrepreneur is an individual who initiates and manages an entrepreneurial venture designed to incorporate green processes and products from the onset. From this definition, provide your understanding of the terms Green entrepreneurship and Green entrepreneur.

During the last three decades, significant interest and research have focused on comprehending the Green growth determinants. A significant issue has eluded policymakers and economists' attention: the ultimate introduction of Green commodities and technologies by Green entrepreneurs in the marketplace. They are economic actors who are instrumental for turning ideas into practice through the transformation of prototypes into viable products. Contrastingly, many policy mechanisms that attempt to spearhead Green growth in sub-Saharan Africa mainly aim to identify technological innovations that can mitigate human effect on the environment and tackle global environmental problems, including biodiversity loss, global warming, and land degradation. In the next section, the focus will be on the important aspects of Green entrepreneurship and its applicability in sub-Saharan Africa.

11.3 Fundamentals of Green Entrepreneurship

11.3.1 Creative Destruction

According to Gibbs (2009), Green entrepreneurship is based on the fundamental concept introduced by Schumpeter about creative destruction. It has the ability to disrupt the current economic model that contributes to the destruction of the planet's resources without substituting or replenishing them naturally. Green entrepreneurship can also become the driving force for a sustainable economic system through consideration of the three fundamental dimensions: society, environment, and economy (see Fig. 11.1).

Green entrepreneurs act as agents of change and contribute to the destruction of conventional production methods, structures as well as consumption models. After destruction, they substitute them with better ecological services and products (Gibbs 2009). Consequently, they can be termed as disruptive innovators.

In detail, provide your understanding of disruptive innovation.

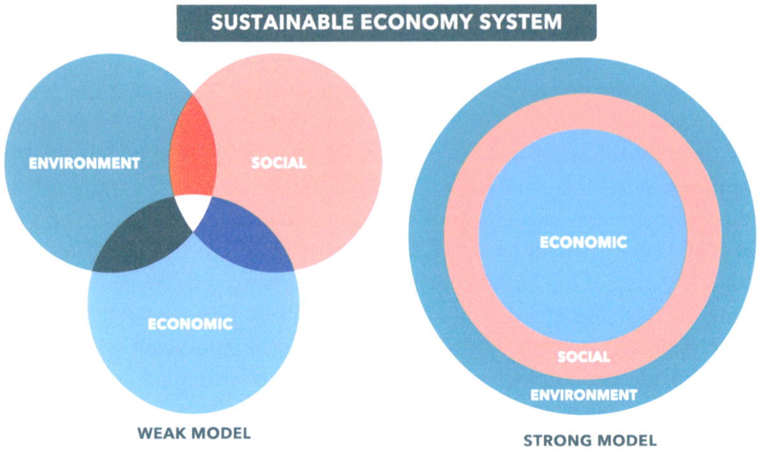

Fig. 11.1 Dimensions of sustainability

11.3.2 Characteristics of Green Entrepreneurship

Many scholars discuss the various intrinsic characteristics of Green entrepreneurs. First, Allen and Malin (2008) believe that they behave like entrepreneurs by coming up with new ideas and taking high risks. Second, Green entrepreneurs have robust internal motivation and sensitivity to environmental issues. Third, they tackle social or environmental issues, and their activities positively affect the natural environment. Finally, eco-entrepreneurs for a more sustainable future and use their environmental and social values to contribute towards it. Consequently, the entrepreneur plays a pivotal role in actualizing the objectives of Green entrepreneurship.

Are there any more characteristics of Green entrepreneurs that you know?

11.3.3 Green Economy

The other fundamental aspect of Green entrepreneurship is the Green economy. The United Nations Environment Program provides a useful definition for the Green economy. It defines the Green economy as economic growth accompanied by the emission of low carbon and pollution, better resource efficiency, and sustained ecosystem services (Allen and Malin 2008). Allen and Malin (2008) argue that a Green economy can be important for tackling significant developmental and economic problems. They include economic growth support, climate change mitigation and adaptation, job creation, promoting the Sustainable Development Goals and eradication of poverty. The Green economy idea has similarities with sustainable development in balancing social, economic as well as environmental concepts (Jones 2017). The focus is mainly on the economic and ecological aspects. According to Jones (2017), the core focus of a Green economy is to develop innovations and policies that allow society to garner value annually while conserving natural resources that sustain people's lives.

Nevertheless, there are numerous markets for the Green economy. The major markets are sustainable water management, clean energy, Green services, sustainable land use, waste management, sustainable transport, sustainable manufacturing activities, and building energy efficiency. Green jobs are more diverse than Green markets in case one adopts a broader definition. They are jobs with skills and functions for the measurement, prevention, and amelioration of adverse effect on the environment. Some examples of Green jobs are water quality technicians, environmental lawyers, and supervisors in eco-industrial plants.

What are the main factors to consider in order to achieve a Green economy?

11.3.4 Green Innovation

Furthermore, another critical element of Green entrepreneurship is Green innovations. Chen et al. (2006, p. 534) define Green innovation as "hardware or software; innovation that is related to Green products or processes, including the innovation in technologies that are involved in energy-saving, pollution-prevention, waste recycling, Green product designs, or corporate environmental management." In other words, it is the development of processes and products that play a crucial role in sustainable development through the commercial application of knowledge to elicit ecological improvements. Entrepreneurs acknowledge that innovation is key to job creation, economic transformation, and environmental sustainability. However, for the case of sub-Saharan Africa, it requires a holistic change in the organization of business, the design of cities, means of living, and the movement of people and goods. Input is required from numerous stakeholders for these changes to occur. Innovation is instrumental not only for the development of ideas but also for Green growth. Conceptually, Green growth implies facilitating economic growth while making sure that the ecosystem continues providing services and resources for life sustenance. The increased concern regarding the viability of economic growth patterns in sub-Saharan Africa bolsters the need for a Greener growth model. Green innovation can be instrumental in preventing natural capital depletion in this region.

Also, innovation and creative destruction can contribute to the development of new ideas and business models, hence leading to the creation of new markets and jobs. Consequently, Green innovation is the foundation for sustainable growth.

International linkages and technologies affect Green innovation by enhancing co-operation in various policy areas, including environmental regulation. It is also crucial for the development of global standards on environmental regulations, trade policy, as well as energy technologies. Technologies, such as Information Communication Technologies and Green innovation, can enhance the environment and address climate change. Green technologies such as nanotechnology and biotechnology in Africa could become "general purpose technologies," because they can be used in various industries, like the Green economy. Overall, this section has determined that the fundamentals of Green entrepreneurship are the Green entrepreneur, Green markets, Green innovations, creative destruction, and Green jobs. However, all these elements can be summarized under environment, economy, and society. The next section examines the evolution of Green entrepreneurship.

Track the definition of Green innovation. What has been consistent in the definition? Identify aspects that have changed.

11.4 The Evolution of Green Entrepreneurship

Jones (2017) provides a holistic and comprehensive evolution of Green entrepreneurship. The economy and the environment have always been at odds. During the industrial revolution, economic production was significantly influenced by steel, iron, coal, and large-scale production. Many entrepreneurial organizations and inventors spearheaded this transition. Businesses were part of the problem of environmental degradation rather than the solution. Despite this, specific individuals were determined and motivated by the desire to promote environmental sustainability through business. Various for-profit businesses during the 19th and early 20th centuries tried to counteract the negative effects of rapid industrialization on the environment (Parry 2012). The other attempt was the intrinsic motivation for "Greenness" by entrepreneurs who understood the significance of environmental, ethical, and health concerns instead of profit-maximization.

To what extent was the movement successful in bringing about Greenness? Use examples to explain your answer.

For example, Rudolph Steiner, a polymath, is among the Green entrepreneurs who encouraged environmental sustainability in current Croatia (Holt 2011). Functioning outside the economic and cultural majority, Steiner advanced his "Anthroposophy" ideology. This ideology emphasized the interdependence between the cultural, legal as well as economic domains. He staunchly advocated for vegetarianism and researched the medicinal value of plants. To accomplish his mission of holistic medicine and health, he developed "biodynamic" agricultural methods (Jones 2017). His motivation for venturing in business was to develop an ethical business culture and fund cultural institutions to propagate his beliefs. Steiner contributed to the development of a Green market by influencing biodynamic farming movements in Europe, particularly Germany.

Poul la Cour, a Danish engineer, was also among the early wind entrepreneurs who tried to solve the scarcity of rural energy supply. She was concerned about the effects of DDT pesticide on health. Other conventional entrepreneurs found themselves, accidentally, becoming Green entrepreneurs. For instance, Edward Buxton and Frederick Selous in East Africa played a crucial role in the development of the eco-tourism industry. These entrepreneurs tried to reimagine trade-offs between economic and environmental sustainability. They tried to find ways for aligning the two concepts but found it challenging to convince investors, consumers, and the government about their radical ideas. During the 1960s, religious and cultural systems increasingly inspired the early entrepreneurs. For example, Theosophy, Buddhism, and Hinduism inspired Steiner's entrepreneurial work as well as the second environmentalism wave. Anecdotal evidence from the Global South also shows models of Green entrepreneurship. For instance, in India, Hamdard began in 1906 to bring Unani medicine into the market. Unani medicine is a traditional system that is practiced in India. Unani medicinal practices entail the use of herbal remedies, alternative therapies, and dietary practices to prevent and treat disease.

Identify three other prominent wind entrepreneurs who were instrumental in the development of the Green movement.

Solar entrepreneurship was the next development in Green entrepreneurship as anticipated financial returns along the socio-environmental and ethical principles attracted entrepreneurs. It is defined as the taking of risks in solar power in order to reap profits. During the first phase of renewables, many early entrepreneurs needed capital as well as production networks. They also experienced stiff competition from fossil fuel organizations that had just emerged. It was challenging to access financial capital, emanating from numerous gas organizations, such as Mobil and Exxon that had varying strategic considerations in the 1970s. In addition, there was little support for Green innovations. For example, Ronald Reagan, President of USA from 1980 to 1988, favored fossil fuels and even ordered the removal of solar panels from the White House's roof. Jones (2017) further explores the critical aspect of the effect of Green entrepreneurship.

Name two more challenges that such organizations experienced.

He traced its evolution during the 1980s in terms of the costs associated with scaling up businesses. In addition, increased scale may be characterized by altering business practices as well as ownership. The growth of start-ups leads to an increased requirement for more finance, particularly because Green businesses find it hard making profits. This is partly because negative externalities are not priced. The evolution further suggests that large companies with resources for driving Green innovation became increasingly involved in sustainability. For instance, large conglomerates, like Unilever, L'Oreal, and PepsiCo, acquired Green start-ups, such as Ben & Jerry's, The Body Shop, and Naked Juice, respectively (Jones 2017). Today, the deterioration of the environment has spearheaded a global movement towards Green consumerism and Green entrepreneurship. There is a widespread belief that Green entrepreneurship should increasingly be adopted to reverse the deteriorating state of the environment.

Is this the right step/initiative?

In addition, gas and oil corporations, such as Chevron and Exxon, as well as technological and industrial organizations, including American corporations such as Wal-Mart and Facebook, have continually reiterated their increased commitment to sustainability despite President Donald Trump withdrawing from the Paris Climate Agreement of 2015.

Perhaps, Africa is among the continents that have been late adopters of the concept of Green entrepreneurship. In sub-Saharan Africa, policymakers have embraced economic transformation as well as industrialization as ways for accelerating inclusive growth. They also acknowledge the importance of economic transformation in delivering Green growth. Many Green entrepreneurs have been acknowledged in the region because of the crucial role they have played in promoting sustainable development. In the absence of Green entrepreneurship policies, there is a risk that environmental stresses attributed to industrialization will destabilize growth.

Describe some of these environmental stresses.

11.5 Current Trends: Green Ideas, Innovations and Businesses in Africa

Whereas there is increased agreement about the importance of economic transformation, there has been relatively little change in Africa. Many African countries still adhere to the typical historical path, and many of their employment opportunities as well as output are in the agricultural sector. In sub-Saharan Africa, a larger proportion of employment is in the agricultural sector. Green initiatives remain unexploited by many countries in sub-Saharan Africa. Despite this, there has been an increased effort for Green entrepreneurs to venture into Green industries.

11.5.1 Architecture

11.5.1.1 Sandbag Houses, Freedom Park South Africa

Architecture is among the most dominant Green entrepreneurship inventions in sub-Saharan Africa. Eco-architecture has found new mechanisms for reducing costs, gaining insights from the traditional building methods, conserving energy, and promoting reuse while remaining relevant. There are various examples of Green architecture, and South Africa is among the leading states in sub-Saharan Africa. The buildings use natural ventilation and cooling systems, thus promoting sustainable use of energy resources. Sandbag Houses found in Freedom Park, Cape Town, South Africa (see Fig. 11.2) are among the most renowned Green architectures in Africa. The objective of a sandbag is the conservation of money and resources. It uses cheap local materials, thus reducing the cost of transportation (Udoma 2014). The assistance of future residents enables the house to minimize its costs further. The house is further constructed through the use of Eco-Beams system that substitutes brick-and-mortar with sandbags. It is a robust, safe, and less expensive way for affordable housing delivery.

What are the major challenges/disadvantages that you believe are associated with this architecture?

11.5.1.2 Vissershok School, Durbanville

The second example of Green architecture is Vissershok School, Durbanville, South Africa (see Fig. 11.3). It is built using recycled shipping containers and acts as a classroom during the morning hours and a library during the afternoon. The

Fig. 11.2 Sandbag House, *Source*: Smithsonian Cooper-Hewitt (2011)

Fig. 11.3 Visserhok School, Durbanville, *Source*: Tsai Design Studio (n.d.)

shelters have a large roof that protects the container from sunlight while the gap provides ventilation and minimizes heat gain. Windows on each side also act as cross ventilation. In addition, the stepped seating provides space for children to consume lunch and serves as an amphitheater for assemblies. The school has also erected a Green wall that acts as a vegetable garden.

How can Africa embrace such buildings in school environments?

11.5.1.3 Eastgate Centre, Zimbabwe

Similarly, the Eastgate Centre in Harare, Zimbabwe, is mainly made of concrete and has a ventilation system that functions in the same way as the mounds created by African termites. Similar to biomimicry, it is inspired by the indigenous Zimbabwean masonry. Despite lacking conventional heating or air conditioning, it is regulated for the whole year, thus minimizing the consumption of energy. Compared to a conventional building with a similar size, it uses 10% of energy (Udoma 2014). The building's owners have saved at least $3.5 million due to the deployment of natural ventilation. These benefits also trickle down to tenants who pay 20% less rent compared to occupants within the surrounding houses (Udoma 2014) (refer to Case Study 1.1).

What impedes Africa from constructing buildings like the Eastgate Centre in Harare?

11.5.2 Green Companies and Other Green Ideas in Africa

11.5.2.1 Ndume Limited, Kenya

Ndume Limited is a large Green entrepreneurship company in Kenya that offers specialized inputs. These inputs are required for conservation agriculture, which is an approach for sustainably increasing agricultural yields with methods that conserve rainwater and soil nutrients, while simultaneously minimizing the costs of production. In addition, a Dutch Organization, Ubbink, has erected a solar module manufacturing organization within Naivasha, Kenya. It is the first of its kind in East Africa, and solar modules are manufactured basing on reworked quality cells.

Similarly, Solar Sister in Uganda was created to empower women by giving them economic opportunities (SEED Case Studies 2011). The role of these women is to distribute clean energy products, including household solar kits as well as clean stoves. Today, Tanzania also has the same experiment to enhance the empowerment of its women.

How does Ndume Limited embrace the ideals of Green growth and sustainability?

11.5.2.2 FONERWA, Rwanda

Rwanda also attracts Green growth by creating and operating a national fund for climate change and the environment. Fonerwa has been instrumental for the rollout of various projects and raising the profile of the country to be among the early movers in implementation of a Green economy. The Fonerwa initiative followed the Green Growth and Climate Resilience Strategy in 2011. It has succeeded in funding 18 projects in sustainable energy, waste management, integrated water management, Green planning, and rehabilitation of the environment.

The Rwandan government has become one of the few countries globally with a well-developed environment and national climate change fund. It is founded on the newly embraced Green Growth and Climate Resilience Strategy. Its major purpose is to ensure that there is sustainable financing to provide support to environmental sustainability, climate change resilience, and Green growth. It is the primary mechanism that enables the country to access, program, and monitor extra-budgetary environmental and climate finance. Funds are distributed to the private sector, civil society, government as well as communities for implementation of various projects.

Describe factors that have contributed to the success of the FONERWA program in Rwanda. In light of FONERWA, how can African politics be oriented to enhance Green entrepreneurship?

11.5.2.3 Climate Resilience Green Economy, Ethiopia

In Ethiopia, the leaders' vision to protect natural assets led to the development of a Climate-Resilient Green Economy in 2011 that guides economic development with the aim to become a middle-income country through Green growth. The strategy's pillars are agriculture, power, forestry as well as transport. There is a strong focus on sustainable land management. The strategy entails the adoption of simple technological solutions, the participation of local communities in planning and implementing activities, adopting irrigation and planting crop varieties as well as learning from restoration. The local members of the community offer free labor for some days by engaging in activities like vegetation, construction of check dams, and afforestation. It has successfully led to the restoration of degraded land within the highlands of Tigray and Amhara to productivity, increased the quantity of water available for agriculture, and enhanced climate resilience. It has also been beneficial for behavioral change, including the use of recycled groundwater for irrigating fields and inclusion of marginalized communities within the program.

Find statistical data about the impact of the program on the country.

11.5.2.4 WeCyclers, Nigeria

Furthermore, an adventurous and new eco-spirit has emerged in sub-Saharan Africa. Various young, inspirational Green entrepreneurs who remain committed to ensuring that there is a difference to the continent's sustainable future are leading this movement. They possess an uncanny ability to unearth environmental issues and rather than being bogged down, they find innovative solutions and develop entrepreneurial opportunities. In Nigeria, Bilikiss Adebiyi-Abiola launched WeCyclers in 2012 to solve trash collection (Hawken 2015). She started a viable and sustainable business model that is currently helping impoverished communities within Lagos to eliminate the waste and pollution problem.

WeCyclers enables poor communities to generate income from waste that piles up in the streets. The organization utilizes cargo bicycles for collecting and recycling unmanaged waste. The bicycles pick up various sachets, cans, and plastics from homes. In addition, the residents get points through SMS depending on the number of recyclables they collect. They can redeem these points for basic consumer electronics and essential food items. The company then aggregates the waste to sell to local recyclers. It is a highly effective and low technology, and yet a high impact recycling scheme that benefits all those involved. By 2015, at least 5000 households had signed up, thus leading to the creation of employment for numerous young people (Hawken 2015). There are also plans to roll out this initiative in other Nigerian cities. The start-up can become a global case example for empowering the community to be sustainable, wealthier as well as healthier lives. Discuss five benefits of WeCyclers.

What are some of the challenges the start-up faces, and how can these challenges be addressed?

11.5.2.5 MitiMeth, Niger Delta, Nigeria

Another entrepreneurial venture is that of Achenyo Idachaba, who weaves stubborn weeds into beneficial products. The problem she identified was that the Niger Delta was grappling with the issue of being filled by aquatic weed, particularly water hyacinth. The hyacinth posed a dire environmental issue that wreaked havoc within local communities and ruined the supply of water. The weeds made water stagnant and depleted nutrients that reduced fish populations, thus leading to a significant issue for local communities dependent on subsistence fishing. Achenyo Idachaba's entrepreneurial vision and energy were instrumental in solving this problem. Her entrepreneurial spirit enabled her to identify the issue and see it as an opportunity (Hawken 2015). The business, dubbed MitiMeth, takes the weeds and transforms them into hand-crafted products that appeal to women. These products include jewelry, baskets as well as tableware. Achenyo has solved an environmental issue and turned it into a lucrative business solution. It entails the development of sustainable business, creation of jobs within local communities, and fixing a prevalent issue. She is an excellent illustration of the new generation of young entrepreneurs who are eager to find Green solutions to problems bedeviling communities.

How can the Victorian hyacinth problem be addressed using the same principles in MitiMeth? Identify four benefits of MitiMeth to the community and environment.

11.5.2.6 Afrocentric Bamboo, Ghana

In Ghana, Winnifred Selby is one of the founding members of Afrocentric Bamboo. The company makes bamboo bicycles that create affordable and innovative transport. Similar to numerous sub-Saharan African nations, Ghana has been struggling to find sustainable solutions to an array of socioeconomic issues, including affordable transport, effective means of eradicating poverty, and a lack of jobs. At the age of 15, Winnifred Selby started the company with Bernice Dapaah. The company manufactures and carries out marketing of bicycles that are made from bamboo. The bicycles are strong, affordable, currently USD $100 in the local market, and designed to address the rough roads found in rural Ghana (Hawken 2015). By using a one-piece frame, they are stronger and economically viable. This innovative approach has been instrumental in solving the issues of poverty, unemployment, and affordable transport. Winifred has demonstrated that Green entrepreneurship can be used for improving the economic future of a nation, particularly young people. Arguably, Afrocentric is a distinct Green entrepreneurship that can inspire many across sub-Saharan Africa to venture into sustainable Green business. Africa has many unused resources, including bamboo and water hyacinth prevalent in some lakes like Lake Victoria in Kenya and Uganda. Tapping on such resources could be a platform for creating employment and reducing the high poverty levels.

Using this case, describe how Green entrepreneurship could be the solution to many African problems. What support structures are needed to make this a success?

11.5.2.7 Eco-Post Kenya, Kenya

In Kenya, Eco-Post Kenya has been instrumental for recycling plastic waste into sustainable or environmentally friendly fencing posts. Founded by Lorna Rutto, the company began with a vision for finding solutions to the large quantities of plastic wastes that filled the streets in her local community. Her love for the environment and the desire to address the waste problem in Kenya made her venture into Green entrepreneurship. Eco-Post is a Green business that manufactures durable and sustainable fencing posts with plastic. It offers an effective solution for managing the vast plastic generated in Kenya every day (Njeru et al. 2016). In addition, it creates more than 300 employment opportunities for young individuals and marginalized women (Hawken 2015). The model seeks to create more than 100,000 jobs in the next 15 years. Furthermore, the company has had an impressive eco-footprint by removing more than one million kilograms of plastic waste from the urban slums in Kenya. It has also saved about 250 acres of vulnerable forests (Hawken 2015). Eco-Post is a demonstration of the possibility of creating a sustainable business in sub-Saharan Africa, while simultaneously benefitting the environment and local communities.

Job creation is at the center of Sustainable Development Goals. Does Green entrepreneurship lead to the creation of more jobs than normal entrepreneurship? Discuss.

11.5.2.8 Repurpose Schoolbags, South Africa

Finally, the use of solar light technology in upcycled schoolbags has drastically changed the lives of many schoolchildren in South Africa. The business uses Green innovations to solve societal problems. It also unearths sustainable opportunities positively affecting low-income communities, with a specific focus on women and children. As a Green entrepreneur, Thato Kgthalhanye sought to solve the problem of students finding it difficult to study at home. She witnessed this in Rustenburg and decided to co-found repurpose schoolbags with Rea Ngwane (Hawken 2015). The company makes schoolbags that have been upcycled through the integration of solar technology. It charges at daytime and provides light for students to study after darkness sets in. The 100% recycled bags have changed the lives of numerous young students. In addition, they cause less harm to the environment while providing the required renewable energy light for young learners to study in an area that lacks electric light. Reflective light material also enables the students to see as they walk long distances to attend school. Through Green entrepreneurship, Thato has positioned herself as a change agent in her community.

Highlight how schools can embrace such ventures.

11.5.2.9 The Green Institute Nigeria

The Green Institute is a campus-based initiative that was founded by Adenike Akinsemolu to promote environmental conservation. The initiative is committed to tackling the challenges of environmental sustainability and climate change through innovative academic research, campus action, and mentoring and development of environmentally conscious leaders. The core aim of the initiative is to rally young people across Nigeria to reorient their thinking about the environment and gain knowledge about Green entrepreneurship. As a training program, it is designed to provide training to learners about sustainable practices and environmental sustainability. Trainees are typically placed in sections that are related to their fields of interest and closely supervised. It provides interns with a chance to undergo experiential learning and gain relevant skills that are beneficial for professional development. Sustainability is among the core goals of the University. This is evidenced in the entire campus through the sustainable facilities, micro-wind turbines, solar panels, and electric car charging stations. Students play a pivotal role in the initiative, primarily through their participation, enthusiasm, and input (Hawken 2015). Campus involvement is encouraged through various programs geared towards electricity reduction and promotion of an Energy Reduction Month. Arguably, the initiative is playing a crucial role in spurring change in Nigeria. The major objectives/goals of the Green Institute's Campus Initiative are:

(a) To implement and handle sustainable initiatives in universities.
(b) To monitor and minimize waste emerging from all campus activities and operations.
(c) Instill a culture of sustainability awareness and action in the local community and on campuses.
(d) To offer practical sustainability learning and training opportunities.

The current projects being undertaken by the initiative are highlighted below:

1. **Romance with the Environment**: On Valentine's Day, students come together across various campuses to celebrate love by taking care of the environment.
2. **Aerated Composting**: This teaches students about static aerated composting and waste management operations. Encourages students to work first-hand with sustainability projects on campus and gain valuable skills.
3. **Bike-Friendly Campus**: This strives to make students become more bike-friendly through increased bike parking, bike lanes, bike loans, as well as bike signage in the community (Green Institute 2018).
4. **Clothing Swap and Free Store**: Students are encouraged to take their clothes once every semester and exchange them for something new. This is a great way of cleaning out their closets and apartments as the semester ends.
5. **Documentary Night**: This is an old yet favorite initiative by the campus that allows students to get together to watch and talk about environmental documentaries.
6. **Green Garden**: The garden provides opportunities for planting, maintaining, and harvesting annually.
7. **Other Initiatives**: These include Green energy, outreach and community service, outings, camping, parties, recycled notebooks, ride board, special events, zero-waste events, new projects, waste, and energy and sustainability audits.

In your local community or family, can you identify people who have come up with Green businesses? Identify at least two, and describe how their work has been instrumental in promoting Greenness.

11.6 The Role of Green Entrepreneurship in Achieving the Sustainable Goals

The 17 Sustainable Development Goals (SDGs) are expected to be achieved by 2030 (refer to Chap. 10). Green Entrepreneurship can be instrumental for achieving these goals.

11.6.1 Tackling Poverty and Social Inequality

The first Sustainable Development Goal seeks to eradicate all forms of poverty across the globe, while the tenth SDG aims to end all forms of inequality, including reduction of income inequalities. Green entrepreneurship can be instrumental in creating jobs and equity to empower people. Green growth targets economic growth, minimal environmental effects, and social goals that can be instrumental for achieving the first and tenth SDGs. Many Green growth initiatives target agriculture and natural resources, which are sectors that many of the people in sub-Saharan Africa work and live.

Why is Green entrepreneurship better poised to sustainably address poverty and social inequality in Africa?

For instance, it is estimated that at least 80% of the people in Ethiopia are employed in the agricultural sector (Sanchez et al. 2009). It provides Green entrepreneurship with an opportunity to enhance people's lives, address poverty, improve access to technology as well as increased productivity, thus contribute the attainment of SDGs No. 1, 2, 3, 4, 5, 6, 8, 10, 14, and 15. These goals revolve around the eradication of all forms of poverty, elimination of hunger, food security, good health and well-being for the people, and inclusive and equitable quality education. Green entrepreneurship also plays a crucial role in the achievement of gender equality and empowerment for women and girls. By creating more jobs, it facilitates the reduction of inequalities and promotes decent work as well as economic growth. Embracing Green entrepreneurship is also crucial for the sustainable use of forests and terrestrial ecosystems. Increased access to technology and higher productivity would contribute to the achievement of SDG No. 9.

Social inequality is increasing at an alarming rate in Africa. Which countries (identify at least three for each category) in Africa have (a) the highest social inequality and (b) the lowest social inequality indexes?

11.6.2 Job Creation

The development of a Green economy is instrumental for creating employment opportunities, thus enhancing livelihoods. With an improvement in people's lifestyles, poverty would significantly decline (SDG 1), and people would be able to educate their children. People would also attain good health and well-being, thus contributing to the attainment of SDG 3. Green entrepreneurship is a key solution for addressing unemployment, especially for the most vulnerable groups, including women and the youth. This is because it has comparatively low entry-level requirements. The technical aspects of Green entrepreneurship can also offer an opportunity for those who have an interest in innovative business solutions (Sanchez et al. 2009). Additionally, the creation of new organizations leads to equal access regarding genders in all areas. Whereas there are still hurdles for Green entrepreneurs, stimulating it could offer new and inclusive employment opportunities, thus leading to equal rights.

Track the progress of Green entrepreneurship in creating jobs in Africa. How does Green entrepreneurship promote the development of skills and knowledge?

11.6.3 Other Roles

The other crucial roles of Green entrepreneurship in the attainment of SDGs are the development of new markets, improved public services, better health outcomes, improved management of land resources, better management of water resources, improved environmental quality, and better adaptation to climate change and coping mechanisms. These roles are essential for the attainment of all 17 Sustainable Development Goals.

11.7 Challenges of Building a Green Business

11.7.1 High Capital Costs/Inaccessibility to Finance

Many developers in sub-Saharan Africa lack adequate funds for starting Green businesses. Compared to other businesses, Green entrepreneurs report that their projects often entail high upfront costs (Koester 2016). Capital costs are often barriers for entrepreneurs who would like to venture into Green business. For example, according to a report from the World Green Building Trend 2016, higher perceived start-up costs are the major challenge for the development of Green buildings.

How can this challenge be solved?

11.7.2 Market Awareness

Apart from the high costs, there are low levels of literacy in sub-Saharan Africa. Due to low educational levels, many people do not value Green entrepreneurship. They also do not think "outside the box" and realize that Green entrepreneurship can be a solution for some of their problems. Green entrepreneurship can be obtained if there is market awareness among the sub-Saharan African people.

Identify ways of solving the market awareness challenge in sub-Saharan Africa.

11.7.3 Low Literacy

The low levels of education imply that people lack the foresight, prerequisite skills, and knowledge for venturing into Green entrepreneurship. Many countries have not incorporated the concept of Green entrepreneurship into their national curriculum, an aspect that would have enhanced the potential for more people to venture into this industry.

11.7.4 Lack of Consumer Awareness

Many people in these countries are not aware of Green products. Green entrepreneurs, therefore, fear that their products would not be purchased or used. Therefore, many decide to venture into a non-Green business where they can find a market for their products (Koester 2016). Familiarity with Green products could be beneficial for increasing the number of people who will become receptive to use Green products.

What programs and policies should be put in place in Nigeria to address this challenge?

11.7.5 Limited Policy Support

Many nations lack policies that support Green businesses. Due to the vulnerability of these businesses, it becomes difficult for them to thrive as they are subjected to the same policies as those governing non-Green businesses. Start-ups in sub-Saharan Africa grapple with many challenges, but many of them are related to education, finance, and support systems.

11.8 Solutions and Recommendations

11.8.1 Recognize and Promote Green Businesses

There is a need for the government, investors, and other financiers to focus on value for stakeholders by adequately recognizing the significance of Green businesses in sub-Saharan Africa. Policymakers should also focus on the development of policy conducive to the success of Green businesses as they spearhead the transition to local economic development, tackle environmental and social challenges as well as inspire other small and medium enterprises to make progress (Koester 2016).

Identify two policies in Nigeria that can help speed this up.

11.8.2 Provide Training to Develop Green Business Initiatives

Green enterprises need support to efficiently contribute to the Green economy and become sustainable as they scale up. There is also a need for policies to focus on women-led Green start-ups that are underrepresented despite being important for driving growth, particularly in agricultural economies.

11.8.3 Facilitate Access to Finance

Finance is among the significant challenges for start-ups. Policymakers should negotiate with financial institutions to ensure that they develop approaches and products that consider the situations of Green Small and Medium-sized Enterprises. An example is Hivo's Green Entrepreneurship program that has played a crucial role by collaborating with government agencies, local incubators, business support organizations, and investors. Through this collaboration, the program supports environmentally and socially conscious entrepreneurs. Its additional investment vehicles, Hivos-Triodos Fund, and Hivos Impact Investment offer finance to hasten the growth of unique SMEs.

Highlight other examples in Africa (at least three).

11.8.4 Facilitate Networks and Partnerships

Partnerships are important for accessing and pooling of resources. They should be developed to facilitate joint learning and sharing of knowledge. Policymakers need to consider themselves as indirect stakeholders and understand the views of Green entrepreneurs to tackle the barriers that they face (Koester 2016).

11.8.5 Empowerment

Small business Green entrepreneurs are key players in the creation of jobs and innovation. Consequently, they need to be invited to engage in policy development discussions, including multi-stakeholder platforms. The creation of platforms for dialogue also offers opportunities for these businesses to develop partnerships important for innovation and unlocking opportunities within the market.

Provide two key examples in Africa.

11.8.6 Conducive Regulatory Environment

The government should play a role in the preparation of an attractive business environment for Green businesses. In sub-Saharan Africa, Green entrepreneurs face many barriers to entry compared to more advanced states. To unleash innovation, there is a need for governments to streamline bureaucratic processes, utilize resources at the local government level to efficiently implement policies, and ensure that courts can efficiently enforce land tenure rights.

11.9 Practical Application: Green Business Ideas for Aspiring Green Entrepreneurship

Do you have an interest in starting a Green business? If yes, the following are innovative Green businesses to consider. The ozone layer is under depletion, global warming is rising, and world leaders are increasingly working hard to make sure that we save the planet. There are global conferences conducted regularly and centered on Greenness. We have a responsibility to reduce our energy consumption and encourage society to embrace eco-friendly energy sources. Regarding the protection of our environment, it is time we acknowledge that each of us is a stakeholder. It is because we all participate in non-environmentally friendly activities that lead to the depletion of the ozone layer. For instance, there are large quantities of materials that are decomposable, and when buried in the ground, they harm the ecosystem, animals, and communities. Various advocacy groups participate in encouraging individuals to do their best in order to save our planet. There are numerous campaigns at the local and international level that are geared towards promoting the need to go Green. It is the only alternative to saving the planet from degradation caused by individuals. Aspiring entrepreneurs need to go Green when choosing a business idea because of the current status of resources. Certain Green business ideas can attract support from stakeholders within the industry.

11.9.1 Selling Recyclable Plastic Bottles

The sale of recyclable plastic bottles is a profitable and thriving business that an aspiring entrepreneur can consider. Many products can be produced from recyclable plastic bottles, including cloth buttons, plastic components, toys, plastic ruler, plastic bags, and plastic chairs. It is crucial to carry out feasibility studies and research to ascertain whether the product has high demand prior to creating a company for recycling plastic bottles. Admittedly, there is a high demand for plastic products, and the plastic bottle recycling venture is open and profitable to all. You should talk to manufacturers of soft drinks and bottled water to determine how you can engage with them. Notably, they will be interested if it is cheaper to obtain plastic bottles from a recycling plant rather than plants or manufacturers who produce virgin plastic bottles.

11.9.2 Producing and Selling Biofuels

Selling biofuels is quite profitable as they are quite critical and popular sources of fuel. The advantage they offer is that they are safe and friendly to the ecosystem. Make sure that you carry out a thorough feasibility study and market survey before venturing into this business. It is quite pertinent for determining whether the business suits you.

11.9.3 Selling Recycled Paper Bags and Glass Products

Selling these bags is a lucrative and profitable business that an entrepreneur seeking to start a Green product business can consider. In North American and European countries, retail outlets are encouraged to use recycled paper bags as they are environmentally friendly and quickly decompose after they have been discarded. Similarly, another Green product business one can consider is selling recyclable glass products. Glass cannot decompose, thus allowing Green entrepreneurs to recycle and continue reusing. The other benefit is that it is pocket-friendly and saves energy compared to the production and use of virgin glass products. It is an excellent way of earning income, and in many cases, you will obtain broken glass as raw materials for free.

11.9.4 Open Up a Consignment Store

A consignment store is one of the Green ventures that require little capital. The essence of the store is to encourage recycling and the reuse of products in order to save the energy used during the production of new products. The business entails selling products that have already been sold, including clothes, caps, shoes, kitchen wares, and bags.

11.9.5 Selling Eco-Friendly Children Toys

Selling children toys that are eco-friendly is a thriving and profitable Green venture that aspiring entrepreneurs can embrace in any part of the world. Environmentally friendly toys are ones that easily decompose after being buried in the ground. Carry out an online search, and you will realize that numerous organizations have ventured into the manufacture of environmentally friendly children toys.

11.9.6 Sale of Refurbished Electronics and Electronic-Related Products

Selling refurbished tablets and mobile phones is another profitable Green product business that one can start in any nation globally. Admittedly, there is a huge market for refurbished tablets, pads, and phones because they are less expensive compared to brand new ones. It is worth noting that more energy is consumed in the manufacture of new smartphones compared to those being refurbished. It is less costly and energy-saving when computers, laptops, and electronic gadgets are refurbished. In addition, there is a large market for refurbished products. Green product sales can play a critical role in saving finite resources and the vulnerable environment.

11.9.7 Retailing Herbal Medicine

Retailing herbal medicine is one of the easiest business ventures. It is not only thriving but also a profitable business. You may require a license to start the business. However, it is worth noting that you need sufficient knowledge to convince people about purchasing your products. The advantage is that in some regions like Africa, a substantial number of people still hold the perception the herbal medicines are safer because of their organic nature, and little if any side effects. It is because of this reason that you may make huge profits.

11.9.8 Development of Green Software Applications

The development and retail of Green software applications is a profitable Green business worth venturing into. You can help spread information and knowledge regarding environmental protection through the creation of a mobile app with a theme on environmental protection. The software you develop will assist people in learning and understanding the need to protect the planet. It could also provide and teach mechanisms for protecting the environment. Therefore, in case you are a software developer, and you seek to leverage the Green business value chain, your best alternative is to begin the development of Green software apps. In case you are unaware of what it takes to start the business, you should consider visiting an expert to assist you. Landscaping Business Providing a landscaping business is also an easy way of improving homes while making profits. Landscape architecture entails planning, designing, managing, preserving, and rehabilitating land. The landscaping business, however, requires one to work hard and smart. Any entrepreneur with the intention to start a landscaping business and reap huge profits needs to be creative and work hard to convince customers that he or she can do a great job.

11.10 Individual Matter: Achenyo Idachaba

Idachaba is an entrepreneur born in America, but working in Nigeria (see Fig. 11.4). She was born in 1969 and pursued her education in the US. In 2014, she was awarded the Cartier Women's Initiative Award within sub-Saharan Africa. She gained prominence after moving to Ibadan, Nigeria in 2009 to start an environmental consultancy firm. She loved Nigeria because her parents were born there and had

Fig. 11.4 Achenyo Idachaba during a TED Talk on water hyacinth, *Source:* TED (2015)

spent considerable time visiting while young. During her time in Nigeria, she noticed that water hyacinth, an invasive species, was causing havoc, yet could be harvested as is done in Asia. Liaising with local craftspeople, she started making various products from the dried plants. The organization was called Mitimeth and began playing a crucial role in developing products like table tidies and wastebaskets. The plant is originally from South America, and they create floating mats that are able to reproduce quickly (Pogutz and Martín-Tapia 2017). By 2013, she employed seven staff after the Nigerian government awarded her a grant. Idabacha remains a popular feature on the Cable News Network (CNN) and has more than one million views on her 2015 TED talk. Mitimeth plays a crucial role in employing women, thus reducing the unemployment rate in the country. More importantly, it has been critical in empowering women and imparting skills as well as knowledge to women who previously were idle.

11.11 Future Trends

11.11.1 Africa

African policymakers will continue emphasizing the urge and opportunities for green growth: growth that protects the natural environment of Africa in ways that enhance the welfare of our generations, and which creates new chances for economic development and transformation. Economic transformation will deliver green growth. Sustainable urban agriculture will become a focus for various entrepreneurs.

11.11.2 Asia

The changing consumer behavior and attitudes towards the green market will lay the basis for the advancement of green entrepreneurship. Green entrepreneurs will also become important for fulfilling consumer needs. There will be a surge in the number of green entrepreneurship ventures, particularly the innovation of green products.

11.11.3 Australia

Entrepreneurs will be at the forefront of driving Australia to a green economy through green innovations. There will also be more investment in low-carbon transport and renewable energy. Greener ways of managing waste will be further embraced while architectures will increasingly construct low carbon or green buildings.

11.11.4 Europe

The deployment of micro-data will facilitate better identification of green activities as well as entrepreneurial dynamics. At the micro-level, the information will enhance the consistency of estimates produced and means for answering and developing indices of entrepreneurial performance that focuses on groups.

11.11.5 North America

The continent will experience a surge in green innovations, particularly in the transport and car industry. Green cars will also become even more prevalent as it shifts towards a zero-carbon economy.

11.11.6 South America

Green entrepreneurship will be characterized by the development of green tech-hubs, government-supported green initiatives for the innovative new business, and more interest in green entrepreneurs. Crowd-funding will contribute to the development of green businesses and entrepreneurship. South American countries will increasingly become desirable hubs for green investment.

11.12 Summary

- Green entrepreneurship has rapidly expanded in many world regions but remains unexploited due to various barriers and challenges.
- This chapter has examined the fundamentals of entrepreneurship, evolution, characteristics of Green entrepreneurs, case studies of Green entrepreneurship, challenges, roles of Green entrepreneurship, and provided recommendations for the industry, especially for the case of sub-Saharan Africa.
- Focus has also been placed on the role of Green entrepreneurship in attaining the Sustainable Development Goals.
- Many young people in the region are increasingly adopting Green entrepreneurship.
- Various governments have set up favorable policies to spearhead the process, but various challenges need to be solved for the industry to thrive.
- There is a further need to support small-scale Green entrepreneurs because they have ideas that can change not only their communities but also countries.
- More importantly, the focus should be on uplifting the industry and ensuring that the population understands and embraces it.
- A key insight from this report is that sub-Saharan Africa's Green entrepreneurship potential remains untapped.
- It should be embraced to solve the perennial problems that continue afflicting the region, including high rates of unemployment, poverty, illiteracy, and environmental degradation.
- This chapter has also examined Green entrepreneurship, particularly the nature-inspired solutions to sub-Sahara African problems.
- The fundamentals of Green entrepreneurship that have been comprehensively discussed are the Green economy, creative destruction, characteristics of Green entrepreneurs, and Green innovation.
- This chapter has established that there are numerous case examples of Green entrepreneurship in sub-Saharan Africa.
- The main case studies in this study are architecture, young entrepreneurs, and individual countries.
- Green entrepreneurship plays a pivotal role in the achievement of the SDGs, and the major roles identified in this study are job creation, reduction of poverty, sustainable utilization of resources, and improved training and skills.
- The main challenges associated with building a Green business are market awareness, low literacy, high capital costs, limited policy support, and lack of consumer awareness.
- Solutions and recommendations are provided to solve these challenges and make Green entrepreneurship a worthwhile venture in sub-Saharan Africa.
- Africa will continue engaging in green entrepreneurship but at a higher level/intensity. Sustainable urban agriculture will also become common.

- Informed consumers in Asia will contribute to more efforts in green entrepreneurship, while in Europe, micro-data will lead to the identification of green initiatives and opportunities.
- North and South America will experience the proliferation of green innovations in transportation and crowd-funding for green ventures.
- For Australia, entrepreneurs will play a pivotal role in the shift towards a greener economy.

11.13 Discussion Questions

1. Is Green entrepreneurship a prudent way of managing natural resources? Explain your answer.
2. Describe the six major challenges that make it difficult for Green entrepreneurs to succeed?
3. What are the barriers to entry of entrepreneurs into Green entrepreneurship?
4. Discuss three Green innovations that will change Africa in the next decade.
5. Project the future of Green entrepreneurship in Africa.
6. Using examples, discuss why Green entrepreneurship is critical for enabling Africa to achieve the 17 Sustainable Development Goals.
7. What Green opportunities remain unexploited in Africa?
8. Discuss how Green business opportunities in livestock, land, and infrastructure could overcome unemployment in rural areas of Africa?
9. What are the barriers and opportunities for eco-friendly investments in infrastructure, and how can Green investments in energy, shelter, and transport boost labor productivity?
10. What are the opportunities for and impediments to sustainable growth and development of Small-and-Medium Enterprises in farming as well as agro-processing within the rural areas of Africa?
11. How can Green entrepreneurship foster innovation in Africa?
12. What is the influence of creative destruction on Africa's resources?
13. Create your Green entrepreneurship venture. Why is it a lucrative business?
14. What future trends do you anticipate to be experienced in developing countries regarding green entrepreneurship?

References

Allen, J. C., & Malin, S. (2008). Green entrepreneurship: A method for managing natural resources? *Society and Natural Resources, 21*(9), 828–844.

Berle, G. (1993). *The green entrepreneur: Business opportunities that can save the earth and make you money*. New York: Liberty Hall Press.

Chen, Y. S., Lai, S. B., & Wen, C. T. (2006). The influence of green innovation performance on corporate advantage in Taiwan. *Journal of Business Ethics, 67*(4), 331–339.

Farinelli, F., Bottini, M., Akkoyunlu, S., & Aerni, P. (2011). Green entrepreneurship: The missing link towards a greener economy. *Atdf Journal 8*(3/4), 42–48.

Gibbs, D. (2009). *Sustainability entrepreneurs, ecopreneurs and the development of a sustainable economy*. Sheffield: Greenleaf Publishing Ltd.

Green Institute. (2018). *The Green Campus Initiative (GCI)*. Nigeria. Retrieved April 9, 2018, from http://www.greeninstitute.ng/research-programs/the-green-campus-initiative/.

Hall, B. H., & Rosenberg, N. (2010). *Handbook of the economics of innovation*. Amsterdam: North Holland.

Hawken, M. (January 7, 2015). *7 Young African women entrepreneurs drive a new eco-spirit on the continent*. Lionesses of Africa. Retrieved April 9, 2018, from http://www.lionessesofafrica.com/blog/2015/1/6/7-young-african-women-entre-preneurs-drive-a-new-eco-spirit-on-the-continent/.

Holt, D. (2011). Where are they now? Tracking the longitudinal evolution of environmental businesses from the 1990s. *Business Strategy and the Environment, 20*(4), 238–250.

Jones, G. (2017). *Profits and sustainability: A history of green entrepreneurship*. Oxford: Oxford University Press.

Koester, E. (2016). *Green entrepreneur handbook: The guide to building and growing a green and clean business*. Boca Raton: CRC Press.

Njeru, E., Grey, S., & Kilawe, E. (2016). *Eastern Africa Climate-Smart Agriculture Scoping Study: Ethiopia, Kenya and Uganda*. Addis Ababa: Food and Agriculture Organization.

Parry, S. (2012). Going green: The evolution of micro-business environmental practices. *Business Ethics: A European Review, 21*(2), 220–237.

Pogutz, S., & Martín-Tapia, I. (2017). Let the Environment Guide Our Development, by J. Rockström, TEDGlobal. Retrieved April 9, 2018, from http://www.ted.com/talks/johan_rockstrom_let_the_environment_guide_our_development?/.

Sanchez, P. A., Denning, G. L., & Nziguheba, G. (2009). The African green revolution moves forward. *Food Security, 1*(1), 37–44.

SEED Case Studies. (2011). *Solar sister. Empowering women in Africa through clean energy solutions*. 2011 SEED Winner. Retrieved April 9, 2018, from https://www.seed.uno/images/casestudies/SEED_Case_Study_SolarSister_Uganda.pdf/.

Smithsonian Cooper-Hewitt. (2011). *Sandbag house*. Retrieved July 12, 2018, from https://www.designother90.org/solution/10x10-sandbag-house/.

TED. (2015). *Achenyo Idachaba*. Retrieved June 20, 2018, from https://www.ted.com/-talks/achenyo_idachaba_how_i_turned_a_deadly_plant_into_a_thriving_business/.

Tsai Design Studio. (n.d.). *Vissershok Classroom*. Retrieved October 02, 2019, from http://www.tsaidesignstudio.com/architecture/safmarine-vissershok-classroom/.

Udoma, O. (April 24, 2014). *10 examples of 'Green' Architecture in Africa*. Future Lagos. Retrieved April 10, 2018 from https://www.smartcitiesdive.com/ex/sustainablecitiescollective/10-examples-green-architecture-africa/243566/.

Chapter 12
Sustainability in Science

Abstract Science is a key driver of the numerous applications, technologies, and solutions embraced in the various disciplines. To this end, this chapter seeks to identify the major terms in science and explore the application of sustainability in each discipline. It uses the case study of biomimicry to demonstrate the way. Science can use sustainability in its efforts. An experiment on biomimicry is also conducted to enhance understanding of the topic. More importantly, Elon Musk's individual matter offers insights on the significance of embedding sustainability in science. This chapter is a further contribution to the sustainability debate, and the larger Going Green initiatives. By understanding the various concepts, it can help learners and tutors to embed the sustainability concepts in science. Future trends in the subject are provided to offer a glimpse into the outlook of sustainability in science.

Keywords Biomimicry · Elon Musk · Green chemistry · Physics and engineering · Velcro · Mathematics · Medicine

12.1 Introduction

Sustainability is an important concept in science. Ideally, science plays a major role in innovation and invention. It also plays a critical role in farming, manufacturing, and industries. Consequently, embedding sustainability in science is pertinent to ensuring that production processes consider the values and benefits associated with sustainability. The primary focus of this chapter will be to examine the applications of sustainability in various subjects: both scientific and non-scientific. A holistic approach is undertaken to ensure that all pertinent issues are addressed.

A. A. Akinsemolu, *The Principles of Green and Sustainability Science*,
https://doi.org/10.1007/978-981-15-2493-6_12

12.2 Chemistry

Chemistry is defined as the study of matter, its properties , and the way substances combine to form others. It also studies the interaction of substances with energy. It is worth noting that chemistry is a core part of our day-to-day lives. One amazing fact is that each material consists of matter, including our bodies. Chemistry is also involved in all human activities: preparation of food, cleaning homes, and launching space shuttles. Overall, chemistry is among the physical sciences that enable us to understand our world. There are five major branches of chemistry:

1. analytical chemistry,
2. physical chemistry,
3. organic chemistry,
4. inorganic chemistry, and
5. biochemistry.

Due to the significance of chemistry in our lives, it is crucial to understand its application to sustainability.

Which of these branches of chemistry do you believe are most important? Provide reasons.

12.2.1 Basic Definitions and Concepts

- What is matter? The general definition of matter is that it is anything that has mass and occupies space (Anastas 2003).
- Mass is the measure of an object's inertia and is proportional to weight. Mass is usually measured by comparing the substance of interest to the International Prototype Kilogram, a standard kilogram.
- Weight is defined as the force exerted on a substance.
- Volume is often defined as the measure of the quantity of space that an object occupies.
- The amount of substance is defined as the quantity of matter that is determined directly or indirectly.
- One mole is the number of particles in 12 grams of Carbon-12. The number referred to as Avogadro's constant is $6.02214078^{18} \times 10^{23}$ particles. The mole acts as a bridge between the atom and macroscopic quantities of material that people work with within the laboratory. It enables chemists to weight out quantities of substances, for example, sulfur and iron, in such a way that equal atom numbers are obtained.

- There are twelve concepts or principles that guide green chemistry.

- **Principle 1**—Prevention: This entails restricting the generation of chemical waste or cleaning it up after it has been generated (Anastas 2003; Anastas and Eghbali 2010).

Explain one application of this concept in industries.

- **Principle 2**—Atom economy: This entails synthetic methods trying to optimize the embedment of all materials utilized within the final production process. An example of this is the peroxide process used for the production of hydrazine without cogenerating salt. The net reaction results in the production of one sodium chloride for equivalent product hydrazine, as shown in the equation below:

$$NaOCl + 2\,NH_3 \longrightarrow H_2N\text{-}NH_2 + NaCl + H_2O \quad \text{(Anastas 2003).}$$

- **Principle 3**—Chemical syntheses: The principle emphasizes the need to use less hazardous chemical syntheses. Ideally, it emphasizes that synthetic methods should not use or generate substances that are toxic to the environment or humans (Berger 2007).
- **Principle 4**—Safe chemicals: This principle emphasizes the need to design chemicals that are non-harmful to the environment. Chemical products need to be designed to attain their desired function while ensuring that they are non-toxic.

Identify some of the safe chemicals that are used for soap-making today.

- **Principle 5**—Auxiliaries and solvents: According to this principle, there is a need to avoid auxiliary substances.
- **Principle 6**—Energy efficiency: The design for energy efficiency should be such that energy requirements need to be reduced, and processes should be carried out at ambient pressure and temperature.
- **Principle 7**—Renewable feedstocks: There is a need to prioritize renewable feedstocks because of their sustainability.
- **Principle 8**—Reduce derivatives: There is a need to minimize or avoid unnecessary derivative generation.
- **Principle 9**—Catalysis: This principle emphasizes that catalytic reagents that can be utilized in small quantities need to be prioritized to stoichiometric reagents (the reagents consumed in a reaction) (Berger 2007).

What is the significance of catalysis?

- **Principle 10**—Design for degradation: This principle emphasizes the need for chemical products to be designed in a way that does not lead to environmental pollution.
- **Principle 11**—Real-time analysis: Analytical methods should be further developed to allow real-time control and monitoring prior to the formation of hazardous substances.
- **Principle 12**—Safe chemistry for preventing accidents: This principle emphasizes that when possible, substances in any process and their forms should be selected to reduce risks like fires and explosions.

12.2.2 Application of Sustainability in Chemistry

- Sustainable chemistry or green chemistry focuses on product and process designs that reduce the utilization and generation of hazardous substances.
- Environmental chemistry focuses on the impact of polluting chemicals, while green chemistry explores the effect of chemistry on the environment.
- Sustainable chemistry is defined as a scientific concept that aims to enhance efficient natural resource use for meeting human needs.
- It entails the design, manufacture, and utilization of effective, safe, and environmentally responsive chemical processes as well as products.
- It is also perceived as a process that promotes innovation across sectors to achieve superior performance and value while meeting the objectives of improving the environment and human health (Berger 2007).
- Sustainable chemistry has numerous benefits, including avoidance of toxic and bio-accumulative materials, the use of renewable resources, and reducing the intake of non-renewable resources.
- Other benefits are the reduction of adverse environmental effects of manufacturing and provision of economically competitive technologies that are advantageous to the industry.
- The major objectives of green chemistry are safe designs, and resource efficiency could be pursued in various contexts.
- Chemicals that have less hazardous properties are used to reduce their harm to the environment.
- Chemicals are perceived as sustainable when there are no detectable hazardous properties. This means that they do not cause damages or persist within the environment.
- An example of sustainable chemistry is chemical leasing, which is a business model that focuses on a value-added approach instead of the traditional approach that seeks to increase the volume of sales of chemicals.

- Chemical leasing is crucial for sustainable chemical management intending to minimize the use of chemicals, thus reducing costs and resource consumption (Anastas and Eghbali 2010).
- Some industries where sustainable chemistry has been applied are cleaning or degreasing, casting, staining, chemical synthesis through catalysts, and the food industry.
- In the food industry, applications have mainly focused on water treatment and extraction.
- Additive production is the other application, including technical gases and cooling goods.
- Sustainable chemical production is also evident in white biotechnology—uses biological processes to produce enzymes or bacteria. An example is vitamin B12 production that is the product of white biotechnology.

Can you provide more examples, specifically in Africa, of white biotechnology applications? Provide your own understanding of sustainable chemistry.

12.3 Physics and Engineering

Physics is a branch of science that studies matter, motion, and behavior. It is among the most critical science disciplines with the objective of understanding the way the universe behaves. Engineering is also a branch of science and technology that deals with the design, construction, and use of machines, structures as well as engines. Advances in physics and engineering lead to new technologies, including nuclear physics and electromagnetism, inventions that have transformed modern society. Physics' role in society today is indispensable, and there is a need to study major concepts in physics before exploring its importance for sustainability. Significant attention will be paid to sustainable physics as well as the application of sustainability in physics.

12.3.1 Basic Definitions and Concepts

- Thermodynamics is a branch of physics that studies the relationship between various forms of energy. Conceptually, three laws of thermodynamics govern physics (Davidson et al. 2010).
- The First law of Thermodynamics states that energy is often conserved and can neither be destroyed nor created .
- The Second law of Thermodynamics states that within an isolated system , the total entropy always increases.

- The Third law of Thermodynamics states that "The entropy of a perfect crystal is zero when the temperature of the crystal is equal to absolute zero."
- A fundamental concept of thermodynamics is that we can divide the planet into a system and the surroundings .
- Sustainable engineering is defined as the process geared towards the design or operation of systems in order for them to utilize resources and energy sustainability. In other words, it considers sustainable resource use at a rate that does not limit or compromise the natural environment.
- Materials of engineering refer to the selection of appropriate materials for application.
- Energy Conservation: This entails the use of materials that use less quantity of energy. It is the most crucial aspect of building green.
- Indoor Environmental Quality: The quality of the inner part of the house. In numerous cases, engineers create a healthier environment by avoiding hazardous materials that are found in carpets, finishes, and paint. It also entails the use of appropriate ventilation and sufficient daylighting.

12.3.2 Application of Sustainability in Physics and Engineering

- In physics, sustainability is applied in five major ways: energy and work, thermodynamics, electricity and magnetism, nuclear physics, and electronics.
- In energy and work, the focus is often on foundations for sustainability issues.
- In thermodynamics, applications have been achieved through issues of heat, energy use, and entropy.

Identify some of the Physics and Engineering applications in Africa that embed sustainability.

- Regarding electricity and magnetism, sustainability is applied through commercial power generation and use.
- In nuclear discipline, application is evident in commercial power generation and waste.
- The sub-discipline of eco-physics highlights the way power generations can be environmentally friendly (Davidson et al. 2010).
- Exergy is a phenomenon that adheres to the First and Second Laws of thermodynamics.
- Exergy is crucial for providing a foundation for resource and emissions accounting.

- Some of the applications of exergy are electricity generation, energy productivity within the industry as well as combined power and heat schemes.
- Sustainable energy solution should also have the same characteristics as low-carbon energy solutions. In other words, low-carbon solutions are an application of physics that seek to provide sustainable energy solutions (Davidson et al. 2010).
- Mimicking nature using twenty-first century technology needs skillful engineering.

Is Africa doing enough to ensure that the study of physics enhances sustainability? Explain how Africa can reorient itself to benefit from concepts of sustainable physics. In what ways can sustainability in the physics subject contribute to a better world? Explain how Africa can exploit engineering to ensure that agriculture becomes sustainable.

12.4 Biology

Biology is defined as the science of life. Biologists often study the function, structure, distribution, growth, and origin of living organisms. There are various branches of biology. Biochemistry is a branch of biology that studies material substances that constitute living things. Botany is the study of plants, whereas cellular biology is the study of the basic units of living organisms. Genetics is defined as the study of heredity, whereas ecology studies the way organisms interact with the environment (Clark 2007). Finally, the concept of equilibrium is whereby all living organisms are expected to maintain homeostasis with their environment. The next section examines the way sustainability is applied in Biology.

12.4.1 Basic Definitions and Concepts

- Active Transport: The movement of molecules in and out of partially permeable cell membranes using energy obtained from respiration.
- Adaptation: The process of getting used to an environment or habitat.
- Addiction: This refers to the situation whereby one has taken a substance for some time to the extent that they cannot stop it without experiencing withdrawal symptoms.
- Aerobic Respiration: The respiration process that makes use of oxygen from the atmosphere.
- Biodegradable: A substance that has properties that can be broken down by natural processes.

- Food security: The condition whereby all people have socioeconomic and physical access to enough quantity and nutritious quality food that sustains their food preferences as well as dietary needs.

12.4.2 Application of Sustainability in Biology

- The major applications of sustainability in Biology are diseases, agriculture, population dynamics, fisheries, carbon cycle, and forestry.
- Sustainable agriculture and organic farming are some of the major applications of biology that promote sustainable use of agricultural methods in enhancing food security (Clark 2007).

Food security and sustainability are compatible aspects. Briefly support this statement.

- The Green Revolution in the twentieth century was made of enhancements to crop genetics, herbicides, and biotechnology. All these are biological concepts that were instrumental in revolutionizing agriculture.

What is the status of the Green Revolution?

- Organic agriculture is inherently biological and is the solution to conventional agricultural methods.

Explain how organic agriculture (using examples) supports the concept of sustainability.

- Biological farmers promote sustainability using carbon obtained from green manure, crop residues as well as livestock manures. Green manure is typically created by leaving the uprooted or sown parts of the crop to wither on a field in order for them to act as soil amendment or mulch. Plants used for green manure are usually cover crops predominantly used for this purpose. On the other hand, livestock manure is organic matter obtained from animal feces that can be utilized as organic fertilizer within agriculture. It entails applying herbicides and pesticides to improve soil life.

- Biological farmers also utilize limited tillage practices for preserving healthy soil structures while embedding carbon from the crop residues.

Identify three examples of limited tillage practices.

- The result is a healthy ecosystem that produces better yields at comparatively lower costs.
- Anaerobic digesters are also crucial applications for reducing agriculture's carbon footprint by processing animal manure to produce carbon-based fertilizers.
- Recent advancements in soil microbiology are crucial for improving biological farming systems.

Identify and discuss two of the recent advances in microbiology.

12.5 Medicine

Medicine is defined as the science as well as the practice of diagnosing, treating, and preventing disease. Medicine also encompasses various healthcare practices that prevent and treat illnesses. Some contemporary applications of Medicine are biomedical research, medical technology, and genetics. The provision of medical care is categorized into primary, secondary, and tertiary care. Sustainability in Medicine is important for ensuring that there is a healthy population (Getz 2006). Physicians, assistants, and nurse practitioners are among the health professionals who provide primary medical services. The following points elaborate on the concept of Medicine and applications of sustainability in Medicine.

12.5.1 Basic Definitions and Concepts

- Body Mass Index: This is a way of determining the weight one should have. It is calculated using one's height, weight, and gender to ascertain their ideal weight.
- Calories: A calorie is defined as a unit of energy and is used in relation to food.
- Medical genetics: A branch of Medicine that entails the diagnosis and treatment of hereditary disorders. It involves applying genetics to medical care.
- Medication: Substances used for diagnosing, curing, treating, or preventing disease.

- Radiation therapy: This is a therapy that uses ionizing radiation as part of cancer treatment to kill malignant cells.
- Drug: This refers to medication or any other substance that poses a physiological impact when it is ingested or introduced into the body.
- Treatment happens in the offices of physicians, clinics, nursing homes, and home visits.
- The primary care provider can treat approximately 90% of medical visits.
- Some treatments are chronic and acute illnesses, health education, and preventive care.
- Medical specialists provide auxiliary medical services in clinics, at the local community hospitals, or within their offices.
- Referrals are made for patients who need procedures or expertise carried out by specialists.
- Referrals include inpatient services/ambulatory care, intensive care medicine, physical therapy, and endoscopy units.
- Specialist hospitals or regional centers provide tertiary medical services, including burn treatment and trauma centers, advanced neonatology unit services, as well as high-risk pregnancy (Getz 2006).

12.5.2 Application of Sustainability in Medicine

- ThedaCare is a Wisconsin hospital that has developed a "sustainability leader" position. It is also in position to host the Climate and Health Symposium for educating employees regarding environmental education.
- The Healthier Hospitals Initiative is another application of sustainability in Medicine. It is a national campaign for improving sustainability within the healthcare sector.
- Tenet Health also embeds sustainability initiatives. In its Sustainability Report, there is a sense of system-wide approach for improved health and highlights the way the strategy is embedded at the local level.
- The other application is the shift from the mere treatment of diseases towards wellness and prevention.
- Magee Women's Hospital of The University of Pittsburgh Medical Center, UPMC, is an example that teaches parents regarding the link between human health and the environment. It uses parental classes and the Baby Steps to Green Parenting program.
- The increasing number of chronic diseases associated with diet has made hospitals roll out sustainable food policies.
- Hospitals and healthcare institutions are working to identify impediments to healthier food systems and enhancing nutritious alternatives.
- Fletcher Allen, in Vermont, is a leader in this movement that buys local organic veggies and cheese. This is important because it is an element of sustainable agricultural and medicinal practices that promote sustainable living.

- Medicine has also been at the forefront of extending equipment's life cycle, enhancing efficiency, preventing waste, and utilizing less water and energy. For instance, Bon Secours St. Francis System in South Carolina was the first healthcare institution in the state to get energy designation when it minimized the use of energy by 20% (Buchwald et al. 2007). The hospital developed the Ecological Stewardship Plan that led to annual energy cost savings of approximately $850,000.
- The Catholic Health Association has embedded some environmental improvement initiatives into health-needs assessments of the community.
- The use of medicinal plants is a crucial part of the African traditional healthcare system. It is a sustainable way of tapping on the natural plants for a cure. Some examples are Aloe ferox Mill. (Xanthorrhoeaceae)—bitter aloe or Cape aloe in South Africa and Lesotho, Acacia senegal or gum Arabic in Northern Nigeria, parts of North Africa and West Africa, and Catharanthus roseus (L.) G. Don (Apocynaceae) popularly known as Madagascan periwinkle in Madagascar (Massele et al. 2015).

Identify specific applications of sustainable Medicine in Africa. What are the major impediments towards sustainability in the field of Medicine?

12.6 Mathematics

Mathematics is defined as the science of quantity. Haskell Curry defined it as the science of formal systems, which consists of tokens, symbols, and rules regarding the way tokens could be combined into formulas. An axiom combines tokens that are incorporated within a particular formal system (Clements et al. 2015). Ideally, the study of quantity begins with numbers, consisting of integers and natural numbers, and arithmetical operations. Mathematics is crucial for planning and also plays a crucial role in various subjects. Applying sustainability in mathematics is critical for ensuring that plans and designs are inherently sustainable.

12.6.1 Basic Definitions and Concepts

- Integers are defined as whole numbers. The number theory studies the deeper properties of integers.
- Cardinal numbers describe the size of sets and include aleph numbers that allow the comparison of large sets.
- Trigonometry is a mathematics branch that studies the relationships between angles and sides of triangles.

- Functional analysis studies spaces of functions.
- The chaos theory defines how many systems are unpredictable but still have deterministic behavior.
- Applied mathematics deals with mathematical methods that are utilized in engineering, business, and science.
- Computational mathematics studies the methods for finding solutions for mathematical problems.

12.6.2 Application of Sustainability in Mathematics

- A sustainable math curriculum embeds sustainability into some Statistics and Mathematics course modules.
- Sustainability in Mathematics is explored through quantitative reading skills that include understanding scale, use of conversion factors, use of unit conversions to guide calculations, reading and creation of graphs, making comparisons, comprehending linear and exponential growth.
- Sustainability is also applied through modeling scenarios that include the determination of key variables, assessment of information, spelling out assumptions, performing sensitivity analysis, and interpretation of findings (Clements et al. 2015).
- Perhaps, the best way of understanding sustainability in mathematics is the concept of Education for Sustainable Development. Mathematics education could be used for bridging the gap or inculcating learners with skills necessary for adopting sustainable ways of living.
- For energy sustainability, mathematics contributes in finding more viable and less polluting ways for exploring new energy, enhancing combustion efficiency, developing alternative energy, managing energy grids and networks, and reducing the climate repercussions of energy use.
- Mathematics also plays a crucial role in the sustainability of economic systems and financial markets.

Identify tangible case studies of the application of sustainability in mathematics.

12.7 Computer Science

Computer science is the study of experimentation, theory as well as engineering that act as the foundation for the design and utilization of computers. It is also defined as the practical and scientific approach to computation as well as its applications.

Computational complexity theory explores the main properties of intractable and computational issues (Mankoff et al. 2008). The next section will examine some of the computer science concepts that are related to sustainability. In the end, the focus will be on the application of sustainability in computer science.

12.7.1 Basic Definitions and Concepts

- Computer graphics emphasize real-world visual applications.
- Programming language theory factors in several approaches while describing computation.
- Computer programming carries out investigations on several aspects of the use of complex systems as well as a programming language.
- Human–computer interaction factors in the challenges associated with making computations and computers beneficial, usable as well as universally accessible to people.
- The Institute of Electrical and Electronics Engineers (IEEE) Computer Society suggests that four areas are important to the computer science discipline: data structures and algorithms, computer architecture, the theory of computation, and programming methodology.

12.7.2 Application of Sustainability in Computer Science

- Computational sustainability is a growing computing field that deals with the application of principles of computer science, methods as well as tools to issues of societal and environmental sustainability. In this context, sustainability is the capacity to produce sufficient energy for the world to support the biological systems.
- Computers can process large amounts of data. Consequently, decision-making algorithms often allocate resources founded on real-time data.
- Smart grids also implement renewable resources to control the production of energy. Intelligent transportation systems analyze road conditions and relay data to drivers to enable them make smarter decisions based on real-time traffic data (Mankoff et al. 2008).
- The first application is computer architecture and hardware. The collective energy footprint of computing is growing globally.
- There is a high need for energy-efficient computer systems made from recyclable components that do not cause harm/pollution to the environment and threaten the health of humans.
- The second application is robotics and sensors, whereby computing is increasingly used for monitoring civil infrastructure, including power to grid and bridges.

Discuss the importance of sustainability in robotics and sensors.

- Sensing capabilities include mobile platforms. These consist mainly of aerial robots, aquatic, and terrestrial. They are utilized in disaster response and routine monitoring of various water bodies, including lakes and oceans.

How do sensing capabilities enhance sustainability?

- The third application is the cyber-physical systems whereby computing is incorporated in physical systems, including cars, buildings, and highways to enhance energy efficiency and safety. Human-built systems like airplanes and cars are on robotic systems (Mankoff et al. 2008).

Buildings can be made sustainable through architectural designs. However, this has often been cited as challenging. What are the challenges associated with the development of sustainable buildings?

- The fourth application is intelligent systems whereby the sensing and control capabilities could only be actualized using intelligent software to include artificial intelligence. Artificial intelligence is also crucial for deliberative human decision-making like resource planning (including water usage and wildlife reserves) with machine learning being detrimental supporting methods (Mankoff et al. 2008).
- The fifth application is social computing and networking. In this application, information and computing technology connects individuals, and thus offers avenues for social ties to alter behavior or environmental sustainability. Sustainability applications would include the formation of online recycling cooperatives as well as promoting the buying of ecologically friendly goods based on full life cycle analysis.
- The sixth application is mobile computing which is at the center of intelligent systems, sensors, social computing as well as other areas. Some other applications include the analysis of data that may range from animals and plants to consumer bar codes (Mankoff et al. 2008).

Using real-life examples, explain how mobile computing can be used for promoting sustainability.

- The seventh application is intelligent transportation systems (ITS) that seeks to enhance travel times and safety while reducing greenhouse gas emissions for all the travelers, but focusing primarily on drivers. Intelligent transportation systems has two systems for data collection/relay and data processing.
- Advanced public transportation systems seek to enhance the efficiency and convenience of public transportation for its riders.
- Advanced traffic management systems are crucial for collecting information about road congestion using cameras as well as other sensors.

Provide examples of advanced traffic management systems that are inherently sustainable.

- Vehicles communicate with manufacturers to facilitate remote installation of software updates during the addition of new features or patching of bugs. Such updates are used by Tesla Motors for increasing the performance and efficiency of their cars. These connections provide ITS with a means for accurately collecting information and even relaying it to drivers without the need for infrastructure.

12.8 Agriculture

Agriculture is defined as the tilling or cultivation of land and domestication or breeding of plants as well as animals to provide medicinal plants, fiber as well as other products for sustaining and enhancing life. Farmers can use numerous agricultural practices for making agriculture more sustainable, including those that result in better water and nutrient efficiencies as well as targeted plant protection. As agriculture plays a crucial role in feeding the world population, food production methods should embrace the aspect of sustainability.

12.8.1 Basic Definitions and Concepts

- Agronomy: The science and technology of producing as well as utilizing plants for fuel, fiber, land reclamation, and food.
- Soil conservation: The use of appropriate methods to ensure sustainable soil usage.

- Agroecology: Management of agricultural systems by emphasizing environmental and ecological perspectives. It is closely linked to areas of organic farming, alternative food systems, and sustainable agriculture.
- Arable land: Land that can be plowed and used for cultivation of crops.

12.8.1.1 Application of Sustainability in Agriculture

- One of the applications is the rotation of crops and the adoption of diversity. Planting various crops can have numerous benefits, including healthy soil and better pest control. Some crop diversity practices are intercropping as well as complex multi-year crop rotations. Intercropping is a cropping practice that entails growing two or more crops close together. The commonest objective of intercropping is to produce a higher yield on some piece of land by using ecological processes or resources that would otherwise not be used by one crop.
- Another application is the planting of cover crops. Cover crops, such as hairy vetch or clover, are planted in the course of off-season periods when soils become bare. Such crops provide protection and build soil health through the prevention of erosion, replenishment of soil nutrients, and minimizing the significance of herbicides (Praneetvatakul et al. 2001).
- The reduction of tillage elimination is an important application of sustainability.
- Integrated pest management also uses various methods, including biological and mechanical controls, that can be applied systematically to keep pest populations under control while reducing the deployment of chemical pesticides.
- Another sustainability application is the adoption of agroforestry practices. It entails the combination of shrubs or trees into their operations.
- Furthermore, the management of whole systems and landscapes contributes to sustainability, whereby sustainable farms regard cultivated areas as integral to the farm (Von Wirén 2001).
- An important application is vertical farming which grows crops in a closed stack system . Vertical farming is a promising solution to challenges associated with traditional agriculture. It uses only 25% of the water used by traditional agriculture and allows farmers to produce crops yearly since all environmental factors are controlled (Von Wirén 2001). It also produces better and healthier yields compared to traditional agriculture and adapts to climate change.
- Sustainable home gardening is crucial for sustaining plants and soils while nourishing and sustaining a family, both esthetically and physically.
- Organic gardening also entails the pursuance of the sustainability objective through the conservation of resources.

How can you practice organic gardening at your home?

- Sustainable gardening is among the most crucial and effective sustainability practices that people could follow. Some of its practices and benefits are the improvement of soil, use of native plants, trees, and shrubs for creating beautiful landscapes as well as using all renewable resources provided by nature.
- Other crucial home gardening practices that promote sustainability are compost, maintenance techniques, planting native and climate-appropriate plants, and saving seed.

After going through this chapter, do you believe that you have a role to play in sustainability? Explain if you believe you have a role to play.

Case Study 12.1: Biomimicry

Fig. 12.1 Velcro image

People have relied on nature as an inspiration throughout the history of humanity. By examining plants, animals, and natural processes, we can learn what is functional and non-functional. These observations are helpful for engineers during the design process as well as for inspiring new inventions. There are numerous examples of biomimicry, with Velcro being among them (see Fig. 12.1). Velcro is a product designed to imitate cockleburs that tend to stick to animals and people. The first invention/case study of biomimicry is airplanes, which are modeled after birds. This can be seen through the airplanes' body shapes that resemble birds' wings and the falcon beak. The Morphing airplane wings tend to change shape in accordance with the speed and flight period (Mathews 2011). This is inspired by different varieties of birds that have wings of different shapes. The variety of wing shapes affect the speed of flight. In addition, fish-inspired airplane wing design enables

(continued)

the plane to move through the air easily. Boat hulls are also designed to emulate the shape of fish. Boat hull material mimic shark and dolphin skin membranes. Other examples that highlight biomimicry are:

- Sonar and radar navigation technology as well as medical imaging. These have been inspired by the echolocation abilities of bats.
- Swimsuit and bobsled clothing fabric are made using woven texture and ribbing to minimize drag while simultaneously maintaining movement. This is an imitation of shark's skin.
- Space applications and adhesives for microelectronics have been inspired by the excellent adhesion capabilities of lizards and geckos.
- Walter filters whose design imitates animal cell membranes enable filters to allow certain things to pass through while ensuring that unwanted substances are kept out (Bartl and Lakhtakia 2015).
- Many running shoes have been designed from technology learned during the study of mechanics of animal feet.
- Super robust and waterproof silk fibers are made without toxic chemicals and were inspired by spiders.
- Film utilized for flat panel displays, phone, and PDA lenses mimics the nanostructures observed in the eyes of night-flying moths.

Some of the inventions founded/inspired by plants are:

- Velcro: hook and loop material that is inspired by cockleburs. Velcro is used for many purposes, including bundle fishing poles, computer cord wrap, hanging and organizing tools in the garage, and wrapping electronic wires (Bartl and Lakhtakia 2015).
- Plant-inspired solar cells are inspired by the process of photosynthesis, especially the capture of energy from sunlight.
- Wind-propelled planetary rover design maximizes drag and imitates tumbleweed. It is used in wind turbines to enhance function.

These case examples suggest that when biomimicry is carried out appropriately, it brings out the benefits associated with the source of inspiration. Janine Benyus, in her book titled "Biomimicry: Innovation Inspired by Nature," insinuates that biomimicry can be successfully applied to the arena of environmentally friendly design. The example of waterproof silk production by spiders is an example. The silk is a tough material and is often produced in the water at 25 °C without any high heat, chemicals, or petroleum oil. From the above cases, it is apparent that biomimicry could be deployed as a model for engineering designs that are beneficial for solving human issues. With environmental concerns, biomimicry can provide suggestions for how industrial designs could be more sustainable and right for various cultures as well as climates.

12.9 Practical Application: Understanding the Concept of Biomimicry

Pre-activity Assessment

Define the term "biomimicry" by breaking it down to enable participants to guess its meaning. Explain the meaning of "bio" and "mimicry." After the class agrees, ask volunteers to provide examples.

Materials Required

Paper, pencil, ruler, colored pencils/markers.

Preparation for Activity

1. Review the various biomimicry inventions.
2. Learners should be divided into pairs.
3. Students should identify things that they have as common interests.
4. Learners agree on any common interests for the design topic area.
5. Learners have 15 min for brainstorming to identify possible ideas for designs using the biomimicry of animals.
6. Learners should try to mention any animals that remind them about the topic. During the brainstorming, they should not air any negative comments, stay focused on the topic, record all ideas, and build on other people's ideas.
7. Required materials (rulers, markers, paper, and colored pencils to students) should be availed for them to draw their new biomimicry products.
8. After completing the design, each of the teams should make a list of special features found in their designs and the animals that inspired the features.
9. Mount the design features and drawing onto a construction paper.
10. Students should role play engineering organizations and present their designs to the class.

The completed designs should then be posted in the school resource center or classroom to enable others learn about them.

Tips

In case students find it challenging to come up with a design idea, they can be assisted. They can also be assigned a common design topic like playground toys or sporting equipment. After each team has presented, let the group vote for the best design.

Assessment Activity Embedded Assessment

Learners should identify features in their design that are inspired by nature. Let them be specific regarding the kind of plant or animal they are mimicking. Let them also describe the inspiration (animal or plant characteristics). Determine whether it is biomimicry. Give examples of biomimicry by including those that are and those that are not. Learners should then vote whether they think the mentioned designs entail biomimicry. Some of the examples that can be used by learners are as shown in Table 12.1. "No" means that the material, part, or invention does not fall under

Table 12.1 Biomimicry activity embedded assessment

Item	No (not biomimicry)	Yes (biomimicry)
Airplane wing		Yes, after bird wings
iPod	No	
Sonar navigation		Yes, after bats
Hall of submarines		Yes, after shark and dolphin skins
Computer printer	No	
Chair's soft cushion	No	
Solar cell		Yes, after leaves
Hard coatings for car windshields		Yes, after abalone mussels' mother of pearl coating

biomimicry. The final column is regarding the plant or animal that the invention has mimicked.

Apart from learning from nature's plants and animals, students can gain insight from its cycles and processes. Ask them to think of the numerous natural closed-loop cycles, including the hydrogen cycle, water cycles, and others that are models that recycle continuously, thus providing long-term sustainability (Lakhtakia and Martín-Palma 2013). Ask them to think of a way that could enable people to do something in a better way by imitating a natural process. After this simple classroom activity, students should have a basic understanding of biomimicry.

12.10 Individual Matter: Elon Musk

Elon Reeve Musk was born on June 28, 1971 (see Fig. 12.2). He is an American business investor and engineer. Currently, he is the founder and designer of Space X, CEO of Tesla, Inc., and co-founder and CEO of Neuralink. The Forbes List of the World's Most Powerful People ranked him position 21 in December 2016. Musk was born in South Africa and taught himself computer programming at a young age. He attended Queen's University in Canada. He moved to the University of Pennsylvania where he graduated with an economics degree from the Wharton School. In 1995, he started a Ph.D. in applied material sciences and physics at Stanford University. However, he dropped out after 2 days to undertake an entrepreneurial career. Subsequently, he co-founded Zip2 that was acquired by Compaq in 1999 (Gibbs 2014). He then founded X.com that merged with Confinity in 2000 to become PayPal. In 2002, he founded SpaceX, where he is the lead designer and CEO. He is also the co-founder of Tesla Inc. in 2003.

Furthermore, he is the inspiration behind the creation of SolarCity . This is a solar energy organization that is currently a subsidiary of Tesla. In 2015, he co-founded

Fig. 12.2 Elon Musk, *Source:* HDQ Walls (2018)

OpenAI . This is a non-profit research organization that aims to enhance friendly artificial intelligence. In July of 2016, he was a co-founder of Neuralink, which is a neurotechnology organization that focuses on the development of brain–computer interfaces. In December, he founded the Boring Company. His contribution to sustainability, especially the fields of physics, science, mathematics, and chemistry, are well known. The objectives of SolarCity, Tesla, and SpaceX revolve around the vision he had for changing humanity and the world. His objectives include the reduction of global warming using sustainable energy production as well as consumption. It also entails minimizing the vulnerability of human extinction through the establishment of life on Mars.

The life of Elon Musk has been described as "The Real Life of Iron Man" because of his sustainable vision. Musk is significantly changing energy, automotive, and aerospace industries with sustainable solutions that could not have been envisioned by past incumbents. The Stanford University dropout has the perception that his academic background in computer science and physics provided him with insight on the way to deal with counterintuitive situations. He is unique because of his philanthropic approach to business. This is evidenced in Tesla's mission that states, "to accelerate the world's transition to sustainable transport." He has set a foundation for a business entity that exceeds the traditional automotive manufacturer. He is the "transitioner" of the shift from gas-powered to electric vehicles (Sheth et al. 2016). The launch of Tesla Motors has been instrumental in proving that electric cars are viable for the future. The bold statement reaffirms that Tesla understands that technological innovations are tools for achieving a sustainable future. With such a confident stature, he is on track to realizing his vision for a sustainable future. Arguably, he plays a crucial role in changing the world and setting the stage for a sustainable future.

12.11 Future Trends

12.11.1 Africa

The continent will increase its investments in science and research and development will raise economic productivity by enhancing innovation and the development of start-ups, jobs, and SMEs. Science will remain crucial for reducing poverty and disease and could lead to the generation of knowledge. Increased investment in sustainability science will significantly affect the relationship between water, food, health, and climate change. In addition, science will increasingly embrace the concept of biomimicry for driving sustainability.

12.11.2 Asia

In the future, the research agenda will be driven by sustainability science. It will also offer insight on the best way forward in addressing development challenges in Asian countries. Asia is also poised to tackle the major challenges that impede it from attaining sustainability. Due to its contribution to pollution, the continent will have many cities that model biomimicry or nature. Finally, more buildings in this continent will be constructed mimicking on nature principles.

12.11.3 Australia

Sustainable innovation will continue driving biomimicry. Sustainability science will become a major driver of innovation and economic progress. Also, sustainability science will become a priority in many mitigation measures to ensure that there is viable production and development.

12.11.4 Europe

Sustainability science will contribute to the development of a better Lisbon strategy. The Lisbon strategy is a 10-year strategy that seeks to make the European Council the most dynamic and competitive in the world. It is a commitment by EU governments to focus their efforts on one objective of bringing socioeconomic and environmental renewal within the EU (Marcu 2006). Europe could also offer examples of good practice of using science and technology to develop social networks and capital. It may also offer critical insights for policy-making on the way of developing new social and ecological systems that are more resilient. Furthermore, sustainability science can result in the development of interdependencies and offer a transdisciplinary approach that enhances joint solutions within a societal context that could be implemented to become more productive.

12.11.5 North America

Sustainability science will be responsive to societal needs while preserving the planet's life-support systems. It will also support the transition to sustainability by developing and utilizing new knowledge, approaches, and tools, learning from experiences, enhancing public understanding as well as inspiring new ethics via participation. Whereas it may not find viable solutions to the persistent unsustainability issues, it will play a crucial role in their mitigation or to minimize their worst consequences by coming up with a new way of framing and approaching them (holistic, integrative, and open).

12.11.6 South America

Researches will increasingly embrace sustainability science to help in solving some of the challenges afflicting the continent. Many of the innovations will revolve around the critical concept of sustainability to ensure that they do not cause deleterious harm to the fragile environment.

12.12 Summary

- Sustainability has been embedded in various subjects.
- Sustainability in Chemistry is achieved through the design of products and processes, chemical leasing, degreasing, staining, water treatment, extraction, additive production, and sustainable chemical production.
- In Physics and Engineering, the application of sustainability is evident in energy and work, magnetism and electricity, electronics, nuclear physics, and thermodynamics.
- In Biology, such applications are evident in diseases, agriculture, fisheries, forestry, population dynamics, and the carbon cycle.
- In Medicine, real-world examples/applications are ThedCare, The Healthier Hospitals Initiative, Tenet Health, Magee Women's Hospital of UPMC, Fletcher Allen in Vermont, Bon Secours St. Francis System in South Carolina, and the Catholic Health Association.
- In Mathematics, sustainability is achieved by modeling scenarios, information assessment, and performance of sensitivity analysis.

- Education for sustainable development is the other concept that promotes the understanding of sustainability in mathematics.
- Mathematics also contributes to reducing polluting ways for energy exploration and the development of alternative energy.
- Computer Science promotes sustainability through computational sustainability, use of smart grids, computer architecture, robotics, and sensors, sensing capabilities, cyber-physical systems, intelligent systems, social networking and computing, mobile computing, intelligent transportation systems, advanced public transportation systems, and advanced traffic management systems.
- Vertical farming, home gardening, crop rotation, embracing diversity, planting cover crops, less tillage, integrated pest management, and agroforestry are cases of sustainability in agriculture.
- Elon Musk is a key contributor to the sustainability concept, especially engineering and physics, by being a co-founder of Tesla and SolarCity.
- Tesla Motors has been crucial for proving that electric cars will be the cars of the future and instrumental for attaining a sustainable future.
- In Africa, the future will be characterized by increased investment in research and development, whereas in Asia, sustainability science will be at the center of the research agenda.
- In Europe, sustainability science could lead to an improved Lisbon strategy while in the Americas, it will support the transition to sustainability.
- Australia will be characterized by sustainable innovation, driving biomimicry.

12.13 Discussion Questions

1. Discuss ways in which sustainability can be incorporated into chemical education.
2. Why has computer science not been quite successful in embedding the concept of sustainability?
3. What is the progress of the African chemical industry on some Sustainable Development Goals?
4. Mathematics has often been heralded as a crucial determinant of sustainability. Explain the truthfulness of this statement.
5. Do you believe that we can enhance agricultural production to feed the expanding population?
6. How can we enhance farm economics?
7. Explain the role of biology in reducing climate effects and minimizing nitrogen runoff that creates dead zones within oceans.
8. What is the role of biology in the creation of nutrient-rich foods?
9. Explain the role of biological agriculture in ensuring world food security.
10. Explain how vertical farming promotes sustainability.
11. Using the highlighted home gardening practices, explain how you would design your garden to ensure that it is sustainable.

12. Why is agriculture an important factor in achieving sustainability in Africa?
13. Identify three benefits of biomimicry.
14. Provide two examples of applications of biomimicry in Sub-Saharan Africa.
15. How can the ideas of Elon Musk be emulated in Africa?
16. Discuss how the emulation of SolarCity can make Africa more sustainable in energy production?

References

Anastas, P. T. (2003). Meeting the challenges to sustainability through green chemistry. *Green Chemistry*, *5*(2), G29–G34.

Anastas, P., & Eghbali, N. (2010). Green chemistry: principles and practice. *Chemical Society Reviews*, *39*(1), 301–312.

Bartl, M. H., & Lakhtakia, A. (2015). The artificial beetle, or a brief manifesto for engineered biomimicry. In *Bioinspiration, Biomimetics, and Bioreplication 2015* (Vol. 9429, p. 94290B). International Society for Optics and Photonics.

Berger, R. G. (Ed.). (2007). *Flavours and fragrances: Chemistry, bioprocessing and sustainability*. Berlin: Springer Science & Business Media.

Buchwald, H., Cowan, G. S., & Pories, W. J. (2007). *Surgical management of obesity*. Amsterdam: Elsevier Health Sciences.

Clark, W. C. (2007). Sustainability science: A room of its own. *The Proceedings of the National Academy of Sciences (PNAS)*, *104*(6), 1737–1738.

Clements, D. H., Sarama, J., Wolfe, C. B., & Spitler, M. E. (2015). Sustainability of a scale-up intervention in early mathematics: A longitudinal evaluation of implementation fidelity. *Early Education and Development*, *26*(3), 427–449.

Davidson, C. I., Hendrickson, C. T., Matthews, H. S., Bridges, M. W., Allen, D. T., Murphy, C. F., et al. (2010). Preparing future engineers for challenges of the 21st century: Sustainable engineering. *Journal of Cleaner Production*, *18*(7), 698–701.

Getz, L. (2006). *Sustainable and responsible preventive medicine: Conceptualising ethical dilemmas arising from clinical implementation of advancing medical technology*. Oslo: Det medisinske fakultet.

Gibbs, S. (2014). Elon Musk: Artificial intelligence is our biggest existential threat. *The Guardian*. https://www.theguardian.com/technology/2014/oct/27/elon-musk-artificial-intelligence-ai-biggest-existential-threat/.

HDQ Walls. (2018). *Elon Musk*. Retrieved April 16, 2019, from https://hdqwalls.com/-download/1/elon-musk-4k-rq.jpg/.

Lakhtakia, A., & Martín-Palma, R. J. (Eds.). (2013). *Engineered biomimicry*. Oxford: Newnes, Elsevier.

Mankoff, J., Kravets, R., & Blevis, E. (2008). Some computer science issues in creating a sustainable world. *Computer*, *41*(8), 102–105.

Marcu, M. (2006). *The Lisbon strategy*. Institute of Economic Forecasting. http://www.ipe.ro/rjef/rjef1_06/rjef1_06_6.pdf/.

Massele, A., Burger, J., Katende-Kyenda, N. L., Kalemeera, F., Kenaope, T., Kibuule, D., et al. (2015). Outcome of the first medicines utilization research in Africa group meeting to promote sustainable and rational medicine use in Africa. *Expert Review of Pharmacoeconomics & Outcomes Research*, *15*(6), 885–888.

Mathews, F. (2011). Towards a deeper philosophy of biomimicry. *Organization & Environment*, *24*(4), 364–387.

Praneetvatakul, S., Janekarnkij, P., Potchanasin, C., & Prayoonwong, K. (2001). Assessing the sustainability of agriculture: A case of Mae Chaem Catchment, northern Thailand. *Environment International*, *27*(2–3), 103–109.
Sheth, A., Anantharam, P., & Henson, C. (2016). Semantic, cognitive, and perceptual computing: Paradigms that shape human experience. *Computer*, *49*(3), 64–72.
Von Wirén-Lehr, S. (2001). Sustainability in agriculture—an evaluation of principal goal-oriented concepts to close the gap between theory and practice. *Agriculture, Ecosystems & Environment*, *84*(2), 115–129.

Chapter 13
The Future: Achieving Sustainability

Abstract There has been significant progress in the quest for sustainability in various world parts, particularly the developed region. Despite developing areas struggling to achieve some of the set targets, they have also made some progress. With the Sustainable Development Goals due in 2030, the next vision will be Agenda 2063. This closing chapter seeks to consolidate the roles that microorganisms, plants, and animals will play in a sustainable future. Focus is also on the role that businesses, educational institutions, governments, religious groups, and citizens will play in such a future. Using the case study of Capsula Mundi, experiment on how to make biogas and individual matter of Jeffrey Sachs, the chapter makes a critical contribution towards understanding how the future of sustainability will look like. Future projections of sustainability are provided at the end to offer insight on what the society should anticipate in the various continents.

Keywords Microorganisms · Corporate Social Responsibility · Jeffrey Sachs · Animals · Plants

13.1 Overview

Sustainability is crucial for the survival of living organisms because it ensures that resources are used in a viable and amicable way. Rapid international urbanization, population growth, declining resources, and degradation of the environment are presenting challenges and opportunities to businesses as well as policymakers globally. The planet is at a critical stage, and humans cannot afford to continue along the same path. The earth's ecosystems are fragile. Humans face problems such as diminished supplies of non-renewable resources, extinction of species, increased signs of global warming, and high incidences of environmental contamination. These are key indicators that there is a need for a shift from contemporary modes of production to sustainable means.

A. A. Akinsemolu, *The Principles of Green and Sustainability Science*,
https://doi.org/10.1007/978-981-15-2493-6_13

Identify some real-world examples of Africa's fragile ecosystems.

Cities globally are responding to urbanism through the promotion of sustainable development solutions via innovative approaches for the reduction of congestion and pollution. The challenges are numerous, but these also present great opportunities for transforming the world with innovative thinking and efficient technology. Today, the world is at a crucial and defining moment where the interdependency of people, society, and the environment are essential for survival. There is a need for innovation, communication, respect, as well as cultural diversity to make sure that we achieve sustainability in many areas. However, achieving sustainability will continue to be difficult if it is left for only certain stakeholders. There is a need to acknowledge the roles of various stakeholders and institutions in the future viability of sustainability. In other words, a sustainable future can only be achieved through an iterative, holistic, and integral process. It is within this pretext that this chapter seeks to examine the roles of various stakeholders, organisms, and institutions in achieving sustainability. We will explore the roles that microorganisms, plants, and animals play in the sustainability of the ecosystem. Citizens, governments, and businesses will form the foundation for the role of various stakeholders. Religious groups and educational institutions will act as organizations that play a crucial role in achieving sustainability.

13.2 Role of Microorganisms

Microorganisms play a crucial role on our planet. Many of the tasks performed are geared towards making the planet habitable and sustainable (see Fig. 13.1). To limit the focus of this subsection, the 17 SDGs will be used to show the crucial way microorganisms assist in achieving sustainability. The first role is the generation of income. Microorganisms are crucial for the production of Gobar Gas and traditional food in Nigeria, including Iru, Ogi, and vinegar, that are crucial for generating income as well as providing food (Khan et al. 2007).

Briefly expound on how microorganisms are crucial for income generation. How can Nigeria improve its potentiality in income generation using microorganisms?

The second role is related to the second SDG of achieving zero hunger. Fermented pickles, milk, cheese, soy sauce, yogurt as well as bread are important foods that are directly or indirectly associated with microorganisms. In the same

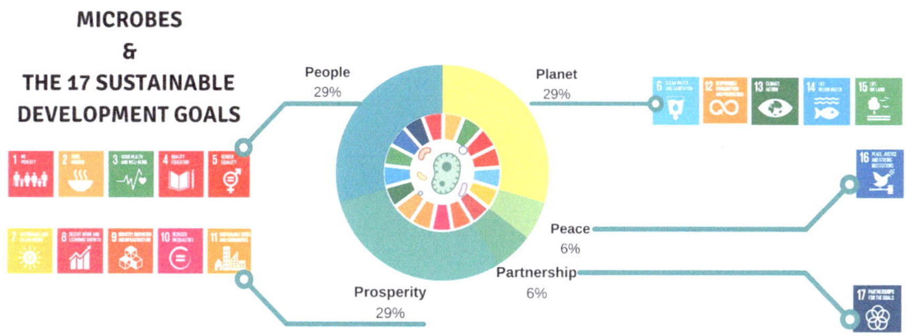

Fig. 13.1 Role of microbes in sustainability, *Source:* Akinsemolu (2018)

breadth, microorganisms play a pivotal role in soil fertility. For instance, symbiotic fungi and other microorganisms contribute to soil fertility. Fungi are critical as decomposers within the soil food web because they help in the conversion of hard-to-digest organic material into forms used by other organisms (Khan et al. 2007). They bind soil particles together, thus helping in the creation of stable aggregates that assist in increasing water infiltration as well as the water holding capacity of the soil.

Enumerate four fungi that help in improving soil fertility.

The fourth role is contribution to good health and well-being. Prebiotics and probiotics are made from microorganisms. Penicillin is also a microorganism that is crucial for the development of new drugs. Other microorganisms also help in preventing diseases, such as gastroenteritis, and prevent infections that may be associated with pollution, including Bacillus and Candida.

What is Candida, and how does it affect the human body?

Microorganisms also play a crucial role in teaching, research, and innovation. This includes the built environment, molecular biology, and education about infant immunity (Akinsemolu 2018). They also act as renewable energy sources for studying rural communities, including biogas and Microbial Fuel Cell.

Microorganisms further play a role in gender equality. A large proportion of women graduate from fields dealing with microorganisms and obtain employment in other areas. Promotion of education among women can enhance the well-being of families, children, and society. Table 13.1 shows the Sustainable Development Goal and the corresponding role of microorganisms.

Table 13.1 The roles of microorganisms in achieving the sustainable development goals

Sustainable development goal	Role of microorganisms
Goal 6: Clean Water and Sanitation	Control of poor-sanitation related diseases
	Production of Rotavirus vaccine, vitamin A and improvement of water sources
	Regulating microbial load
Goal 7: Affordable and Clean Energy	Generation of electricity
	Use organic matter to generate usable electricity
	Use waste products to produce bioenergy
Goal 8: Economic Growth	Research and development
Goal 9: Industry, Innovation and Infrastructure	Biofertilizers
	Microbial products
Goal 10: Reduced Inequalities	Agricultural and land management practices
Goal 11: Sustainable Cities and Communities	Solid waste management
Goal 12: Responsible Consumption and Production	Environmental education
	Bioremediation
	Responsible production using microbial cellulose
Goal 13: Climate Action	Biogeochemical cycles
	Biofuel production
	Nitrogen fertilizers
Goal 14: Life Below Water	Aquaculture
Goal 15: Life on Land	Ubiquitous functions including the sustenance of livelihoods
Goal 16: Peace, Justice, and Strong Ecosystems	Combating bioterrorism
Goal 17: Global Partnerships for Sustainable Development	Capacity building on aspects related to trade

Source: Adopted from Akinsemolu (2018)

13.3 Role of Plants

Plants contribute to the alleviation of poverty. They act as a source of medicine, construction materials, financial income, and food. People living in poverty, especially those living in rural areas in developing states, rely on plants to provide these functions. Wild plants shield communities against famine and support livelihoods (Eickhout et al. 2006). They also play a crucial role in ending hunger, achieving food security, and improving nutrition, while simultaneously promoting sustainable agriculture. With plants acting as the foundation for food systems globally, access to their diversity is an important factor in making sure that there are better nutrition and sustainable agriculture. Botanic gardens play a crucial role in sustainability by encouraging sustainable and equitable trade through effective management of cultivation programs. They also deploy training and educational programs for promoting the sustainable harvesting of natural resources.

A substantial body of evidence suggests that transgenic crops can contribute significantly to the three major pillars of sustainability: environmental, social, and economical (refer to Chap. 12 for more information). Managing insect-resistant and herbicide-tolerant crops to reduce the vulnerability of pests and weeds is crucial for sustainability. It is generally acceptable to consider transgenic crops as one of the tools that is crucial for meeting the aspirations for sustainable global development. Transgenic crops have genes that are inserted artificially. Other plant genes are taken and transferred to a particular plant or could transfer from various species to specific plant/crop.

Identify some of the transgenic crops and areas where they are found in abundance in Africa.

Whereas plant diversity could enhance well-being through poverty alleviation opportunities and ensuring that there is food security, it also plays a significant role in enhancing healthy lives. Consequently, plants act as a foundation for the sustenance of life by providing material and non-material benefits. Without adequate management, there would be a crisis on the planet since humans would be denied access to the vital services provided by plants.

What do you think is the role of home-gardening in sustainable development?

13.4 Role of Animals

Animal products play a crucial role in the attainment of food security. First, they are a crucial source of balanced and bioavailable micronutrients that include vitamin B-12 and iron. Moderate intake of such products could play a crucial role in attaining a nutritionally balanced diet, particularly within the developing world. Also, farm animals are crucial for the development of a sustainable agricultural system, particularly for smallholders who make up many of the world's farms (Reynolds et al. 2015). Farm animals are a major source of high-quality food.

Provides examples of these foods.

Such food enhances the nutritional status of people, while improved efficiencies, through enhanced management technologies and genetic selection, can save vast resources and significantly minimize the carbon footprint associated with animal production. The conversion of forages can enhance grasslands' biodiversity and health. For example, in Brazil's Cerrado region, pasture losses were reversed through the conversion of more tropical forests into pasture. Notwithstanding, sustainable farm animal production will also need a holistic understanding of their effect on the environment. There is also a need to understand the particular role of animal products within the diet.

It is worth noting that solving issues and overcoming the bottlenecks for sustainable food security are not alternative approaches since food security is crucial to national security and global stability. Feeding the expanding population is, therefore, the major challenge in this century. Animals will continue playing a crucial role in the feeding of this population (Reynolds et al. 2015). In summary, farm animals will contribute to sustainability through the provision of manure for fertilizer and on-farm power. They also offer economic diversification, particularly for smallholders.

Africa has found it challenging to maintain sustainable animal production. What are the major challenges? Which countries have tried to embed sustainable animal production policies in their agricultural practices? Have they been successful? Provide measures on how sustainable animal production can be achieved in the African continent.

13.5 Role of Citizens

The central role of citizens in sustainability is engaging in responsible production and consumption. This is crucial for reducing food losses along supply chains. To achieve this objective, citizens can modify their habits by not overbuying products. It is crucial to plan meals and utilize shopping lists when making purchases. People can also freeze fresh products as well as leftovers in case they do not have the opportunity to eat them before going bad (Kasemir et al. 2003). Responsible consumption also entails eating less fish, poultry, and meat. The production of meat consumes more resources compared to the production of plants. Another responsible way of consumption is buying and eating seasonal produce that originates from local growers. Responsible consumption and production also entail recycling, reusing, and preventing the use of unnecessary resources. Citizens can donate things they do not use, including books, furniture, clothes, and food, and thereby contribute to other objectives that seek to alleviate poverty, end hunger, and minimize inequalities.

What is the role of citizens in achieving the Sustainable Development Goals? (Identify and briefly discuss four major roles).

As citizens, we have a critical role and should reconsider our lifestyles. Taking responsibility further entails undertaking the crucial steps and personal actions to create a more responsible community and business. There is a need to change our attitudes to enhance environmental quality, economic health, and achieve social equity. Therefore, we should begin taking small steps that bring about positive changes and inspire others to drive our society towards sustainability. If we, as citizens, feel individually engaged and dedicated to sustainability, we can lay the foundation for embracing change at a societal level.

13.6 Role of Government

The government plays a crucial role in enhancing sustainability. It protects the environment and individuals through the various legislations that it implements. During the regulatory process, governments are well positioned to enhance eco-friendly practices that are naturally sustainable. For example, in the US, the federal government has played a crucial role in enhancing awareness regarding sustainability issues. The promulgation of eco-friendly regulations also results in the attainment of sustainability objectives. A government can also promote the utilization of best practices to ensure that there is an effective and efficient use of resources, especially since they are usually used to advance corporate objectives (Laszlo and Zhexembayeva 2017). Some of the best practices include recycling of e-equipment, the development of complex systems for renewable energy development, reduction of toxic contaminants in the use of materials, and encouraging the use of certain fuels other than hydrocarbons. Rwanda is among the countries that developed measures to ensure that eco-friendly practices are embedded in government activities.

The government is also critical in advancing sustainability by encouraging sustainable actions that can be rolled out easily. In building a sustainable economy, the roles of the government entail funding basic science that is required for renewable energy and technology, using the tax system for steering capital towards investment in sustainable technologies and businesses, investing in sustainable infrastructure, and minimization of ecosystem destruction. The government also has a role of measuring the progress of the society towards sustainability through the development and maintenance of sustainability metrics.

Do you think governments have done enough to promote sustainability in Africa?

13.7 Role of Businesses

During the last few years, the definition of sustainability has shifted from environmental stewardship to incorporate social equity and economy. Organizations that have an interest in sustainability measure their performance using the triple bottom line that is made up of social, environmental, and economic criteria. The increased interest of investors in business sustainability resulted in the development of Dow Jones Sustainability Indexes in 1999 that compare the performance of the leading business globally based on the triple bottom line. Unless multinational organizations sustainably conduct their businesses, they risk losing community support, alienating clients, and making suppliers lose business (Slaper and Hall 2011). More importantly, there is a need for sustainable business to ensure that the available resources are used prudently and promote wellness. Perhaps, the best way of achieving this is through the deployment of the concept.

Explain the contribution of Dow Jones to sustainability. As communities, what can we do to ensure that businesses become responsible to us and the environment?

13.7.1 Corporate Social Responsibility (CSR)

Corporate Social Responsibility (CSR) refers to the initiatives undertaken by a company to conduct assessments and take responsibility for its impacts on the environment and social well-being. This term usually describes efforts that go beyond what the authorities and environmental lobby groups require. It could also entail incurring short-term costs that may not immediately offer financial returns to the business, but rather enhance positive socio-environmental change. This is crucial for achieving sustainability because it focuses on the most salient needs of the firm. The success story of Unilever is sufficiently publicized, but it is difficult identifying with a business of such magnitude. The positive aspect is that even small businesses benefit when prioritizing CSR with small changes. Whereas profits could be the ultimate objective of businesses, responsible businesses can attract more investors, minimize their vulnerabilities, and tackle stakeholder concerns. There is more interest in corporations demonstrating CSR today.

13.7.1.1 Examples of Corporate Social Responsibility (CSR) Organizations

As shown in Table 13.2, numerous organizations have successfully integrated CSR into their programs. General Electric (GE) is one of the CSR organizations whose GE Foundation, committed towards the transformation of communities and shaping tomorrow's diverse workforce, contributed about $88 million to educational and community programs in 2016 (Castelló and Lozano 2009). Its Developing Healthcare Programs often focus on access to healthcare globally for communities. Deloitte is committed to spurring environmental sustainability and societal change. Many of its employees donate time towards pro bono work. International Business

Table 13.2 Examples of Corporate Social Responsibility by companies

Company name	CSR activities
Walt Disney Company	Supports communities by instilling hope, happiness, and comfort to children and families
	VoluntEARS program encourages the employees to donate their time
Google	Supports various initiatives to attain social objectives
	Uses data to unearth racial injustice and improve economic mobility
Bayerische Motoren Werke (BMW)	Has a wide range of social sustainability programs, including wellness, education, and intercultural understanding
Dell	Rolled out the 2020 Legacy of Good Plan as its commitment to foster human progress through environmental sustainability, addressing community challenges, and hiring diversity
Ingvar Kamprad Elmtaryd Agunnaryd (IKEA)	Established the IKEA Foundation that focuses on Circle of Prosperity, which funds health, education, and sustainable income for communities in need
Starbucks	Sensitive to social responsibility and committed to strengthening communities. For example, Starbucks provides one million coffee trees to farmers
Ben & Jerry's	Ben & Jerry's Foundation is responsible for 7.5% of the organization's pretax profits that are allocated to philanthropy. Every year, the foundation awards this amount for funding social change, sustainability, and community action

Source: Adopted from Idowu et al. (2013)

Machines (IBM) is committed to corporate citizenship. It supports various efforts in disaster relief, education, and global health. Zappos' major values are Embrace and Drive Change. Its charitable group called Zappos for Good works with other organizations to donate goods like shoes and books. TOMS was established as a platform for sustainable giving. Its One for One program supports various services, social entrepreneurship, and partners with other charitable organizations. Apple also takes global responsibility seriously. It was named by Greenpeace as the greenest tech company globally for three consecutive years. Table 13.2 provides further information on companies that have embedded CSR into their programs.

In Africa, Unilever has embedded CSR through the Sustainable Living Plan, while Exim Bank Ltd. is famous for its Management Development and Leadership Program. African Air Rescue (AAR) Uganda has on the other hand, been committed to environmental excellence because of its Trees for Health Project, while Sarova Hotels Resorts and Game Lodges have successfully implemented the Sarova Green Initiative. Serengeti Breweries Ltd is known for its Responsible Drinking Campaign, while Unilever Tea Kenya is famous for rainforest alliance certification. Other successful organizations are:

(a) Standard Chartered Bank—Seeing is Believing Initiative.
(b) Bedi Investments Ltd—Feed Projects.
(c) Kuoni Private Safaris East Africa—Child Protection Initiative.
(d) Vodacom Tanzania Ltd—M-Pesa Woman Empowerment Program.
(e) Tigo Tanzania—Tigo Lapdesk Initiative.

> What other African organizations are famous for their CSR programs? Identify the companies and the associated CSR programs.

13.7.1.2 How to Incorporate Corporate Social Responsibility (CSR)

What is the first thing that an entrepreneur should engage in case they want to embed CSR into their business model? There are various steps you can take to ensure that you embed CSR into your programs. First, it is crucial to offer pro bono services to staff while unleashing their expertise. The second way is hosting a fitness challenge to nurture wellness habits and good health among employees. Organizations should deploy fitness tracking applications like UP to track the workers' fitness. The third way could be racing for a cause that aligns with employees. The other crucial ways for incorporating CSR are crowd-funding as a team, volunteering as a group, promoting educational programs in the community, hosting a donation drive, encouraging board service, providing time off for employees to volunteer, and examining the organization to determine what is important.

13.8 Role of Educational Institutions

Educational institutions play a crucial role in inculcating knowledge, values, experience, skills, and knowledge that are relevant for achieving sustainability. These institutions also contribute to the attainment of the SDGs by supporting the implementation of all these goals. This is achieved through research, organizational governance, external leadership, teaching and learning, operations, and culture (see Table 13.3). There are five major roles that educational institutions can play in enhancing sustainability:

1. Assisting in the design of sustainability-based policies.
2. Encouraging sustainability-oriented research.
3. Nurturing and protecting sustainability businesses, including high-tech startup organizations located close to university research programs.
4. Sustainability-based training of future leaders through experiential learning.
5. Enhancing multi-stakeholder engagement.
6. Promoting the understanding of human relations and enhancing creativity to solve world issues that may impede sustainability.

How do you think our institution can help on this front? Explain how you can participate in such a process.

An example of an institution that is playing a crucial role in sustainability is the Green School in Bali. Green School educates for sustainability using community-integrated entrepreneurial learning in a natural environment that does not have any walls. The school's holistic student-guided approach is a source of inspiration and empowerment for green leaders. Green School is a structure that has been constructed using sustainable materials, including alanglang, a local bamboo. Located on a sustainable campus along the Ayung River, it was founded by environmentalists and designers. The innovative bamboo structures depict the architectural potentialities of a sustainably developed institution. Its seventy five buildings are powered by alternative energy sources, including bio-diesel, solar panels, and a hydro-powered vortex generator. The number of such schools should increase for the world to become more sustainable.

The second institution that has played a crucial role in environmental sustainability is Ashesi University. Founded by Ghanaian Patrick Awuah, the university is committed to environmental sustainability (Dias et al. 2005). From its walls to the water running on its taps and stones, the university has embraced the best environmental practices.

Finally, The Earth Institute at Columbia University strives to mobilize education, public policy, and the sciences for achieving a sustainable Earth. Through the

Table 13.3 Summary of the various roles of stakeholders in achieving the SDGs

	Microbes	Plants	Animals	Citizens	Government	Businesses	Educational institution	Religious groups
Agriculture	Produce Gobar Gas and food	Food source	Food source	Responsible farming	Sustainable agriculture	Support farmers	Research	Sustainable agriculture
Economy	Research and development	Alleviate poverty	Food security	Consuming and producing sustainable goods	Sustainable consumption Green entrepreneurs	Support communities	Teach values and knowledge	Sustainable consumption
Building and construction	Inspire building construction	Materials	Inspire building designs	Using eco-friendly products	Enforce sustainable building codes	Eco-friendly construction	Teach values and knowledge	Advocate for sustainable cities and construction
Environment	Solid waste management Biofuel production	Sustain life Sequester carbon	Sustainable agriculture	Eco-tourism Reducing carbon footprint	Sustainable environmental standards	CSR Sustainable Production practices	Training and research	Environmental awareness
Manufacture	Biofertilizers and microbial products	Source of materials	Source of materials	Sustainable resources	Sustainable production and manufacture standards	Sustainable manufacturing practices Eco-friendly production	Training and research	Promote sustainable behavior
Energy	Generate electricity and bioenergy	Convert light into chemical energy Sequester carbon	Diversify economy	Sustainable energy	Promote renewable energy sources	Reduce non-renewable energy use Install technologies that conserve energy use	Research Education and knowledge transfer	Spearhead renewable energy use
Health	Prebiotics and probiotics Drug development	Source of medicine	Nutrition	Reducing pollutants released	Sustainable health policies Awareness on food nutrition	Support sustainable initiatives	Training	Promote healthy living

leadership of Professor Sachs, the institute significantly contributes to the necessary knowledge for tackling the challenges of the twenty-first century. With curricula for 24 degrees and an active fellowship program, it is also educating new leaders to ensure that they become professionals in sustainable development. Such institutions and leaders need to devise innovative strategies for protecting the future of our vulnerable planet.

13.9 Role of Religious Groups

Religion plays a crucial role in conservatism, environmental stewardship, and responsible behavior. These are crucial for the adoption of sustainable activities. The three major functions of religion are campaign and intermediation, enhancing responsible behavior, and the dissemination of worldviews and values that empower environmental action and attitudes. Many religious groups are disturbed by the increasing trends of climate change (Minton et al. 2015). They express the desire to reform and adopt environmentally friendly practices. Also, they advocate for more sustainable behaviors and the development of pro-environmental policies that include the use of cleaner energy (see Table 13.3). Many Western Christians have demonstrated that they are committed to enhancing sustainability by engaging in ethical and responsible behaviors.

What has been the main challenge that deters Christians from environmental conservation?

Various initiatives that focus on engaging the local congregation are evident. For example, the Shrinking the Footprint campaign conducted by the Church of England aims to minimize CO_2 emissions. Similarly, Islam has theological and ethical underpinnings with harmony values, stewardship, and environmentalism. Muslim communities also tend to associate themselves with better environmentalism. According to the Quran, the natural order was put in place by Allah. It further provides insight that humans were granted the responsibility to become the custodians of the Earth. This is also evident in the Bible, thus suggesting that religion is a crucial way of enhancing sustainability. Some of the Islamic projects that promote sustainability through environmentalism are Muslim Khatri Association and the Muslim Green Guide to Reducing Climate Change (Hope and Jones 2014). The guide offers advice on the way humans can live sustainably. With this in mind, religion could play an indispensable role in promoting sustainability.

How can Muslims take advantage of their teachings to enhance sustainable development?

Religion can also foster sustainability through the following ways:

- Enhancing respect for various beliefs and faiths.
- Strengthening the reduction of corruption by enabling faith-based ethics.
- Engendering peace through the diffusion of religious tensions, thus minimizing conflict and violence related to religion.
- Propagating broader freedoms.
- Development of the economy.
- Alleviating overregulation associated with various trends including coercive blasphemy laws, and increasing trust among employees.

Case Study 13.1: Capsula Mundi

Fig. 13.2 Capsula Mundi, *Source:* Capsula Mundi (n.d.)

Capsula Mundi is a cultural project that envisions a unique approach to the thinking about death. Made of biodegradable material, Capsula Mundi is an egg-shaped pod that facilitates the burial of the dead (see Fig. 13.2). The project will have ashes held in urns with a small-egg shape, whereas the bodies will be placed in a fetal position. It will then be buried similar to a seed in the soil. During lifetime, a person will be expected to choose a tree. This tree will be planted on top of the Capsula Mundi and act as a memorial for the departed. It will be a legacy for the continuity of their life and posterity while driving the future of our planet forward (Rashmi et al. 2015). Friends and family will continue caring for the tree. In this project, it is anticipated that cemeteries will have a new look rather than the cold grey landscape seen today. They will grow to become vibrant woodlands. Arguably, such projects will play a crucial role in the sustainability of the planet because they embed eco-friendly practices.

13.10 Individual Matter: Jeffrey Sachs

Sachs is a famous professor of economics and leader in sustainable development (see Fig. 13.3). His monthly newspaper columns make appearances in at least 100 nations. Sachs is also a co-recipient of the 2015 Blue Planet Prize, a prize in recognition for his contribution in environmental leadership and sustainability. Time Magazine also named him number two among the 100 most influential leaders in the world. The New York Times also referred to him as "probably the most important economist in the world." Currently, he is the Director of the Centre for Sustainable Development at Columbia University. He is also a Special Advisor to the UN Secretary-General Antonio Guterres on Sustainable Development Goals. Previously, he has advised the United Nations Secretary-General Ban Ki-moon on the Millennium Development Goals and Sustainable Development Goals (Sachs 2012). He also advised Kofi Annan while he was still UN Secretary-General on the Millennium Development Goals. Currently, he serves as Director of the UN Sustainable Development Solutions Network that is led by the UN Secretary-General. He is a Founder and Chair of SDG US, a Non-Government Organization initiative to enhance Sustainable Development Goal ideas in the US. The work he has carried out in these fields has made him visit more than 125 nations.

Furthermore, Sachs is the author and editor of numerous books, some of which are New York Times bestsellers: The Price of Civilization (2011), The End of Poverty (2005) and Common Wealth: Economics Crowded Planet (2008).

Fig. 13.3 Jeffrey Sachs during the 2011 World Economic Forum on East Asia in Cologny, Switzerland, *Source:* Columbia University (2011)

Many consider Professor Sachs as among the leading experts on issues of global macroeconomics, economic development, and the quest to end poverty.

Since 1995, he has been actively involved in Africa's quest to end poverty. He has done much work in more than 24 African nations and has advised many leaders at various Africa Union (AU) summits. During the 1990s, he worked with the Clinton Administration to come up with the African Growth and Opportunity. Also, he has engaged many African leaders to develop smallholder agriculture and fight disease burdens by strengthening the primary health systems. Currently, he is an advisor to various African governments, including Tanzania, Nigeria, Malawi, Ethiopia, Ghana, Senegal, Rwanda, Mali, Uganda, and Malawi. He directed the Millennium Villages Project in 10 African states that covers at least 500,000 individuals. Sachs has also been successful in advocating for the need to raise agricultural production, minimizing stunted growth of children, and reducing child mortality rates. Sachs contends that in case the rich nations offer substantial investment to rural and agricultural areas within developing nations, the poor will live better lives and have a brighter future. He also supports the Anti-Hunger Programme initiated by the Food and Agricultural Organization.

Mr. Sachs warned that it costs money to increase productivity in agriculture, health, and education and to reach the UN Millennium development goals. Without added assistance from the rich countries, we will not make any progress. We also require leadership from poor and rich countries, from industry, non-governmental organizations, and scientists. The guiding principles for these issues are integrated in rural development to attain the Sustainable Development Goals in Nigeria and Mali. Many other African states are also applying them to provide support to anti-poverty programs. In close collaboration with the Islamic Development Bank, he has spearheaded the scaling up of programs regarding integrated rural development as well as sustainable agriculture among member states of the Bank. One of these projects supports the pastoralist communities in the Horn of Africa. The nations that participate in the project are Kenya, Uganda, South Sudan, Djibouti, and Ethiopia.

Finally, as Director of the Earth Institute of Columbia University (2002–2016), he led an organization of more than 850 professionals to support sustainable development. He is a consistent advocate for the expansion and development of university education to support sustainable development. He was also instrumental in the introduction of the Ph.D. program in Sustainable Development, the first of its kind in the US. The Earth Institute has also guided the adoption and advancement of sustainable development. As a leading research institute, it plays a crucial role in sustainable development with Sachs at the helm. Arguably, he has played a critical role in advancing the issue of sustainable development.

13.11 Future Trends

13.11.1 Africa

Opportunities provided by the green economy could be exploited for their benefits. Africa could also use climate change as an opportunity for developing new industries across the region. Towards this end, there will be a significant shift towards wind and solar energy. With Kenya having put up the largest wind energy plant in the continent, other governments could follow suit. Green entrepreneurship will become more familiar with more support from the government.

13.11.2 Asia

Asia could collaborate with global organizations as they pursue the SDGs. It can also be projected that Asia will increase its efforts in green infrastructure, clean energy and climate friendly activities to reach its targets. China could put in mechanisms for ensuring that the cancer villages caused by pollution are not only reduced but also mechanisms developed to reverse the trend by addressing the root causes. The region could begin decoupling the use of energy and GDP to achieve energy efficiency. The national energy policies could be aligned with SDG 7. China will be among the first nations to build a smart city using artificial intelligence and renewable energy. The country will also take steps to prevent pollution while simultaneously speeding up its transition to clean energy. Also, changes within the urban environment in the Middle East will be fueled by an international evolution in the fuel, transportation, and construction industries. Autonomous driving will also become common in the Arab world.

13.11.3 Australia

There could be a remarkable improvement in sustainability performance. Australia will have to make hard choices for it to achieve its sustainable development targets by 2030. The adoption of renewable energy could significantly increase in this region.

13.11.4 Europe

There will be continued mainstreaming of the sustainable development goals in European Union (EU) policies. These goals will continue inspiring political decision-making and guide future growth strategies. The region will increasingly prioritize external action while simultaneously consolidating the current sustainability targets. It will also promote its socio-environmental and governance standards more robustly through multilateral negotiations and trade agreements.

13.11.5 North America

Variable energy sources could offer about half of the electricity demand. However, this could only be achieved with the appropriate policies and mechanisms in place. Further investment in green transportation could lead to a reduction in the amount of harmful gases released into the atmosphere. Sustainable urban agriculture will also continue being vital for producing the regions' food.

13.11.6 South America

The continent will continue relying on the green economy for driving its growth, particularly through further green industrialization. Solar and wind energy will also continue being prioritized in government plans, while the private sector will increase its investment in renewable energy. Also, the continent could continue with its efforts geared towards mainstreaming SDGs into its policies.

13.12 Summary

- Microorganisms enhance sustainability by generating income, providing nutritious food, improving soil fertility, contributing to good health, teaching, research, innovation, acting as sources of renewable energy, and promoting gender equality through the enhancement of the well-being of the family.
- Plants play the role of poverty alleviation, acting as a source of medicine and food, ending hunger, achieving food security, supporting livelihoods, enhancing nutrition, and sustainable agriculture. Botanic gardens promote sustainable harvesting of natural resources.
- The roles of animals in sustainability are availing balanced micronutrients, source of high-quality food, economic diversification, and food security.
- Citizens engage in responsible production and consumption that is critical for minimizing food losses. Environmental stewardship by citizens is also crucial for achieving sustainability.
- Governments protect the environment using various legislations. They also embed ecologically, friendly practices. The other roles played by governments towards sustainability are research and development, funding basic science,

measuring progress, providing incentives, maximizing resource efficiency, and spurring private investment.
- Regarding businesses, CSR is crucial for ensuring that the environmental and societal pillars are considered during decision-making.
- Educational institutions inculcate knowledge, experience, values, and skills relevant to attaining sustainability. The other roles are sustainability-oriented research, protection of sustainability businesses, sustainability-based training, and improving multi-stakeholder engagement.
- Finally, religious groups play a crucial role in conservation, responsible environmental behavior, economic development, and alleviation of overregulation.
- Professor Sachs has played a critical role in advancing sustainable development at Earth Institute and globally. His contribution to sustainability in the African continent is impeccable.
- Africa will exploit the opportunities provided by the green economy, while Europe will increasingly mainstream the SDGs into their policies, plans, and programs.
- Asia will increase the partnership and collaboration efforts with other regions to meet its targets as the Americas increasingly rely on the variable and renewable energy sources for meeting their energy demands.
- In Australia, renewable energy will be increasingly embraced.

13.13 Discussion Questions

1. How can religion be aligned to increase its role in sustainability?
2. With increased skepticism about religion, will it play a crucial role in enhancing sustainability?
3. The provision of ecosystem services is slowly declining. What do you think would happen in a world characterized by non-functional ecosystems?
4. Identify plants that will be crucial for driving the sustainability agenda in two decades.
5. The problem of educational institutions has been the failure to lead in research. Explain four benefits of educational research in achieving sustainable development.
6. How should sustainability be embedded in learning programs?
7. Many African nations find it difficult achieving sustainability. Identify and discuss four challenges responsible for this trend.
8. Discuss five businesses in Nigeria that you believe have the best CSR programs.
9. How can organizations successfully incorporate CSR into their policies?
10. How can Capsula Mundi related projects be integrated in Africa?
11. Describe six key challenges that prevent sustainability from being actualized.

References

Akinsemolu, A. A. (2018). The role of microorganisms in achieving the sustainable development goals. *Journal of Cleaner Production, 182*, 139–155.

Capsula Mundi. (n.d.). *What is Capsula Mundi?* Retrieved December 02, 2018, from https://www.capsulamundi.it/en/project/.

Castelló, I., & Lozano, J. (2009). From risk management to citizenship corporate social responsibility: Analysis of strategic drivers of change. *Corporate Governance: The International Journal of Business in Society, 9*(4), 373–385.

Columbia University. (2011). *Jeffrey Sachs*. Retrieved June 4, 2018, from https://up-load.wikimedia.org/wikipedia/commons/8/88/Jeffrey_D._Sachs_-_World_-Economic_Forum_on_East_Asia_2011.jpg/.

Dias, M. B., Mills-Tettey, G. A., & Nanayakkara, T. (2005, April). Robotics, education, and sustainable development. In *Proceedings of the 2005 IEEE International Conference on Robotics and Automation, 2005. ICRA 2005* (pp. 4248–4253). Barcelona: IEEE.

Eickhout, B., Bouwman, A. V., & Van Zeijts, H. (2006). The role of nitrogen in world food production and environmental sustainability. *Agriculture, Ecosystems & Environment, 116*(1–2), 4–14.

Hope, A. L., & Jones, C. R. (2014). The impact of religious faith on attitudes to environmental issues and Carbon Capture and Storage (CCS) technologies: A mixed methods study. *Technology in Society, 38*, 48–59.

Idowu, S. O., Capaldi, N., Zu, L., & Gupta, A. D. (2013). *Encyclopedia of corporate social responsibility* (Vol. 21). New York: Springer.

Kasemir, B., Jäger, J., Jaeger, C. C., & Gardner, M. T. (Eds.). (2003). *Public participation in sustainability science: A handbook*. Cambridge: Cambridge University Press.

Khan, M. S., Zaidi, A., & Wani, P. A. (2007). Role of phosphate-solubilizing microorganisms in sustainable agriculture—A review. *Agronomy for Sustainable Development, 27*(1), 29–43.

Laszlo, C., & Zhexembayeva, N. (2017). Embedded sustainability. In *Embedded sustainability* (pp. 116–140). Abingdon: Routledge.

Minton, E. A., Kahle, L. R., & Kim, C. H. (2015). Religion and motives for sustainable behaviors: A cross-cultural comparison and contrast. *Journal of Business Research, 68*(9), 1937–1944.

Rashmi, A. S., Namratha, V., & Sahithi, P. (2015). Capsula Mundi: An organic burial pod. *European Journal of Advances in Engineering and Technology, 2*(8), 49–53.

Reynolds, L. P., Wulster-Radcliffe, M. C., Aaron, D. K., & Davis, T. A. (2015). Importance of animals in agricultural sustainability and food security, 2. *The Journal of Nutrition, 145*(7), 1377–1379.

Sachs, J. D. (2012). From millennium development goals to sustainable development goals. *The Lancet, 379*(9832), 2206–2211.

Slaper, T. F., & Hall, T. J. (2011). The triple bottom line: What is it and how does it work. *Indiana Business Review, 86*(1), 4–8.

Glossary

Acidification This is a reduction in pH.

Acidulation This can be attained through the addition of sulfuric acid to decompose large oil particles that elude from the equalization tank into smaller particles.

Additive production This is an emerging transformative approach to industrial production that enables the creation of lighter, flexible, stronger parts, and systems in order to improve resource efficiency and other sustainability practices.

Aerosols These are tiny particles released by trucks and cars, which may slow down the process of global warming.

Agroforestry An agricultural practice that integrates growing of trees and crop food or animal rearing on the same piece of land.

Alien species An alien species is a species introduced outside its normal distribution.

Anthropogenic activities These are human activities that have an influence on the environment.

Aquaculture Cultivation of aquatic organisms under controlled conditions.

Aquifers These are underground layer permeable rocks that bear water.

Artificial intelligence Intelligence exhibited by machine.

Biofuels They make use of edible feedstock, including corn, cellulosic ethanol can be produced through the use of raw materials, such as grass and wood.

Biogeochemical cycles This is the pathway by which essential elements are being transferred across the Earth.

Biomass energy This is produced through the burning of timber, wood, and municipal, agricultural, and landfill waste.

Biomimicry This is the imitation of nature to proffer sustainable solutions to human problems.

Biomining This utilizes microorganisms in the recovery of metals, also offers the possibility of mining with minimal disruption to land.

A. A. Akinsemolu, *The Principles of Green and Sustainability Science*,
https://doi.org/10.1007/978-981-15-2493-6

Bioremediation Bioremediation is a technique used in waste management that entails using organisms to neutralize or eliminate pollutants from contaminated sites. It can also be defined as the treatment that uses natural organisms to decompose hazardous substances into non-toxic or less toxic ones.

Bioterrorism This is an act of terrorism that involves the intentional release of viruses, bacteria, or other germs that can cause disease to living organisms.

Capitalist economy A system in which individuals or businesses own capital goods.

Carbon footprint This is to identify the aspects of human activities that contribute most to environmental degradation.

Carbon sequestration The process of removing carbon dioxide from the atmosphere.

Carbon sinks These are artificial or natural reservoirs that can store and accumulate materials with carbon compounds for an extended period.

Carrying capacity This refers to the maximum number of individual organisms that a specific environment can sustain during a particular period, and is defined in terms of population, species, time, and space.

Categorical imperative Each individual should act on universalized principles or the principles that can be rationally and universally accepted.

Cellulosic ethanol It is a second-generation biofuel manufactured through the conversion of vegetation that is not suitable for human consumption into ethanol.

Clarifier This is a built settling tank for continuous removal of solid mechanically.

Climate change Climate change refers to prolonged weather events that bring about changes in climatic conditions and weather.

Climate regime This is a structure put in place by governing bodies to regulate the interaction between human activities and climate systems.

Climate variation This is an unusual change in the average climatic condition of a place.

Computational complexity theory This explores the main properties of intractable and computational issues.

Computational sustainability This is a growing computing field that deals with the application of principles of computer science, methods as well as tools to issues of societal and environmental sustainability.

Computer programming This carries out investigations on several aspects of the use of complex systems as well as a programming language.

Congenital Physical aberration or abnormality from birth.

Dammed To hold back for retention or diversion.

Deforestation This is indiscriminate destruction of the forest to make the land available for other uses.

Denitrification This refers to the conversion of nitrates into nitrogen gas.

Desertification This minimizes the arable land for farmers and cattle to graze.

Domino effect This is the effect in which an event sets off a chain for more similar events.

Dow Jones Sustainability Index This compares the performance of the leading business globally based on the triple bottom line.

Early Warning Systems This is a policy setup to detect and reduce the impact of disasters.

Ecofeminism It contends that capitalistic societies have resulted in a disastrous impact on the environment and the people.

Ecological balance This is the equilibrium formed through the harmonious co-existence of organisms, including flora and fauna.

Ecology This is the study of the relationship between organisms and the environment. It can also be defined as the study of the abundance and distribution of living things and the way they are influenced by their interactions with their environment.

Economic sustainability This refers to the capacity industries hold in order to continue their activities or processes in an indefinitely way for an indefinite period.

Economically sustainable An economy growth that supports the social, environmental, and cultural stability of a community.

Eco-physics This highlights the way power generations can be environmentally friendly.

Ecosystem services These are the numerous benefits that humans freely accrue from the natural environment and appropriately functioning ecosystems, some of which are aquatic, forest, and grassland ecosystems.

Education for sustainable development This is a type of education that promotes knowledge, skills, attitudes, and values necessary for environmental, social, and economic sustainability. It also promotes participatory teaching and cognitive learning methods to empower learners to collaborate, change behavior, and take measurable action for environmental, social, and economic development.

Endangered species These are animals that are categorized to likely go extinct in the nearest future.

Enkai Also refers to as Ngai, is believed to be a higher spirit by the Maasai.

Environmental degradation This is the deterioration of the environment as a result of exhaustion of natural resources.

Environmental ethics This is an area of philosophy that examines the relationship between humans and the environment from an ethical and moral perspective.

Exergy This is a phenomenon that adheres to the first and second laws of thermodynamics.

Extinction Extinction is depicted as the termination of an organism or various organisms.

Flow equalization This is the process involved in controlling the flow rate in a wastewater treatment system. This is also involved in attaining normal flow conditions through a wastewater treatment system.

Gasification This is the decomposition of organic waste materials by exposing waste to low volumes of oxygen and extremely high temperatures.

Gender equality This is a state of equal access to resources, opportunities, and privileges regardless of gender.

Genetic modification This refers to the use of biotechnology to manipulate the genes of an organism.

Genetically modified organisms These are organisms whose genetic makeup has been altered through genetic engineering.

Global warming Global warming is the slow rise in the average temperature of the Earth over a period of time.

Going Green This is defined as the adoption of environmentally friendly lifestyles through activities such as recycling, reusing, reducing, or minimizing.

Green energy This refers to environmentally friendly energy.

Green growth This describes an economic growth that utilizes natural resources sustainably.

Green industrialization This offers a flexible and practical approach for attaining measurable progress across the environmental as well as economic pillars while considering the social repercussions of greening.

Green innovation This includes all environmentally friendly innovations.

Green transportation The use of vehicles that run on alternative fuels and promotes the use of minimal energy.

Greenhouse gas emissions This involves the emission of greenhouse gases (GHG) that contribute to global warming.

Human-induced Instigated by human.

Hybrid cars These are cars that depend on electricity and are designed such that they automatically recharge their batteries through energy conversion.

Hydrogen fuels Hydrogen fuels are made of fuel cell stacks, and each has an anode, cathode as well as a proton exchange membrane.

Hydrosphere This is the total mass of water found on earth.

Incineration Incineration is the combustion of solid and organic waste in order to convert them into gaseous products.

Industrial revolution The process of change from agricultural dominated society to industrial and manufacturing society.

Infection This is the invasion of the host's body by disease-causing organisms.

Infrared light This is a type of electromagnetic radiation that is invisible to human eyes but can be felt as heat.

Integrated pest management An ecosystem-based approach involved in the control of pest.

Inter-generational equity Deals with equity between people of a particular generation and the next generation.

Intra-generational equity Deals with equity between people of the same generation.

Invasive species This is a given species that establishes itself in a new environment.

Irradiated waste These are wastes containing radioactive substances.

Least developed countries These are low-income countries with difficulty meeting up with sustainability.

Lithosphere The lithosphere consists of a crust and supports the biosphere.

Marker-assisted breeding This refers to indirect selection process whereby a trait is chosen based on a marker (could be biochemical or morphological) linked to a trait of interest (including disease resistance, quality, and disease resistance).

Mass transit systems Public means of transportation.

Microbes Tiny living organisms that can only be seen with the aid of a microscope.

Microbial fuel cell This is a bio-electrochemical device that is detrimental for harnessing respiring microbes for the conversion of organic substrates into electrical energy.

Ocean acidification This refers to the ongoing reduction in pH of the Earth's oceans, attributed to the uptake of carbon dioxide (CO2) absorbed from the atmosphere.

Organic agriculture Organic agriculture has the same principles as sustainable agriculture, including natural plant fertilization, ecological balance, natural pest management, biodiversity, sustainability, soil integrity, and natural pest management.

Photovoltaic cells This is a device that converts solar energy to electricity.

Plutonium reactors These are reactors containing plutonium isotopes used in the creation of atomic bombs.

Pre-settling basins The sewage moves through large tanks called pre-settling basins during the primary sedimentation stage.

Product lifecycle A product lifecycle is the sequence of stages that a new product undergoes during its life span.

Programming language This is a set of rules setup to instruct a computer.

Protozoa These are free-living or parasitic unicellular eukaryotes that feed on organic matter. They can be found in moist environment.

Recombinant DNA Joining of two different DNAs and inserting into a new host.

Recycling Recycling enables the reuse of materials that are found in unwanted materials for making new items.

Replenish Renew.

Risk avoidance This entails avoiding an activity that could pose a risk.

Risk treatment Risk treatment entails identification of the range of alternatives for treating risks, evaluating alternatives, preparation of risk treatment plans, and implementation of the plans.

Sedimentation The process whereby silt, sand as well as other particulates settle at the bottom.

Sludge Wastewater treatment plants utilize bacteria which consumes organic residue then turns it into sludge.

Smart grid This is an electricity network that utilizes diverse digital technology, including renewable energy and energy-efficient resources, smart meters, and appliances to reduce the energy cost and maximize the reliability of the energy supply chain, ultimately meeting the future energy demand.

Social sustainability This involves meeting the current needs of humans, societies, and cultures without compromising the ability of future generations to meet their needs.

Socioeconomic equity This defines the quality of being where individuals have equal economic and social status.

Solar glass Solar glass is a transparent glass panel with the ability to harvest solar energy.

Space-based solar power This is geared towards the collection of solar power within outer space. These systems are instrumental in converting sunlight into microwaves.

Sustainable agriculture This is defined as the production of animal or plant products, including fiber and food, using farm methods that conserve and protect the environment, human communities as well as the animal welfare.

Sustainable gardening Its practices and benefits are the improvement of soil, the use of native plants, trees, and shrubs for creating beautiful landscapes as well as using all renewable resources provided by nature.

Sustainable infrastructure This involves developing roads, buildings, energy, and water infrastructure with due regard to economic, social, and environmental sustainability.

The International Union for Conservation of Nature (IUCN) This is an organization that focuses on nature conservation and the sustainable use of natural resources.

Theory of human motivation This states that individuals are motivated to achieve specific needs at a specific pace.

Thermal infrared radiation This is the heat transferred by electromagnetic waves.

Thorium nuclear reactors These reactors make use of thorium nuclear reactors rather than plutonium for nuclear fission.

Tragedy of the commons This describes a situation where individual users of a resource act independently out of self-interest. This behavior contradicts that of common good users since it depletes and spoils that resource.

Undernutrition This is the inability to have access to sufficient energy and nutrients that are necessary to maintain good health.

Urban sprawl This is an uncontrolled growth in an urban area.

Urbanism This is the study of the way inhabitants of a place relate with the human-made environment.

Urbanization This is a population shift as a result of rural-urban migration.

Utopian society This describes an ideal and perfect society that is unrealistic.

Vertical farming This involves the production of medicinal plants and food in stacked layers as well as inclined surfaces.

Virtue ethics This focuses on what makes one a good person and the meaning of living a good human life.

Vulnerable species A vulnerable species is one which has been categorized as likely to be endangered by the International Union for Conservation of Nature due to unfavorable circumstances that threaten its habitat or livelihood.

Wangari Maathai She was a Kenyan environmental activist and founder of the Green Belt Movement.

Waste reduction It is the first process of waste management and entails reducing the quantity of waste generated.

Water reclamation This enables wastewater to be utilized for other purposes.

Wet prairies These are native wetland grassland flooded by streams and surrounded by lake margins and isolated depressions.

Index

A. A. Akinsemolu, *The Principles of Green and Sustainability Science*,
https://doi.org/10.1007/978-981-15-2493-6

Q

MIX
Papier aus verantwortungsvollen Quellen
Paper from responsible sources
FSC® C105338

If you have any concerns about our products,
you can contact us on
ProductSafety@springernature.com

In case Publisher is established outside the EU,
the EU authorized representative is:
Springer Nature Customer Service Center GmbH
Europaplatz 3, 69115 Heidelberg, Germany

Printed by Libri Plureos GmbH
in Hamburg, Germany